"COMPREHENSIVE AND ACCESSIBLE...A MUST HAVE for anyone interested in flexagons. The authors combine scholarly analysis with infectious enthusiasm. Instructions are included for dozens of examples that are easy to make and fun to explore."

Paul Jackson, *award-winning origami artist, paper engineer, teacher, and author*

"HIGHLY RECOMMENDED! A lucid and comprehensive treatment of every aspect of flexagons. Beginners will find a gentle way into a fascinating subject, and experienced flexers will find themselves learning something new on nearly every page of this beautifully illustrated book."

Jason Rosenhouse, *professor of mathematics, James Madison University, author of Games for Your Mind: The History and Future of Logic Puzzles*

"WOW! THE POSSIBILITIES JUST KEEP GOING AND GOING. I had no idea there was so much variety in flexagons! Quite an impressive book. This book goes into amazing depth in the incredible variety of flexagons. There's something for everyone: if you just want to make cool gadgets, there's clear instructions (and downloadable templates) for a wide variety of shapes, but for those who want to go deep, the mathematics of flexagons are also presented, rich and rewarding. Be forewarned, though: they're addictive."

Robert J. Lang, *physicist, virtuoso origami artist, and master of origami mathematics, theory, and real-world applications*

The Secret World of Flexagons

The hexaflexagon is a folded paper strip of colored triangles that has long delighted people with how it "magically" changes its appearance when "flexed". This hands-on, comprehensive book goes beyond the hexaflexagon, the standard version of this folded puzzle, exponentially expanding the barely explored field of flexagons as it brings new options and fresh insights to light.

- Learn over a dozen different flexes, and make dozens of different flexagons with the aid of step-by-step illustrated directions and templates to copy and print.
- Delve into the internal structure of flexagons and discover a universal way to describe and predict their behavior.
- Learn how to create your own custom flexagons with a special computer program.
- Understand how flexagons are connected to group theory, computer science, and topology.
- Have fun decorating flexagons and make flexagon books, puzzles, pop-ups, mazes, and more.

Written in a clear, easy-to-understand, and conversational style and enhanced with challenges and tips to broaden your flexagon skills and spark creativity, *The Secret World of Flexagons: Fascinating Folded Paper Puzzles* is a must for flexagon enthusiasts, teachers, students, libraries, mathematicians, and everyone who loves to solve a good puzzle.

AK Peters/CRC Recreational Mathematics Series

Series Editors

Robert Fathauer
Snezana Lawrence
Jun Mitani
Colm Mulcahy
Peter Winkler
Carolyn Yackel

For more information about this series please visit: https://www.routledge.com/AK-PetersCRC-Recreational-Mathematics-Series/book-series/RECMATH

The Secret World of Flexagons

Fascinating Folded Paper Puzzles

Scott Sherman, Yossi Elran, and Ann Schwartz

CRC Press
Taylor & Francis Group
Boca Raton London New York

CRC Press is an imprint of the
Taylor & Francis Group, an **informa** business

AN A K PETERS BOOK

Designed cover image: Scott Sherman and Susan Brown

First edition published 2025
by CRC Press
2385 NW Executive Center Drive, Suite 320, Boca Raton FL 33431

and by CRC Press
4 Park Square, Milton Park, Abingdon, Oxon, OX14 4RN

CRC Press is an imprint of Taylor & Francis Group, LLC

ISBN: 9781032560472 (hbk)
ISBN: 9781032553832 (pbk)
ISBN: 9781003433538 (ebk)

DOI: 10.1201/ 9781003433538

Typeset in Avenir
by Deanta Global Publishing Services, Chennai, India

To Les Pook
and past, present, and future
flexagon enthusiasts everywhere

Contents

Contents

Acknowledgements

First off, the authors thank the late Martin Gardner for reintroducing flexagons to the world. Without him, this book could not and would not have been written. In the same vein, we thank the Gathering 4 Gardner organization and its biennial gatherings that bring recreational mathematicians and others together. It was at one of these gatherings that the authors met and began collaborating.

We are grateful to the four flexagon pioneers, Arthur Stone, John Tukey, Bryant Tuckerman, and Richard Feynman, and all those who explored flexagons before us, especially Les Pook, Anthony S. Conrad, Daniel K. Hartline, Paul Jackson, Robin Moseley, Bruce McLean, Jean Pedersen, and Peter Hilton. And we are grateful to Vi Hart whose flexagon videos have reached millions of viewers and ushered flexagon awareness into the twenty-first century.

The authors also thank the Davidson Institute, the educational arm of the Weizmann Institute of Science in Rehovot, Israel, for featuring flexagons at its Recreational Math, Puzzles and Games Conference in 2015, and Neil Shore for sponsoring the Neil Shore Flexagon Workshop in 2019, also at the Davidson Institute, which brought the three coauthors together and sparked the idea for this book.

And speaking of the book, Scott thanks his wife, Susan Brown, for many of the photos and the cover design. The authors thank Audrey Nassar for her creative and fun tetraflexagons and Ela Schwartz for her ingenious hexaflexagon puzzles, which they generously permitted us to use. Ann also thanks her older sister, Martha, for showing her how to make a hexaflexagon so many decades ago.

On the publishing side, we are grateful to Paula Breen for her astute negotiating skills and agent Mel Parker who enthusiastically recommended her, as well as the folks at CRC Press: Callum Fraser, for wanting a book about flexagons and leading this recreational mathematics line; Mansi Kabra, for her hand-holding and elastic deadlines; and Charlotte Byrnes for her expert copyediting. And, lest we forget, Colm Mulcahy for connecting us to CRC Press in the first place.

Most of all, we thank our families for their patient understanding and tolerance toward our flexagon obsession, and for their loving support.

About the Authors

Scott Sherman is a leading software designer focused on building tools for seeing and understanding information, from computer-aided design to diagramming to data visualization. He was a researcher and software engineering architect at Tableau, a data visualization company, where he built the first version of many essential tools including dashboards and highlighting. At Microsoft, he designed and built the layout engine for Microsoft Office's SmartArt feature. Scott has been fascinated with flexagons since he first learned about them from Martin Gardner's writings and has figured out techniques for generating a huge variety of new flexagons and flexes. He has also applied his software skills to create programs for exploring flexagon dynamics. He has given talks on various topics in recreational mathematics, including flexagon theory, at the Gathering 4 Gardner and Georgia Southern University. With Dr. Elran and Ann Schwartz, he co-led the 2019 Neil Shore Flexagon Workshop at the Davidson Institute, the educational arm of the Weizmann Institute of Science in Rehovot, Israel. He also is the author of an online, interactive book called *Explorable Flexagons*.

Yossi Elran is a British-Israeli recreational mathematician at the Davidson Institute of Science Education, the educational arm of the Weizmann Institute of Science in Rehovot, Israel, and teaches various courses at the Western Galilee College in Israel. He holds a PhD in theoretical quantum chemistry and has done postdoctoral research on decoherence, one of the main challenges of quantum computing. Yossi frequently journeys nationally and globally, delivering engaging and captivating presentations to diverse audiences ranging from schools and universities to festivals and other events on various subjects, including recreational math, astronomy, quantum mechanics, the history and philosophy of science, and creativity. He has written many papers, is the author of *Lewis Carroll's Cats and Rats and Other Puzzles with Interesting Tails* and *Archimedes' Stomach and Other Puzzles You'll Love to Digest*, and coauthor of *The Paper Puzzle Book*. He is the creator of four popular FutureLearn massive open online courses on recreational math, two of them on flexagons, and two TED-Ed puzzle videos that have reached nearly 20 million views. Yossi is a longtime member of the Gathering 4 Gardner Foundation, whose mission is to stimulate curiosity and the playful exchange of ideas and critical thinking in recreational math, magic, science, literature, and puzzles to preserve and extend the legacy of writer and polymath Martin Gardner.

Ann Schwartz has created more than a dozen new flexagons. She has presented a new flexagon at the biennial Gathering 4 Gardner, which she has attended since 2006, and has given flexagon presentations and workshops at the National Museum of Mathematics (MoMath) in New York City and that museum's MOVES conferences. In 2015 she was the guest speaker at the Recreational Math, Puzzles and Games Conference at the Davidson Institute of Science Education, the educational arm of the Weizmann Institute of Science in Rehovot, Israel; and in 2017 gave the only flexagon presentation at the Seventh International Meeting on Origami in Science, Mathematics and Education (7OSME) in Oxford. With Scott Sherman and Dr. Yossi Elran, Ann co-led the Neil Shore Flexagon Workshop at the Davidson Institute in 2019. She coauthored "The Hexa-Dodeca-Flexagon," a chapter in *Homage to a Pied Puzzler* (2009), and the chapter "Should We Call Them Flexa-Bands?" that was published in *The Mathematics of Various Entertaining Subjects*, Volume 3.

Introduction

Flexagons are fascinating, origami-like puzzles folded from strips of paper that you flex in various ways to reveal previously hidden portions of the flexagon. The huge variety of different flexagons range from simple and elegant ones to others that can be surprisingly tricky to find all the secrets tucked away inside.

Flexagons sit at the intersection of paper folding, puzzles, and mathematics. You could choose to enjoy the pleasure of using your hands to flip through a kaleidoscope of arrangements. Or come up with interesting ways to decorate them. Or explore what's hiding inside, possibly discovering new flexes and new types of flexagons. Or challenge yourself to figure out how to "solve" a flexagon in the same way you might solve a Rubik's Cube. Or investigate the mathematics that can untangle how the twisted strips of paper actually work. In this book, we touch on each of those areas, leaving it up to you which of these most interest you.

The original discovery of flexagons was a happy accident. In 1939, Arthur Stone traveled from England to attend Princeton University in the US. Due to the difference in standard paper sizes between the two countries, he needed to trim off one edge of each sheet to fit them in his binder. After experimenting with different ways of folding those strips of paper, he stumbled upon a particularly interesting arrangement, which he called a **flexagon**, short for *flexible polygon*. He and his friends formed the Flexagon Committee to explore these mysterious objects. The other members of the committee were Bryant Tuckerman, Richard Feynman, and John Tukey, each of whom went on to have distinguished careers in mathematics or physics. While playing with flexagons might not always lead to such careers, there's certainly correlation between experimental play and science. It is believed that Feynman even used some of the techniques he developed for flexagons when he later explored concepts in quantum physics!

Arthur Stone called his original discovery the **trihexaflexagon**: *tri* for the three faces you cycle between as you flex, and *hexa* because it folded into a hexagon. The trihexaflexagon is folded from a straight strip of regular triangles, where every side and every angle are the same. Later on in the book you'll learn how to construct and flex this and the other flexagons shown in this introduction, as well as many more.

But the story of flexagons probably would have come to a screeching halt right there if Stone hadn't generalized this idea, realizing something interesting could happen if he doubled the strip. He called this second discovery the **hexahexaflexagon**, a hexaflexagon that can flex between six different faces. Furthermore, he realized he could keep doubling the length of the strip, making ever more complex flexagons.

The Flexagon Committee also explored flexagons made from squares. They called these **tetraflexagons**, since the faces are divided into four squares and *tetra* means four.

Popular science and mathematics writer Martin Gardner first introduced flexagons to a broader audience in a 1956 *Scientific American* article, where he described Arthur Stone's hexaflexagons.

The flexagons proved so popular that that particular issue became the best-selling issue of *Scientific American* for decades. In addition, many readers sent letters describing all the entertaining ways they decorated their flexagons, including making greeting cards out of them. *Scientific American* soon gave Gardner his own monthly column to discuss other topics in recreational mathematics. His column, "Mathematical Games", continued into the 1980s and was very influential for a whole generation of budding mathematicians. One of his later columns described tetraflexagons.

But there was still a lot more to be discovered. Can these examples be generalized? If you can make a hexaflexagon with 3 or 6 faces, what about 12 or 5 or 100? If you can make a flexagon by folding up a strip of regular triangles, what about right triangles or scalene triangles? If you can make a flexagon out of triangles or squares, what about pentagons or hexagons or octagons? If you can reveal new faces by flexing it one way, are there other ways to flex it that might behave differently? Can you describe how the internal structure of a flexagon changes as you flex it in order to make predictions?

The answer to all those questions is yes. We'll give examples for each of those answers throughout the book. Those answers help us uncover a huge variety of flexagons, some of which are shown in figure 0.1. Note that we're not offering a complete guide to all the possible things you could do. There are still plenty of questions left to answer that begin with, "What if instead I try …?"

FIG. 0.1 A few of the intriguing flexagons shown in this book.

This book is arranged in five parts.

Part One starts with an introduction to making a hexaflexagon and pinch-flexing it, followed by a variety of generalizations. Later chapters add more hidden faces, change the shapes of the triangles and the flexagon, and vary how the pinch flex itself is performed. The part finishes with flexagons made from squares, pentagons, and hexagons.

Part Two focuses on flexagons through the lens of different ways you can flex them. It treats a flex as the fundamental building block for understanding and creating more-complex flexagons. Each chapter starts with the simplest flexagon that supports a new flex, shows variations, and builds more interesting flexagons from there. It then breaks flexes into pieces, showing how you can combine those pieces in different ways to discover new flexes.

Part Three gives you tools and notation for describing your flexagon explorations and for creating new flexagons with custom behavior. This includes diagrams of how flexes change a flexagon, notation for sequences of flexes, a way to describe the internal structure of flexagons, and atomic flex theory for describing the smallest building blocks of flexes. You can use these tools to analyze various types of cycles and sequences of flexes. For those who are interested, it also dives into some of the mathematical theory behind flexagons, including group theory and a demonstration that flex sequences are Turing complete, meaning they can emulate a computer.

Part Four has chapters that cover individual types of flexagons, trying out a variety of flexes you can use to explore them. It describes Flexagon Inspector, a tool for creating and exploring your own custom flexagons.

Part Five provides a sampling of some of the creative ways to play with flexagons. You can draw designs on them that get rearranged in interesting ways as you flex, make books that require flexing in order to read, or create tricky puzzles and mazes on them.

You don't have to read this book in sequential order, nor do you have to read every chapter. Start with the first three chapters of Part One on the pinch flex to get familiar with the terminology and the basics of creating and flexing a regular hexaflexagon. The rest of the initial section covers different generalizations of these concepts. The chapters in Part Two are fairly independent, describing a variety of interesting flexes you can do as well as giving examples of different types of flexagons. Some chapters in Part Three build on knowledge from other chapters, leading to a deeper understanding of how flexagons work. Part Four references flexes and techniques that have been previously described, and finishes by describing the Flexagon Inspector, a useful app for creating and exploring custom flexagons. Part Five is a potpourri of different creative ideas that can be explored in any order.

While some of these topics have been published before, such as the basics of hexaflexagons and tetraflexagons, most of this book includes new material appearing in print for the first time. The new content includes lots of novel flexes and flexagon shapes, cohesive flex notation, innovative flex diagrams, atomic flex theory, new definitions of *flexagon* and *flex*, original mathematical results in group theory and topology, and a variety of playful ways to decorate flexagons and use them for puzzles.

This book is intended to be hands-on, conveying the magic of flexagons through making and flexing them. You will need scissors and tape (preferably clear tape). Ideally, you have access to a computer and printer to print out the paper patterns that you can fold to make a flexagon. (Printable versions of the templates in the book are available online in our Flexagon Templates app; see chapter 1.) Be prepared to cut, fold, and tape to make a lot of different flexagons.

Above all, flexagons are fun to play with. Treat this book as a collection of starting points for your own explorations and experiments.

Happy flexing!
Scott, Yossi, and Ann

PART ONE

Symmetric Flexes

The first few chapters explore a variety of shapes and types of flexagons using **symmetric flexes**, flexes that change a flexagon in a way that's rotationally symmetric. This means that if you put a pencil in the middle of the flexagon and spun it to the correct angle, the flexagon would look identical.

DOI: 10.1201/9781003433538-1

Chapter 1

The Pinch Flex

To introduce you to the world of making and flexing flexagons, we'll start with the **pinch flex**, abbreviated as P, on a **tri-hexaflexagon**. This hexaflexagon starts as a flat hexagon as shown in figure 1.1. Just as a coin has two sides – heads and tails – this hexagon has two visible sides, which we call **faces**. The tri prefix indicates that it actually contains *three* different faces, one of which starts out hidden. The hexa prefix indicates that there are six triangles on a face. First, we'll step through how to make the flexagon, then we'll show how to use the pinch flex to reveal the concealed face.

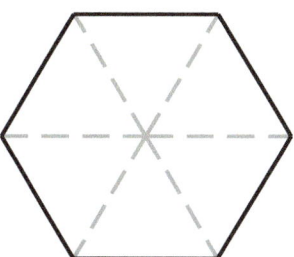

FIG. 1.1 One face of a tri-hexaflexagon.

A flexagon **template** is an unfolded flexagon. Each triangle in the template is called a **leaf**. Each leaf has two **leaf-faces**, opposite surfaces of the leaf, which have separate labels. In figure 1.1, the leaves are **regular triangles**, meaning that all edges of the triangles are the same length and all angles are equal. A **hinge** is an edge between two leaves that can bend and fold.

Print or draw the tri-hexaflexagon template in figure 1.2 onto a piece of paper. One way to get a copy of the template is to type https://loki3.github.io/flex/templates.html into the address bar of your browser to bring you to our Flexagon Templates app and print out the image from that page. Once you have the template on paper, cut it out along the four outside edges. Then, to make it easier to fold up into the proper shape, prepare the template by folding along the dotted lines between leaves – creasing both ways – then unfold. The larger number on each leaf is for the front face of the leaf while the smaller number is for its other face. Copy the small numbers onto the back

DOI: 10.1201/9781003433538-2

of the corresponding leaf. Note that the first and last leaves in the template are marked with a star (*) to indicate the edges that need to be taped together after folding.

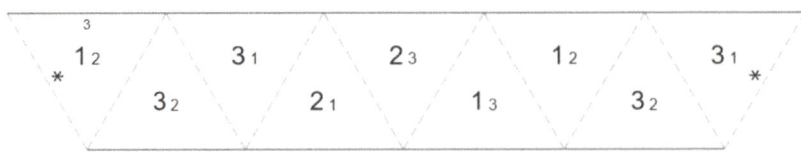

FIG. 1.2 The tri-hexaflexagon template.

To make a working tri-hexaflexagon from the template, follow these steps while referring to figure 1.3:

1. Fold the leftmost pair of 3's face to face.
2. Turn the template over.
3. Fold the visible pair of 3's face to face.
4. Fold the final pair of 3's face to face.
5. Tape together the 1's that were on the ends of the template along the edges marked with stars.

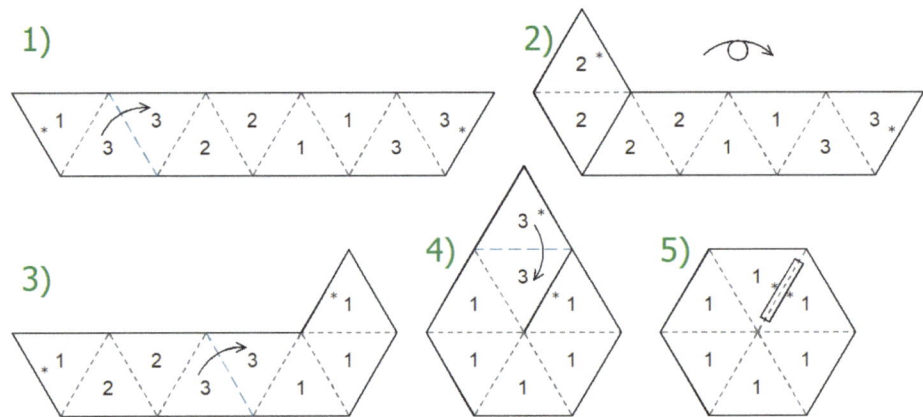

FIG. 1.3 Folding the tri-hexaflexagon from its template.

Now that you've created your tri-hexaflexagon, you should see one face where all six leaves are labeled with 1's and, if you turn it over, a second face with all leaves labeled with 2's. To pinch-flex, start with face 1 in front and face 2 in back. Follow these steps, while referring to figure 1.4:

1. Find a **thumbhole** under one of the leaves, a gap where you can slide your thumb. Note that some leaves won't have a thumbhole.
2. Position one hand at the hinge with the thumbhole and your other hand two hinges to the right, as shown in the picture.
3. With each hand, pinch the leaves together backward. The hinges will form three **mountain folds** pointed toward you and three **valley folds** pointed away from you.
4. Open the flexagon from the center to reveal face 3, which was previously hidden. You have now done one pinch flex.

5. Rotate both hands one hinge to the right.
6. With each hand, pinch the leaves together backward across the hinges, as you did in step 3.
7. Open the flexagon from the center to reveal face 2. You have now done a second pinch flex.

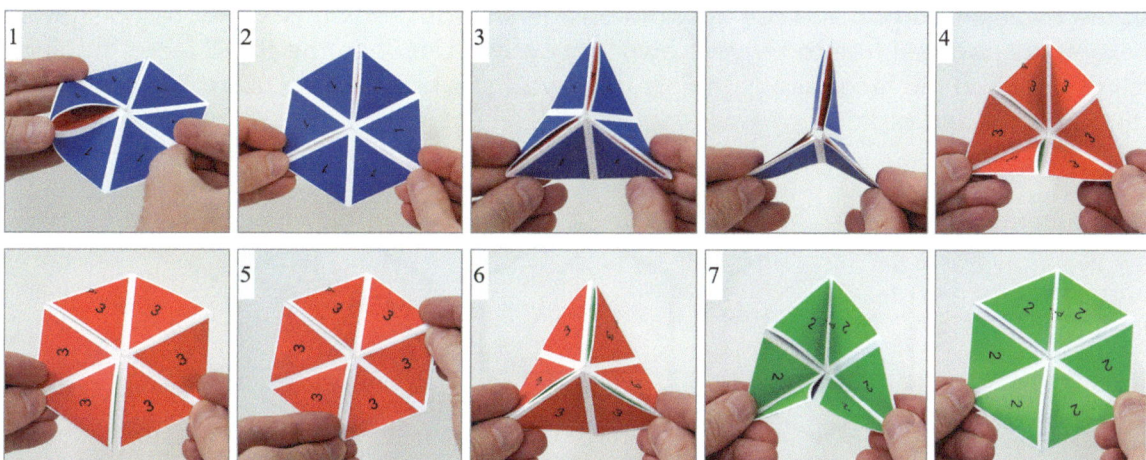

FIG. 1.4 Pinch-flexing the tri-hexaflexagon. In the pictures, face 1 is colored blue, face 2 is green, and face 3 is red.

If you repeat steps 5–7, you will find yourself back where you started, with face 1 in front and face 2 in back. You can then travel between the three faces forever. With practice, you can make the entire series of flexes very smooth. A sequence of flexes that starts and ends with the same arrangement is called a **cycle**.

If somewhere along the way you mistakenly position your hands at the wrong hinges, the flexagon won't open from the center. To rectify this, flatten your flexagon again, move your hands one hinge to the right or left, and try again.

Take a moment to examine your tri-hexaflexagon. Its overall shape is hexagonal. The entire arrangement of all the leaves is called a **state**. Each face is made up of six triangles. Notice how the thickness of the flexagon alternates as you traverse the triangles with your fingers. Three triangles have two layers of paper, which alternate with triangles that only have one layer. Each triangular stack of leaves is called a **pat**. Using flexagon terms, we say that there are six pats that alternate between one leaf and two leaves thick. There are three adjacent pairs of pats with the same structure, rotated around the flexagon. Equivalent, rotationally symmetrical sets of pats on a face are called **sectors**. If we closely examine the tri-hexaflexagon, we can see it has three sectors, where each sector has two pats with a total of three leaves in each sector. Figure 1.5 illustrates these terms.

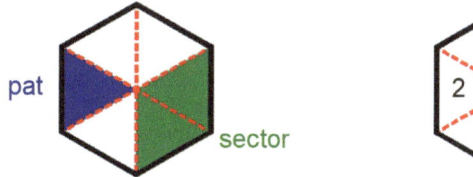

FIG. 1.5 A pat and a sector (left) and the number of leaves per pat (right) of a tri-hexaflexagon.

Sector is a useful concept as we use the pinch flex on flexagons with more faces. In these cases, the number of leaves in a sector is equal to the number of faces of the flexagon. For example, the tri-hexaflexagon has three leaves per sector and a total of three faces.

Now that you have a working flexagon and have learned some useful terms, take some time to experiment with how the leaves rotate and change position as you flex. A fun way to do this is to draw on the tri-hexaflexagon and see how the design changes as you flex. Figure 1.6 shows several simple examples, but feel free to try your own. Try cycling between the three faces, turning it over, and cycling between the faces again. In Part Five, we'll see how we can use this behavior to create interesting designs and puzzles.

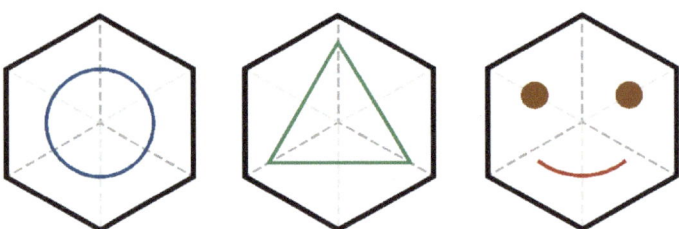

FIG. 1.6 Simple patterns to draw on a hexaflexagon. How do they change as you perform flexes? The darker dashed lines indicate hinges with thumbholes, folds you can slip your finger into. The lighter dashed lines show hinges that are solid paper.

We have created our first flexagon, a tri-hexaflexagon with one hidden face, and practiced flexing it with the pinch flex. We also covered some terminology for basic parts of a flexagon.

Chapter 2

General Tips

Folding templates

This book contains a large number of flexagon templates that all follow a consistent style. In this chapter, we'll explain how to fold the templates so we don't need to include instructions next to every template.

The templates are available online in the Flexagon Templates app at https://loki3.github.io/flex/templates.html. At the top of the web page is a dropdown list where you can choose the template you want by the figure number in this book. For example, the tri-hexaflexagon we saw in the first chapter is under figure 1.2.

The basic steps to make a flexagon from a template are as follows:

- Make a copy of the template onto a piece of paper and cut it out.
- Pre-crease along the dotted lines to prepare the hinges.
- Copy each of the small numbers to the back surface of the corresponding leaf.
- Find the highest numbers and fold adjacent pairs of those numbers face to face.
- Repeat until only 1's and 2's are visible.
- Tape together the two edges of the final hinge, marked with stars.
- Fold back and forth across all the hinges to prepare it for flexing.

Figure 2.1 shows the meanings of the various pieces of a template. Most templates have an additional small number or piece of text on them to distinguish them from other templates. For example, figure 1.2 has a small 3 to indicate that it has three faces.

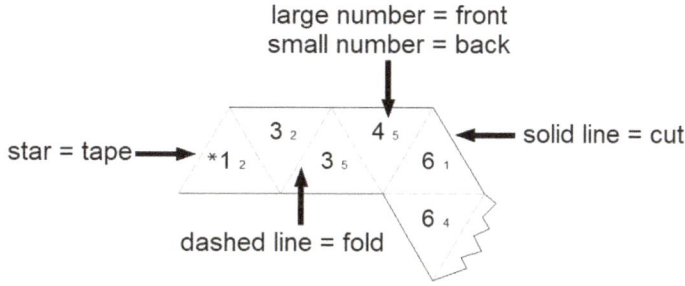

FIG. 2.1 Meanings of template symbols.

DOI: 10.1201/9781003433538-3

Here are some additional details on each of those steps.

Copy: Print or copy the template.

Cut: Cut out along the outside edge of the template using scissors or a craft knife. While your cutting doesn't need to be perfect, it's best to trim off a shape a bit too much rather than too little, since excess paper can interfere with flexing.

   The solid lines in the template indicate where you should cut the paper, while the dashed lines indicate hinges where you should fold. Notice, however, that the first and last leaves in the template have dashed lines that don't join to another leaf. These edges are marked with a star (*) and are the ones that you will tape together after folding everything, creating the final hinge in the completed flexagon. Therefore, you should cut along those edges when cutting out your flexagon. See figure 2.1 as a reference.

Pre-crease: Once you have cut out the template, fold along every dotted line – creasing both ways – then unfold it. Pre-creasing all the hinges makes the flexagon easier to fold and flex.

Copy numbers: Both faces of every leaf will have a label. Since these illustrations only show one side of the template, both labels are shown on the same side. The large number is the label for the front face of the leaf. Copy the small number to the back face of the leaf. For example, if a leaf says 1 $_2$, this indicates that the label for the front side is 1 and the label for the other side is 2. Copy the 2 onto the back of the leaf. Repeat for all leaves. It's useful to underline 6's and 9's so you can tell them apart. With some templates, you will end up attaching the front of one leaf to the back of another, so you may wish to copy the stars to the back face before folding to make it easier to find the edges that need to be taped.

Fold and repeat: Now find the highest numbers that are visible. Fold together all the adjacent pairs of this highest number. Keep repeating with the next highest numbers until the only visible numbers are 1 and 2. If the pats are thick, using a paper clip to temporarily hold together the stack of triangles can be useful to help keep things together.

Tape: Tape together the first and last edges, which should each be marked with a star (*). Sometimes it may be easier to add tape to an edge before you fold it up, especially if the leaf ends up in the middle of a pat. Another approach is to put a paper clip on that leaf.

Pre-crease all hinges: Fold back and forth along each hinge to help prepare it for flexing. You may find it useful to repeat this after the first few flexes until all the hinges bend easily.

Flexing

There are lots of different ways to flex various flexagons. This book will cover some flexes, but there are many others so feel free to experiment. It's ok if you sometimes get lost or end up with a tangle. Take the tape off or cut a hinge, fold it back together, retape, and try again. Draw on your flexagon and see how it changes as you flex. Note how the leaves change as you flex. Play around with them and see what you find.

   The numbers on the templates often serve two different purposes: to show how to fold the template into a flexagon and to help you track how everything changes as you flex. Sometimes the same labels can elegantly serve both purposes. Other times, it may be important to individually label every leaf in order to see exactly how the leaves move around.

   The text uses a slash (/) to separate the top and bottom numbers on a leaf or pat. For example, 1/2 indicates that 1 is on the front and 2 is on the back.

   It's a good idea to practice each new flex several times until you're comfortable with it.

Double-sided color

Most of the templates that appear in this book are designed to be printed out in black and white on a single side of a piece of paper, which is also the default view in the Flexagon Templates app. This is the simplest way to print a template and works on almost any printer with minimal effort.

However, it can be nicer to have all the leaves colored. Besides looking more interesting, you may find it easier to tell the difference between colors versus just numbers. This is even more important once we get to flexes that mix everything up. While the book usually refers to the numbers on the leaves, many of the illustrations also include color as extra information for keeping track of how things are changing. The illustrations in the book follow a consistent number/color scheme – 1-blue, 2-green, 3-red, 4-purple, 5-yellow, 6-brown, and 7-gray – as shown in figure 2.2.

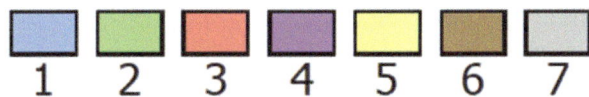

FIG. 2.2 Number/color scheme used in the book.

The most basic method for making colored templates is to color them yourself using crayons, colored pencils, markers, paint, etc. If you use ink or paint, you may need to be careful about the color bleeding through the paper. You may wish to use a slightly thicker paper.

Or, if you have access to a color printer that can print on both sides of the paper, you can use the Flexagon Templates app to generate the two sides of the template in color. If you look at the app again with figure 1.2 selected, you should see three options for showing faces: *front & back*, just *front*, and just *back*. The default option, *front & back*, shows the front and back numbers in black and white. If you instead pick *front*, it will show the numbers and colors for just the front face. Likewise, picking *back* shows the numbers and colors for just the back face, which has been flipped horizontally (top to bottom). On many computer systems, you can right-click on each image to save it as an image file. You can then use software such as Adobe Acrobat Reader, Microsoft Word or PowerPoint, or Google Docs to put each image on a separate page of a document. Make sure each image is centered on its page and sized appropriately. When you print your document, find the option for printing on both sides of the page. With the proper alignment, you should be able to print out a double-sided color template.

Summary

The templates used throughout this book are explained in detail in this chapter. We provide practical advice for how best to construct flexagons from these templates.

Chapter 3

More Faces

While the tri-hexaflexagon offers the surprise of a single hidden face, it's even more surprising that you can create flexagons with many more hidden faces. In this chapter, we'll explore hexaflexagons with four, five, and six faces, and learn how to make flexagons with any number of faces.

Four and five faces

First up is the **tetra-hexaflexagon**, with the tetra prefix indicating that the hexaflexagon has four faces. Figure 3.1 shows its template.

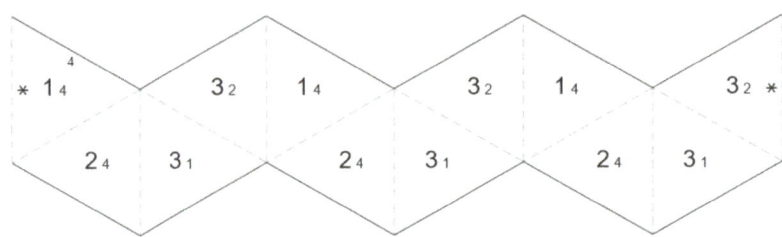

FIG. 3.1 Template for the tetra-hexaflexagon.

After cutting out the template, follow the general strategy of pre-creasing it by folding along all the dashed lines then unfolding them again. To prepare for taping together the first and last leaves, you might find it easiest to first put tape along the starred edge of the last leaf, with half of the tape hanging off waiting to attach to the starred edge of the first leaf. Fold together the adjacent pairs of leaves with 4's on them, then the adjacent pairs of 3's. Finish by attaching the tape to the first leaf. You should now have all 1's on one face and all 2's on the opposite face.

Now that you've made your flexagon, you'll notice one key difference from the tri-hexaflexagon: there are two different places where you can pinch-flex from face 1. Pinch-flexing one set of three hinges will lead you to face 3, while pinch-flexing the other set of hinges will lead you to face 4, as illustrated in figure 3.2. Each leads to a different cycle of three states as you continue to pinch-flex. As you explore the flexagons in this chapter, you will find that some faces support two pinch flexes while others only support one.

DOI: 10.1201/9781003433538-4

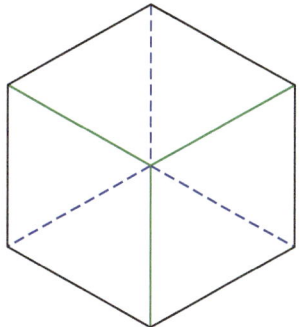

FIG. 3.2 Two different pinch flexes from the same face. Pinch the solid green hinges to reach face 3, or pinch the dashed blue hinges to reach face 4.

Figure 3.3 shows the template for the **penta-hexaflexagon**, with the penta prefix indicating that it has five faces. Note that you will end up attaching the front of the first leaf to the back of the final leaf (or vice versa), so you may wish to copy one of the asterisks to the back face to help you line up the edges that need to be taped. See if you can find all five faces as you pinch-flex.

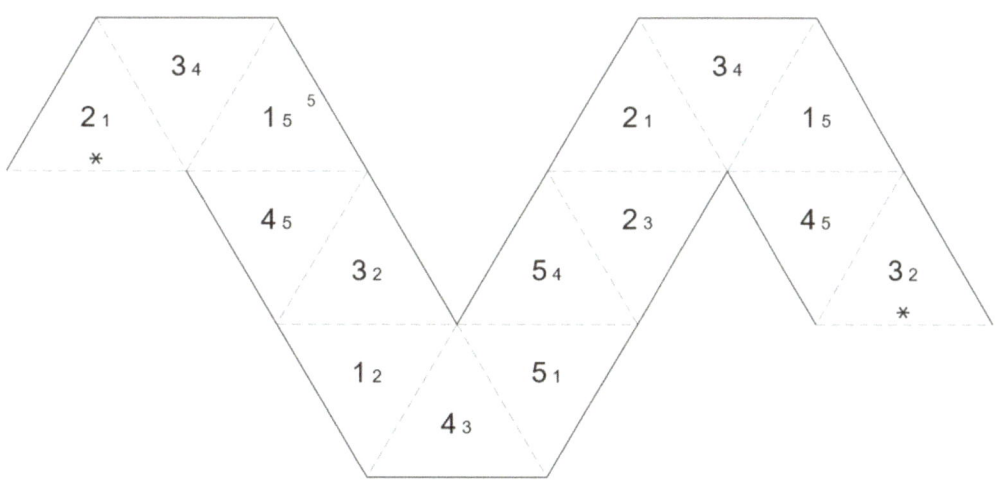

FIG. 3.3 Template for the penta-hexaflexagon.

Six faces

While the tri-hexaflexagon is interesting, it's the straight-strip **hexa-hexaflexagon** that really gets people hooked. It seems so simple and yet has really surprising behavior, and it can be tricky to find all six faces. While each face of the tri-hexaflexagon appears one third of the time, you'll find that three faces of this hexa-hexaflexagon appear more often than the other three. See figure 3.4 for the template.

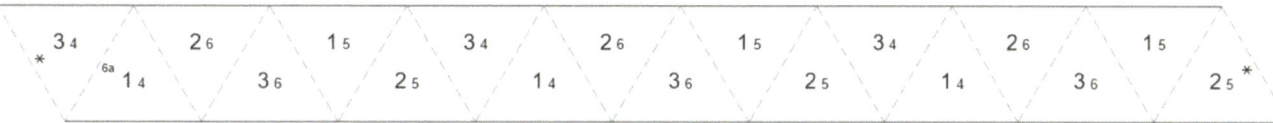

FIG. 3.4 Template for the hexa-hexaflexagon, variation 6a.

While there's only a single template for the three-, four-, and five-faced hexaflexagons, there are three different templates that fold into a hexa-hexaflexagon, one of which we just saw. Each has its own distinct properties and is worth exploring. As you make each one, see if you can figure out how to find all the faces and what pattern they make as you flex between them. We'll refer to the straight template in figure 3.4 as variation 6a. See figure 3.5 for variation 6b and figure 3.6 for variation 6c.

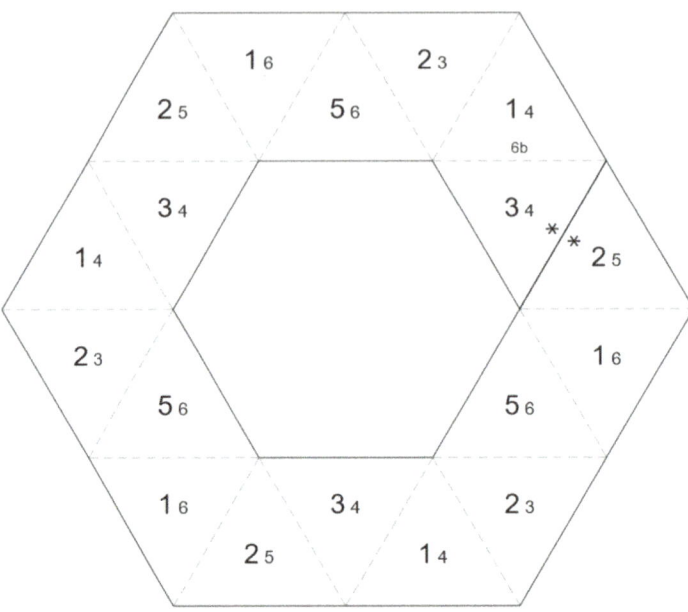

FIG. 3.5 Template for the hexa-hexaflexagon, variation 6b. To cut out the center section, you can either cut along the solid line between the *'s or fold the hexagon in half.

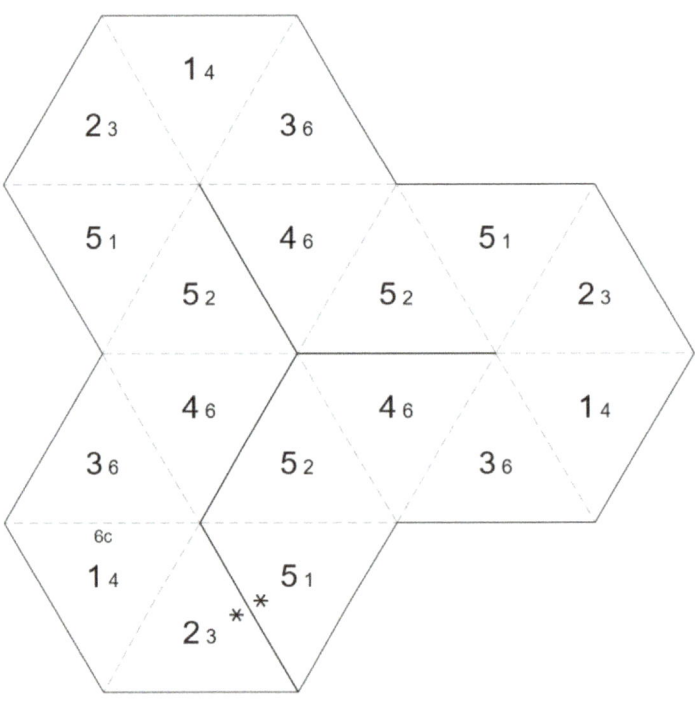

FIG. 3.6 Template for the hexa-hexaflexagon, variation 6c.

Tuckerman Traverse

Back in 1939, Bryant Tuckerman of the Flexagon Committee discovered a general technique to find every face of a flexagon with the least number of pinch flexes, and his flexagon friends dubbed it the **Tuckerman Traverse**. You may wish to see if you can figure it out for yourself before reading the next paragraph.

Recall the instructions for cycling through all the faces of the tri-hexaflexagon: after finishing one pinch flex, rotate both hands by one hinge and pinch-flex again. With the Tuckerman Traverse, you instead first try to do a pinch flex at the same set of hinges. If that doesn't work, shift your hands to the adjacent hinges and pinch-flex there. If you keep following this recipe, you'll visit every face, eventually arriving back where you started.

As you explore the flexagons in this chapter using the Tuckerman Traverse, you may notice that some faces and combinations appear more often than others. What patterns do you notice?

Doubling and reducing templates

Consider again the straight hexa-hexaflexagon template in figure 3.4. When you fold it, you start by folding 6 on 6, then 5 on 5, and then 4 on 4. This essentially "winds up" the template, leaving you with the tri-hexaflexagon template.

We can repeat this process of winding to make even longer templates. Figure 3.7 shows a portion of the template for a straight **dodeca-hexaflexagon**, where dodeca indicates that it has 12 faces. Make three copies of the template and tape the last leaf of the first copy (labeled 6/12) to the first leaf of the second copy (labeled 4/7), then tape the last leaf of the second copy to the first leaf of the third copy. You should now have a long strip of paper with 36 leaves. As you fold up this template, notice that after folding 12 on 12 through 7 on 7, you should have a shorter template that looks like the straight hexa-hexaflexagon template. And, as noted earlier, after folding 6 on 6 through 4 on 4, you have the tri-hexaflexagon template.

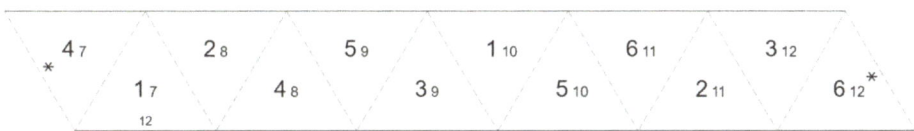

FIG. 3.7 One third of a straight dodeca-hexaflexagon template.

Figure 3.8 doubles the template again, giving you 24 faces. We call this the **icositetra-hexaflexagon** (icositetra means 24). As with the previous template, make three copies and tape them together in a long strip. Make sure that the 12 at the end of one piece is taped to the 7 at the start of the next piece. The strip will be easier to fold and flex if it's at least a couple centimeters wide. Expect that you will end up with a very long strip. You may wish to use paper clips to keep the leaves in the pats neatly stacked as you fold it up.

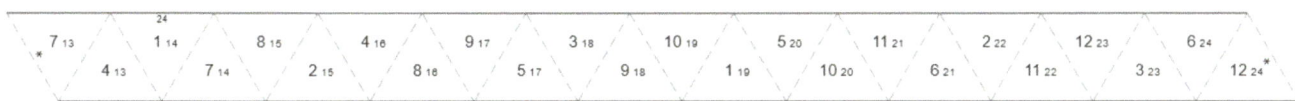

FIG. 3.8 One third of a straight icositetra-hexaflexagon template.

In theory, you can keep doubling the template to double the number of faces. In practice, however, paper models quickly get a lot harder to fold and flex.

Doubling the templates lets us make flexagons with an arbitrarily large number of faces. But we can also reduce the number of faces. To figure out what a shorter template with fewer faces looks like, simply paste together the faces you don't need.

To see how this works, refer again to the template for the straight hexa-hexaflexagon in figure 3.4. If you paste together all of the adjacent 6's, you end up with the template for the penta-hexaflexagon in figure 3.3. If you also paste together all the adjacent 5's, you get the template for the tetra-hexaflexagon in figure 3.1. You can do this with any template to reduce the number of faces.

It should be noted that the numbers or labels assigned to faces are somewhat arbitrary. You may assign them in the order you encounter the faces while flexing or label the template aesthetically. The important part is the pattern of the labels on the template, which can communicate both how to fold the template and where you are when flexing.

Conrad and Hartline (1962, p. 70) computed how many different hexaflexagon templates there are for a given number of faces, as shown in table 3.1.

TABLE 3.1 The number of different templates based on the number of faces in a hexaflexagon.

number of faces	number of different templates
3	1
4	1
5	1
6	3
7	4
8	12
9	27
10	82
11	228
12	733
13	2,282
14	7,528
15	24,834
16	83,898
17	285,357
18	983,244

This series of numbers – 1, 1, 1, 3, 4, 12, 27, 82, … – is sequence A000207 in the On-Line Encyclopedia of Integer Sequences: https://oeis.org/A000207.

Summary

In this chapter, we first made flexagons with additional hidden faces. Then we learned two techniques for creating flexagons with any number of faces: doubling the straight templates to increase the number of faces and pasting together leaves to reduce the number of faces.

Chapter 4

Triangle Tetraflexagon and Octaflexagon

Take another look at the template for the tetra-hexaflexagon in figure 3.1, paying close attention to the number labels and shapes. You may notice that the exact same pattern repeats three times, with the numbers 1/4, 2/4, 3/1, and 3/2 appearing on the same pattern of triangles. If you look at the various templates for the hexa-hexaflexagons in figures 3.4, 3.5, and 3.6, you should likewise see that the same pattern repeats three times in each template. Similarly, we made the full templates from figures 3.7 and 3.8 by taping together three copies of a shorter template.

Now take a look at the tri-hexaflexagon template in figure 1.2. It may not be as obvious, but it also repeats the same pattern three times. The first and last three leaves are labeled 1/2, 3/2, and 3/1. The three leaves in the center may look different because they're labeled 2/1, 2/3, and 1/3, but this is the exact same pattern, except that the front and back numbers are swapped. (Note that flipping the strip over to its back face is the same as swapping the number labels, so 1/2 turns into 2/1.) If you look at the penta-hexaflexagon in figure 3.3, you will see that it behaves in a similar fashion: 2/1, 3/4, 1/5, 4/5, 3/2, followed by the inverted version 1/2, 4/3, 5/1, 5/4, 2/3, followed by the original sequence again.

A natural question to ask is *what happens if you repeat the pattern a different number of times?* We've already seen that repeating a pattern three times gives us a flat hexagon. This is because each copy of the pattern adds two pats, each angle of a regular triangle is 60°, and therefore we end up with $3 \times 2 \times 60° = 360°$, a full circle. Repeating the pattern fewer times won't make a complete circle, while repeating it more times will give us more than a circle. This means that the resulting flexagons won't be flat.

Two repeats

First, let's try repeating the pattern just twice. Because the resulting flexagon has four pats instead of six, we call this a **tetraflexagon**. And since the pats don't make a full circle, the flexagon folds into a shape that looks like a pyramid. Figure 4.1 contains the template for the **tri-tetraflexagon** (*tri* indicates three faces and *tetra* indicates four pats). Flexing this flexagon is a slightly different maneuver than the pinch flex on a hexaflexagon. This flex is called the **pocket flex**, which is basically the tetraflexagon version of the pinch flex. We'll later see that the pocket flex can be used on other flexagons as well.

DOI: 10.1201/9781003433538-5

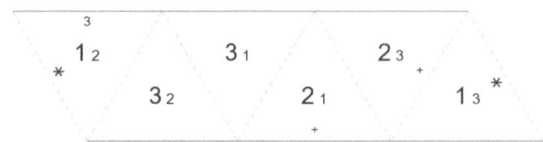

FIG. 4.1 Template for the tri-tetraflexagon.

Fold the template into a pyramid with 1's on the outside and 2's on the inside before taping the first and last leaves together. Follow along with figure 4.2 to flex it to a position with 3's on the inside:

1. Start with 1's on the outside of the pyramid and 2's on the inside.
2. Flatten the pyramid by folding pairs of 2's face-to-face, including folding together the 2's with plus signs (+'s) on them.
3. Close the pyramid to the left and open it again from the right.
4. The inside should now be 3's. The outside is still all 1's.

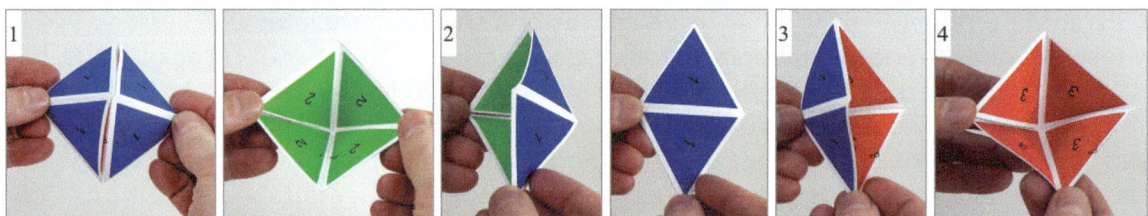

FIG. 4.2 Flexing the tri-tetraflexagon

Practice flexing back and forth between having 2's on the inside and having 3's on the inside. Notice how there are two possible ways to flatten the pyramid – flatten left to right or top to bottom. Step 2 told you to fold the two leaves with plus signs together, but you could have instead folded each plus sign against the adjacent leaf without a plus sign. If you had done that however, it wouldn't have opened back up.

But in this next tetraflexagon, sometimes both ways of flattening the pyramid will open up. This time, we take the pattern from the penta-hexaflexagon and repeat it twice, giving us the **penta-tetraflexagon**.

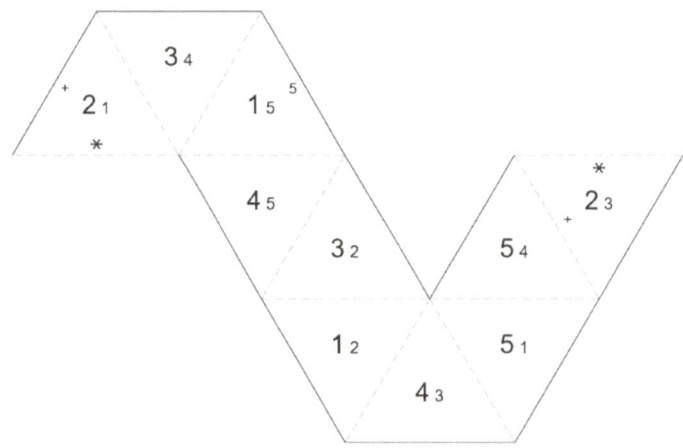

FIG. 4.3 Template for the penta-tetraflexagon.

Fold the penta-tetraflexagon template in figure 4.3 into a pyramid with 1's on the outside and 2's on the inside. Pocket flexes change the inside of the pyramid from 2's to 3's to 4's to 5's, but 1's will always be on the outside. Follow along with figure 4.4 to do this series of flexes:

1. Flatten the pyramid by folding pairs of 2's face-to-face, including the 2's with plus signs on them.
2. Close from the left and open from the right to reveal the 3's.
3. Flatten the pyramid by folding the inside top and bottom faces together.
4. Close from the front and open from the back to reveal the 4's.
5. Flatten the pyramid by folding the inside left and right faces together.
6. Close from the right and open from the left to reveal the 5's.

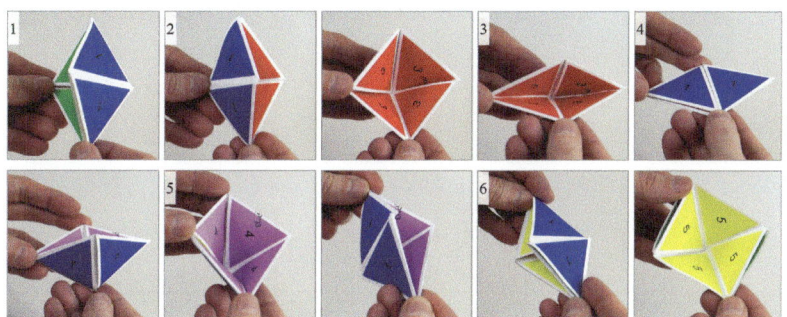

FIG. 4.4 Flexing the penta-tetraflexagon.

Four repeats

Next, let's try repeating the underlying pattern of numbers four times, which creates a **tri-octaflexagon**. We call this an **octaflexagon** because there are eight pats around the center. Even though it has too many regular triangles around the center for it to lie flat, you can still pinch-flex it. The trick is to alternate mountain folds and valley folds so every other hinge is a mountain pointing toward you, while the remaining hinges are valleys pointing away from you.

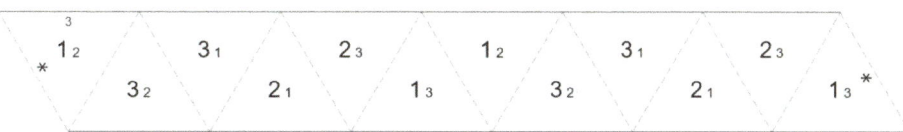

FIG. 4.5 Template for the tri-octaflexagon. Note that it's very similar to the tri-hexaflexagon, but repeats the underlying pattern four times rather than three times.

Fold the tri-octaflexagon template from figure 4.5, then follow along with the pictures in figure 4.6 to pinch-flex it:

1. Start with face 1 in front and face 2 in back. Find the hinge with the stars (*) on both sides of it. Make that hinge into a valley and the adjacent hinges mountains, and continue alternating between mountains and valleys all the way around the flexagon.
2. Pinch the mountains together.
3. Open up the flexagon from the center to reveal face 3.
4. This should naturally give you mountains and valleys that you can pinch-flex again, revealing face 2.

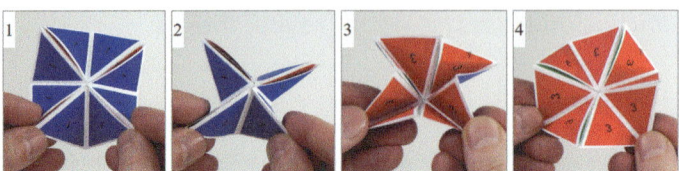

FIG. 4.6 Pinch-flexing the octaflexagon.

Cycling through the three faces of the tri-octaflexagon feels fairly natural because the mountain and valley folds stay the same as you continue to pinch-flex. However, once you have additional faces, you need to occasionally turn all the mountains into valleys and the valleys into mountains in order to pinch-flex to all the faces. Let's try this on the **penta-octaflexagon**.

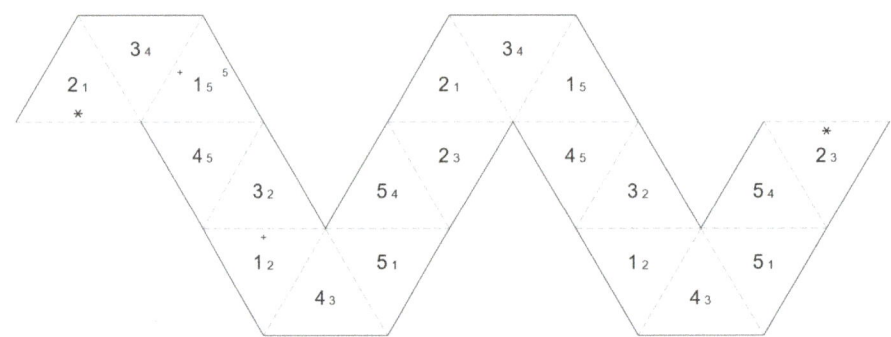

FIG. 4.7 Template for the penta-octaflexagon.

Use the template in figure 4.7 to fold the penta-octaflexagon. Follow along with the pictures in figure 4.8 to find all five faces:

1. Start from the state with 1's on the front and 2's on the back. Find the hinge with the +'s on either side. Make that hinge into a mountain and the hinges on either side valleys, and continue alternating between mountains and valleys all around the flexagon.
2. Open from the center to reveal face 3.
3. Take the valley fold by your right hand and invert it into a mountain fold.
4. Continue around the flexagon, inverting each mountain into a valley and each valley into a mountain.
5. Pinch the mountain folds.
6. Open from the center to reveal face 4.

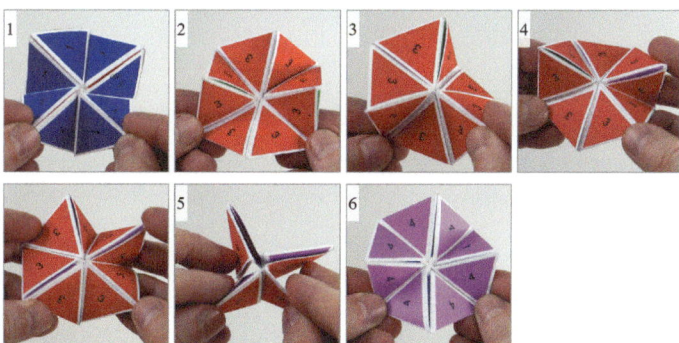

FIG. 4.8 Pinch-flexing to face 4.

Recall that when we do the Tuckerman Traverse on a hexaflexagon, we attempt to do a second pinch flex at the same hinge as the previous pinch flex. If that doesn't work, we shift by one hinge and pinch-flex there.

With the octaflexagon, the way we do a second pinch flex at the same hinge is by reversing the mountains and valleys as we did in steps 3 and 4 just now. As with the Tuckerman Traverse on the hexaflexagon, sometimes we will be able to pinch-flex after we do this and sometimes we won't. If we can't, we need to switch the mountains and valleys back to their original form and try again at the adjacent hinge. On the penta-octaflexagon, following this recipe should eventually lead you to face 5. The entire cycle leads you through every face.

It's also important to note that there are various ways to fold the octaflexagon flat, but you won't see all eight leaves of the face in those cases.

Summary

So far, we've tried three copies of certain leaf patterns in the hexaflexagon, two copies in the tetra-flexagon, and four copies in the octaflexagon. What if we try more than four copies? The pinch flex and mountain-valley inversion both continue to work, though the flexagons get more difficult to handle.

But what if we try just a single copy of one of the patterns? It seems to be completely immobile. Is there any way we can get it to flex? Does trimming off parts of the triangle help? We'll return to this question later in the book.

A key lesson is this: if the sum of the angles of the leaves around the center is 360°, the flexagon will lie flat; if less, it will form a cup; if more, it will make mountains and valleys. We will explore this further in the next chapter.

Chapter 5

Different Triangles

All our flexagons so far have used regular triangles where all the sides and angles are the same. What happens if we try different triangles?

Bronze hexaflexagon

For our first experiment, we'll use a triangle where the angles are 30°, 60°, and 90°. In origami, this is sometimes called a **bronze triangle**, so we call flexagons made from them **bronze flexagons**. Figure 5.1 shows a bronze hexaflexagon in its flat position.

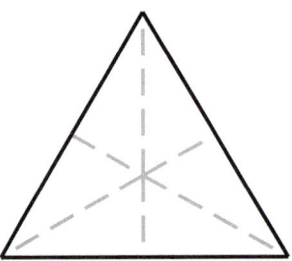

FIG. 5.1 A bronze hexaflexagon in its flat position.

As we did with the regular hexaflexagon, we'll start with three faces. Figure 5.2 shows the template for the bronze tri-hexaflexagon.

DOI: 10.1201/9781003433538-6

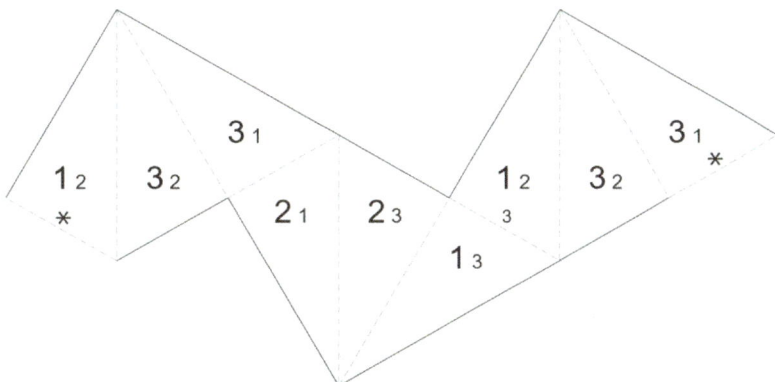

FIG. 5.2 Template for a bronze tri-hexaflexagon.

We'll start with a few observations:

- The leaf numbers follow the same pattern as the regular tri-hexaflexagon: 1/2, 3/2, 3/1, 2/1, 2/3, 1/3, 1/2, 3/2, 3/1.
- Unlike the tri-hexaflexagon template, this template isn't straight. However, as we'll see later, the way the triangles are attached to each other still follows the same pattern.
- Each bronze triangle is attached along two of its edges to a mirror image of itself, with the first and last leaves attached to each other after folding. They need to be mirror images because when we fold and flex pats together, adjacent pats align with each other.

Since the leaves are attached to mirror images, when all six triangles meet in the center of the flexagon, the leaf angles will all be the same, all either 30°, 60°, or 90°. Similar to what we saw in the previous chapter, this means that we may end up with a cup (6 × 30° = 180°), flat (6 × 60° = 360°), or mountain-valley (6 × 90° = 720°) form. Indeed, with the bronze hexaflexagon, we get all three positions from the same flexagon.

Follow along with figure 5.3 to pinch-flex the bronze tri-hexaflexagon:

1. From the flat 1/2 state, with face 1 in the front and face 2 in the back, pinch the corners of the triangle together.
2. Open from the center to reveal face 3 to finish the first pinch flex.
3. Notice that state 3/1, with 3's on the front and 1's on the back, isn't flat. It has three long legs sticking down, with the hinges alternating between mountain folds and valley folds.
4. Shift your hands to a pair of mountain folds and pinch each mountain.
5. Open from the center to reveal face 2 to finish the second pinch flex. This is a cup with 2's on the inside and 3's on the outside. From here, reverse your previous moves to return to state 1/2.
6. Turn it over so face 2 is on top with face 1 in the back.
7. Pinch together the shorter hinges that are in the middle of the triangle's edges rather than the longer ones that go to the corners as you did in step 1.
8. Open from the center to reveal face 3. You now have a cup that looks like step 5, except that the faces have been swapped, with 3's on the inside and 2's on the outside.

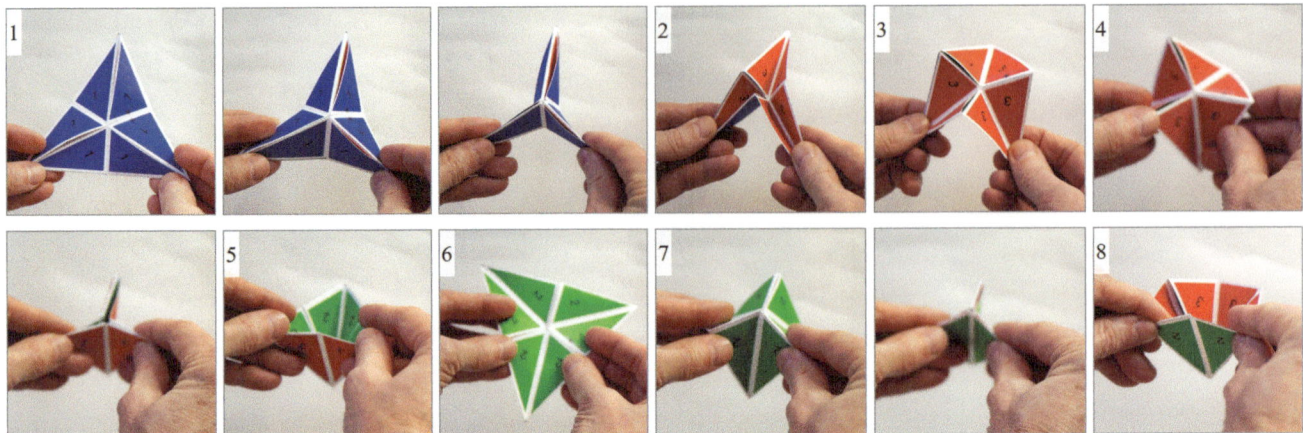

FIG. 5.3 Pinch-flexing the bronze hexaflexagon from a corner or an edge.

Earlier we noted that the regular tri-hexaflexagon template is straight but the bronze tri-hexaflexagon template is not. To see that they follow the same underlying pattern despite this difference, we'll label all the edges in the two templates and compare. On the template for the regular tri-hexaflexagon, label the edge of the first leaf where it connects to the second leaf with an *a*. Going clockwise around the first leaf, label the next two edges *b* and *c*. Label the edges on the second leaf such that the labels align with the labels on the first leaf when you fold them together. Continue in this manner labeling all the leaves as shown in figure 5.4. If you look at the labels for edges where two triangles are connected, you'll get the series *aa, cc, bb, aa, cc, bb, aa, cc,* and *bb*, repeating the pattern *acb* three times. If you fold up the template, you'll find that edges with the same labels are always stacked together at the same hinge – for example, the *a*'s are always aligned with each other. You may find it useful to circle the letters to make them stand out more.

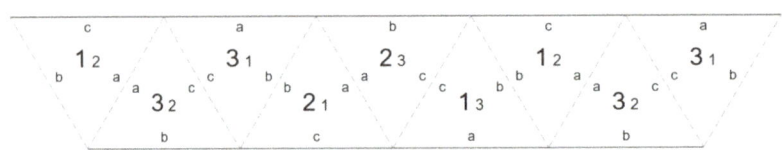

FIG. 5.4 The regular tri-hexaflexagon template with the edges labeled.

Now let's apply this same pattern to the bronze tri-hexaflexagon template, as shown in figure 5.5. Start with the leftmost triangle 1/2, label the edges *a*, *b*, and *c*, and connect a mirror image of the leaf across edge *a*, labeled 3/2, to match figure 5.4. Connect a mirror image of leaf 3/2 across edge *c* to create leaf 3/1. Then connect a mirror image of leaf 3/1 across edge *b*. Continue repeating the same *acb* pattern for every new leaf. The resulting template follows the same pattern as the regular tri-hexaflexagon template, but with a different triangular shape for the leaves. This is how we derived this template in the first place and why it supports the same pinch flex.

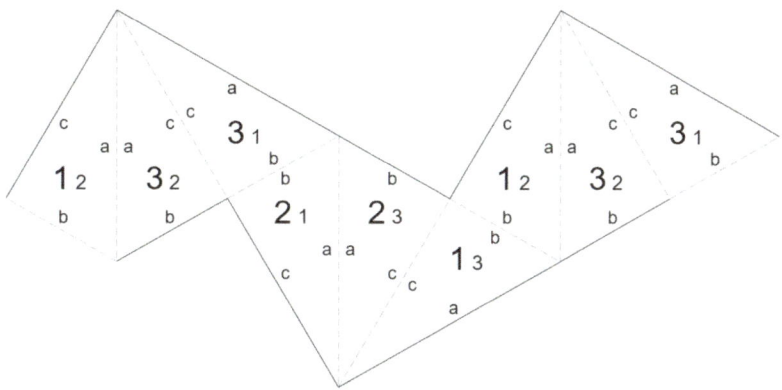

FIG. 5.5 The bronze tri-hexaflexagon template with the edges labeled.

One interesting difference, however, is that every edge of the bronze triangle is a different length. This means that, unlike the regular hexaflexagon, you may get a different template depending on which of the three edges you start reflecting over. Let's see this on the bronze tetra-hexaflexagon, with four faces. When we label the leaf edges of the regular tetra-hexaflexagon from figure 3.1 and apply its pattern to the bronze triangle, we end up with three different templates depending on which of the three edges we start mirroring over. The three templates are shown in figures 5.6, 5.7, and 5.8, which we'll refer to as 4a, 4b, and 4c, respectively. When folded to state 1/2, each of the templates has a different initial shape. The first template folds to the mountain-valley shape, the second one folds flat, and the third folds to a cup.

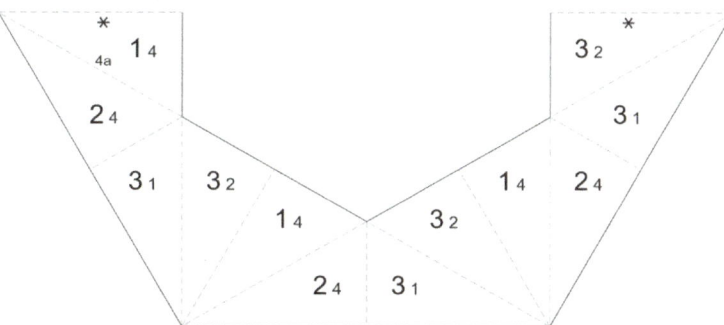

FIG. 5.6 Template for the bronze tetra-hexaflexagon that starts by mirroring over the long edge (variant 4a). State 1/2 is mountain-valley.

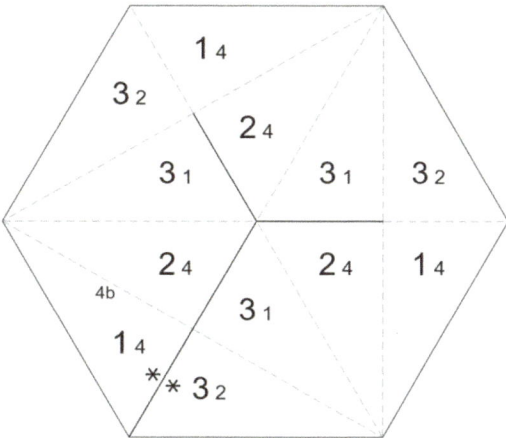

FIG. 5.7 Template for the bronze tetra-hexaflexagon that starts by mirroring over the medium edge (variant 4b). State 1/2 is flat.

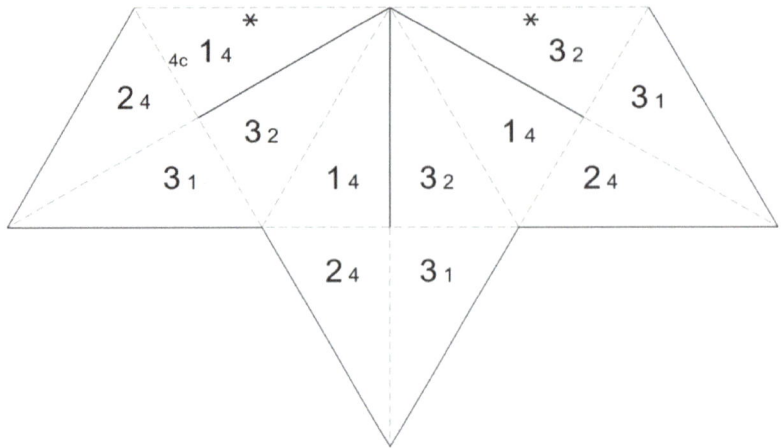

FIG. 5.8 Template for the bronze tetra-hexaflexagon that starts by mirroring over the short edge (variant 4c). State 1/2 is a cup.

Let's explore the 4a variant from figure 5.6. See figure 5.9 for illustrations of the different arrangements.

1. Start from state 1/2. There are actually two possible configurations for this state. If you see state 1/2a, with face 1 on the outside and the legs sticking down, it will flex to a cup with 4 on the inside and 1 on the outside. In this case, reverse the mountain and valley folds so it's in state 1/2b, with the legs sticking up.
2. From state 1/2b, with face 1 on the inside, pinch-flex to state 3/1, which is flat.
3. Pinch-flex from the shorter hinges at the center of the edges of the flexagon to get cup 2/3, with face 2 on the inside and face 3 on the outside.
4. Note that you can't flex again, so undo that flex and turn the flexagon over to state 1/3.
5. Pinch-flexing from the longer hinges at the corners of the triangle brings you to the non-flat state 2/1a.
6. If you pinch-flex from the mountain folds, it will take you to the cup 3/2. Undo that flex to get back to state 2/1a.
7. Reverse the mountain and valley folds to reach state 2/1b.
8. A pinch flex will give you state 4/2, which is flat.
9. Pinch-flex from the hinges at the center of the edges of the flexagon to get cup 1/4.

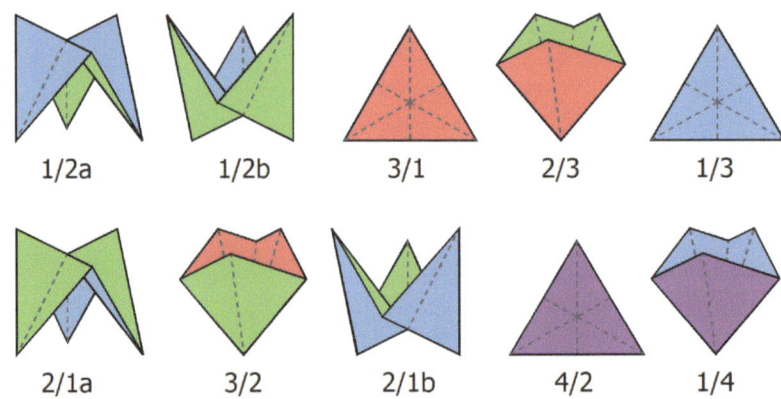

FIG. 5.9 Exploring the 4a version of the bronze tetra-hexaflexagon, with face 1 in blue, face 2 green, face 3 red, and face 4 purple.

The 4c variant from figure 5.8 gives us one additional case to consider. From the initial 1/2 cup with face 1 on the outside of the cup, you can pinch along one set of three hinges to get to the flat state 4/1, or pinch along the other three hinges to get to mountain-valley state 3/1.

What implication do the non-flat states have for the Tuckerman Traverse? Recall that there are two possible places to do a pinch flex on each face. And if you turn over the flexagon, the opposite face also has two possible places to do a pinch flex. However, sometimes a cup may be a dead end, so you need to reverse course, while other times you may be able to open up the cup from a different set of hinges. When the central angle is greater than 360°, you may be able to continue by inverting the mountains and valleys as we saw with the regular octaflexagon.

To experiment further with these options, try the template in figure 5.10 to make a bronze hexa-hexaflexagon. It can be tricky to find all six faces. Keep in mind that each face has two possible sets of hinges where you may be able to do a pinch flex. You may want to keep notes about which sets of faces you've visited and which sets of hinges you've tried.

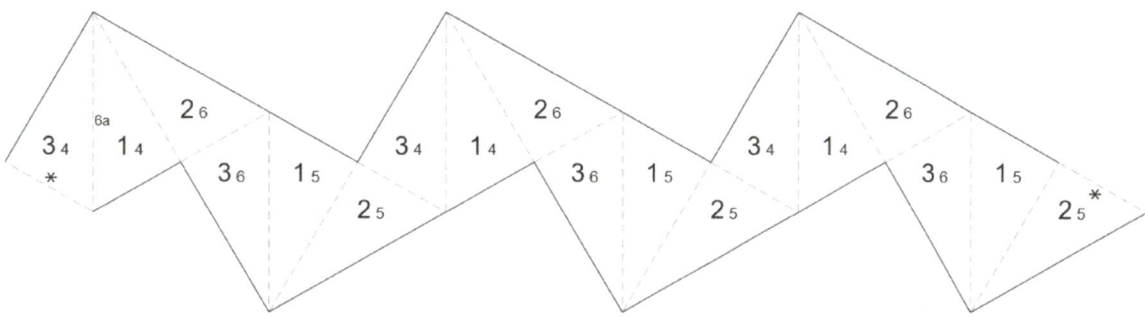

FIG. 5.10 Template for a bronze hexa-hexaflexagon.

Silver octaflexagon

Now let's try a triangle with angles of 45°, 45°, and 90°. In origami, this is sometimes called a **silver triangle**, so we call flexagons made from them **silver flexagons**. One way to arrange a collection of silver triangles into a square is by taking eight of them and putting the 45° angles in the center, which we'll try first. Figure 5.11 shows a silver octaflexagon in its flat position. Note that this position has four hinges at the corners, which we call **corner hinges**, and four hinges at the middle of the edges, which we call **edge hinges**.

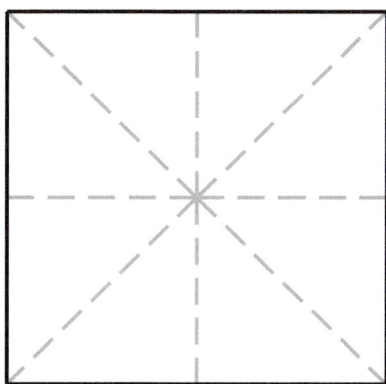

FIG. 5.11 A silver octaflexagon in its flat position.

Figure 5.12 shows the template for the silver tri-octaflexagon. Since two of the three angles of a silver triangle are 45°, two of the three states of the tri-octaflexagon are flat (8 × 45° = 360°), both states 1/2 and 3/1. The third state, 2/3, has 90° angles in the center, so it's a mountain-valley shape (8 × 90° = 720°).

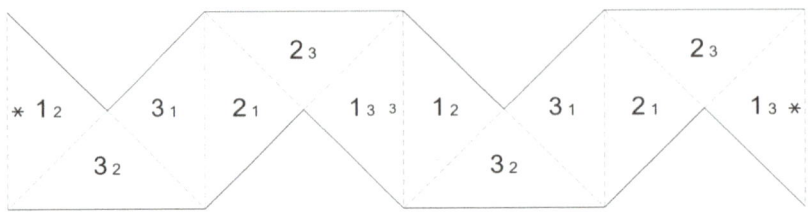

FIG. 5.12 Template for the silver tri-octaflexagon.

One nice attribute of the silver octaflexagon is that you can sometimes pinch from one flat position to another, unlike the bronze hexaflexagon. For example, starting in flat state 1/2, you can pinch-flex the edge hinges to get to state 3/1, which is also flat. However, pinch-flexing from 3/1 will lead you to a mountain-valley state, 2/3. From there, you can pinch-flex back to the flat 1/2 state.

Follow along with the pictures in figure 5.13 to do a series of pinch flexes on the silver tri-octaflexagon. The sequence of two pinch flexes in steps 3–6 is sometimes called the **pass-through flex** because it travels from flat state to flat state, passing through the unwieldy mountain-valley state.

1. Start in state 1/2 with your hands at opposite edge hinges. Pinch at the edge hinges, bringing the four corners of the square together in the back.
2. Open from the middle to reveal face 3, with face 1 in the back.
3. Shift your hands to the corner hinges and pinch, bringing the four edge hinges together in the back.
4. Open from the middle to reveal face 2. This is a mountain-valley state.
5. Bring the four lower spikes together in the back.
6. Open from the center to return to state 1/2.

FIG. 5.13 A sequence of three pinch flexes on the silver tri-octaflexagon.

The silver octaflexagon can be fun because there are lots of different ways to flex it. For example, figure 5.14 shows an alternate way to flex from 1/2 to 3/1.

1. Start in state 1/2 with your hands on opposite corner hinges. Fold the top corner down until the flexagon folds flat, with 2's on the outside.
2. Open the left half of the flexagon from the center front, pulling the flap down to reveal four pats with 3's on the front.
3. Open the right half of the flexagon from the center back, pulling the flap down to reveal four pats with 2's on the front.
4. Fold the two pats in the upper left corner backward until they fold flat in the back.
5. Fold the two pats in the upper right corner forward until they fold flat in the front.
6. Open the flexagon from the top corner of the back, swinging it forward until you see face 3. Face 1 is on the back.

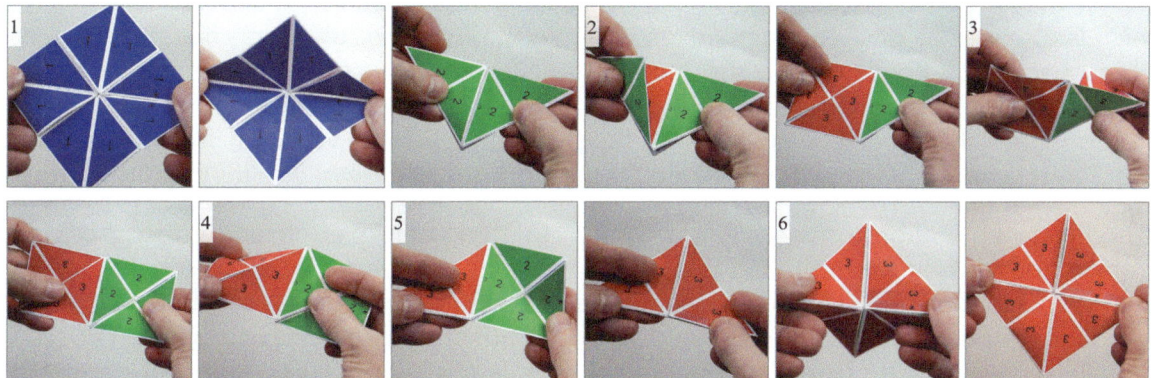

FIG. 5.14 An alternative to the pinch flex on a silver tri-octaflexagon.

For the tetra-octaflexagon, there are two possible templates because the edges of the leaves are two different lengths. See figures 5.15 and 5.16 for the two silver tetra-octaflexagon templates.

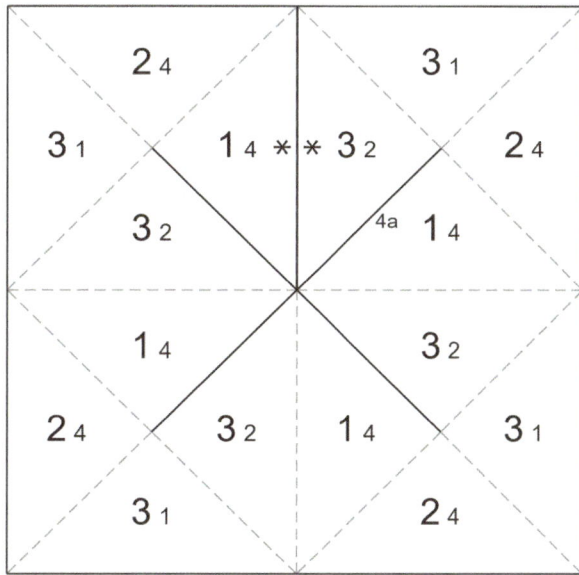

FIG. 5.15 Template for the silver tetra-octaflexagon, variant 4a, which starts by mirroring over a short edge.

FIG. 5.16 Template for the silver tetra-octaflexagon, variant 4b, which starts by mirroring over the long edge.

If you start from state 1/2 on variant 4a, you can pinch-flex either from the four short hinges that meet the middle of the square's edges, or from the four long hinges that meet the corners of the square. If you pinch-flex the short hinges, you get to 4/1, which is flat. If you pinch-flex the long hinges, you reach 3/1, which is mountain-valley.

The mountain-valley states on the silver tetra-octaflexagon are very unwieldy, so you may prefer to skip past these states, instead focusing on the flat states. Earlier, we showed the pass-through flex, which is one way to do two pinch flexes in a row to pass through a non-flat state. Another way to combine two pinch flexes to skip the mountain-valley state is called the ***reverse pass-through flex***, where we reverse the mountains and valleys in the middle of the flex.

To demonstrate the reverse pass-through flex, we'll use the 4b variation from figure 5.16. Start with face 2 in the front and face 1 in the back. Refer to figure 5.17 while following these steps:

1. Position your hands at opposite corners and pinch, bringing all four edge hinges together in the back.
2. Carefully open from the center so that you can see all eight leaves labeled with 3's. You have just pinch-flexed to state 3/2, which consists of jagged mountains and deep valleys.
3. The goal of the next few steps is to turn the four mountains (at edge hinges) into valleys and the four valleys (at corner hinges) into mountains. Start with the valley by your right hand, flattening it and the mountain immediately above it.
4. With your left hand, push forward on the pat that was on the left half of the mountain, which will also move the pat to its left. In addition to flattening another valley, this should reverse the hinge that was previously a mountain, turning it into a valley.
5. In the next few steps, we'll continue to work our way around the flexagon to the left. Push the next pair of pats backward.
6. Pull the next pair of pats to the right.
7. Push the next pair of pats back.
8. Reverse the last remaining mountain, turning it into a valley.
9. Open from the center to reveal face 4.

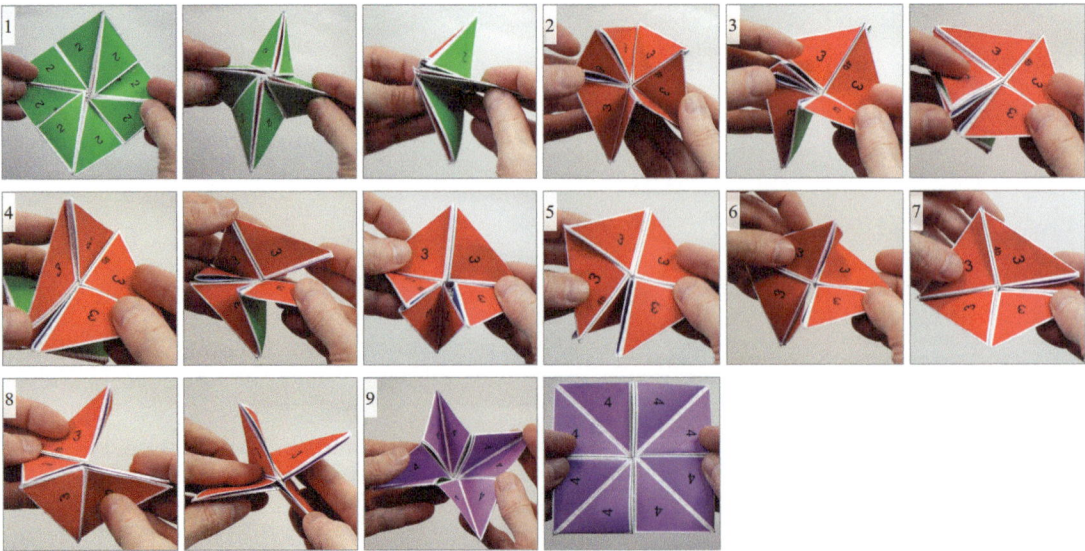

FIG. 5.17 The reverse pass-through flex on a silver tetra-octaflexagon.

Silver tetraflexagon

Another elegant use of the silver triangle is to put four right angles in the center of a flexagon. Since $4 \times 90° = 360°$, we can make a flat silver tetraflexagon as shown in figure 5.18. But when either of the 45° angles is in the center, we get a cup, since $4 \times 45° = 180°$.

FIG. 5.18 A silver tetraflexagon in its flat position.

Figure 5.19 contains a template for a silver penta-tetraflexagon (five faces, four pats per face) that can flex between two different flat states by traveling through multiple cups. Cut out the two pieces of the template and tape the two edges labeled *a* together. Note that the template will overlap itself, which is why it starts out in two pieces. When folded, the flexagon should be a square made up of four right triangles.

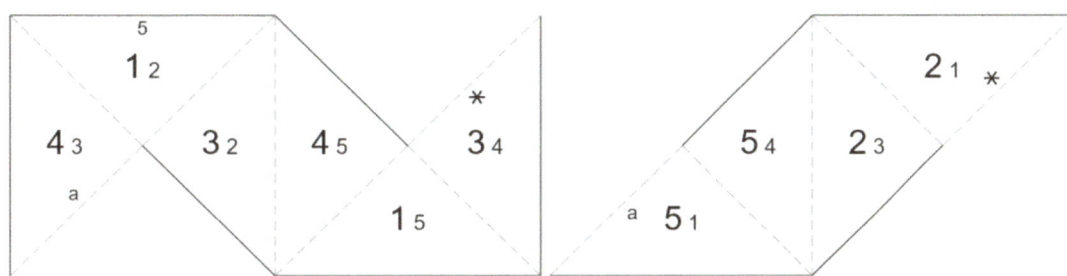

FIG. 5.19 Template for the silver penta-tetraflexagon. Tape together the edges labeled *a* before folding.

To flex from face 1 to face 5, refer to figure 5.20 while following these steps:

1. Start in state 1/2, with a single-leaf pat in the upper left corner and a pat with multiple leaves in the upper right corner. Fold the corner on the right backward until it folds against the opposite corner.
2. Open up the center of the flexagon. Continue to pull the two sides apart so that the top and bottom corners come together.
3. Open up the flexagon from the back corner. Continue to pull the two corners apart so that the two sides come together.
4. Open from the corner that makes a 90° angle to reveal face 5.

FIG. 5.20 Flexing the silver penta-tetraflexagon.

Arbitrary triangles

Our technique for applying a pattern from an existing template to create a new template works with any triangle we want, not just bronze and silver triangles. For example, figure 5.21 contains a template for a tri-hexaflexagon where the angles in the triangles are 40°, 65°, and 75°.

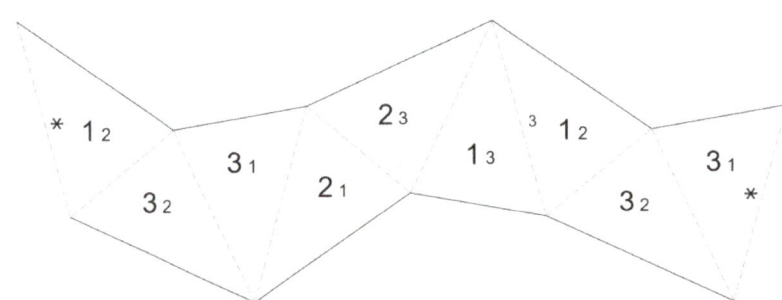

FIG. 5.21 Template for a tri-hexaflexagon made from 40°-65°-75° triangles.

Summary

Throughout this book, we'll see flexagons made from a wide variety of different polygons. Right angles and equal angles make for nice symmetry in a flexagon, so we'll generally choose right triangles, isosceles triangles, or regular polygons, but any set of angles can be used.

The techniques from the last few chapters can be combined to make a large variety of triangle flexagons: doubling the template, pasting together unused faces, repeating the template pattern a different number of times, and changing the shape of the leaves. We'll see many other examples in later chapters.

Chapter 6

Pinch Flex Variations

Up until now when we've done the pinch flex, we've pinched every other hinge all the way around the flexagon before opening it up. This means that we need to have an even number of pats on a face. It also means that all leaves on a face travel together as we flex.

What if we only pinch every third hinge? Or every fourth? Or skip hinges in some other pattern? This causes the faces to get mixed up as we flex. It also opens up the possibility of flexing a flexagon with an odd number of pats.

To name a specific pinch flex variation, we use a series of numbers to indicate how many hinges we step through to get to the next hinge we pinch, moving clockwise. For example, P345 indicates that we pinch a hinge, step three hinges and pinch that hinge, then step four hinges and pinch that hinge. Stepping five hinges returns to the first hinge. After pinching each of those hinges, we then open up the flexagon. The sum of the numbers in the name must match the number of pats in the flexagon, so the P345 flex requires 3 + 4 + 5 = 12 pats. Figure 6.1 shows examples of some pinch flex variations.

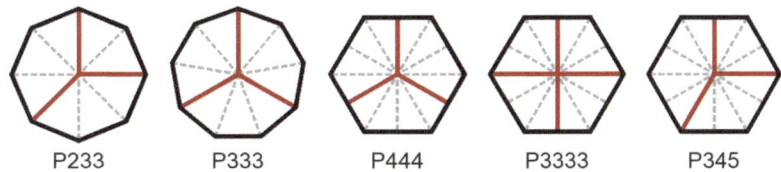

FIG. 6.1 Examples of sets of hinges you can pinch-flex on different flexagons.

We can also apply this naming convention to the pinch itself if we want to be explicit. Since we pinch every second hinge, we use a series of 2's that depends on how many pats the flexagon has. For example, on a hexaflexagon, the pinch is P222 (2 + 2 + 2 = 6), while on the octaflexagon, the pinch is P2222 (2 + 2 + 2 + 2 = 8).

P333

Our first example is P333, pinching every third hinge on an **enneaflexagon** (ennea means nine, so nine pats per face). See figure 6.2 for the simplest flexagon template that supports two P333 flexes. We call the simplest flexagon that supports a given sequence the **minimal flexagon** for that sequence.

DOI: 10.1201/9781003433538-7

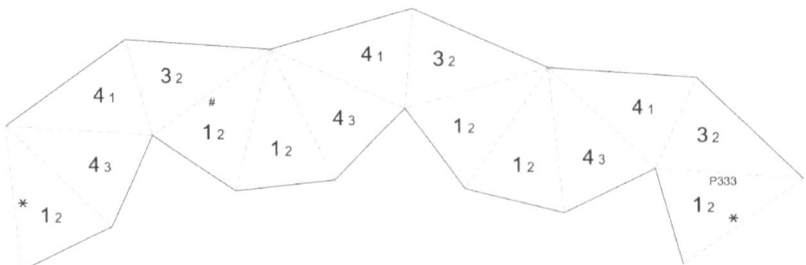

FIG. 6.2 Template for the minimal enneaflexagon for two P333 flexes.

Follow along with figure 6.3 to do two P333 flexes in a row. After folding the template from figure 6.2, start from flat state 1/2.

1. Place your left hand at the hinge marked with #. Place your right hand three hinges to the right.
2. With each hand, pinch the leaves together backward. The hinges will form a mountain pointed toward you, including a third hinge at the top.
3. Open the flexagon from the center to reveal a face with a mix of 3's and 2's and a hole in the middle. The face will be slightly cupped. This completes the first P333.
4. With one hand, snap half of the face toward you.
5. With the other hand, snap the remainder of the face toward you.
6. Pinch the hinges between adjacent pairs of 3's.
7. Open from the center, revealing a flat face with a mix of 4's and 1's.
8. To return to the original state, turn the flexagon over and repeat steps 1–7, where you start by pinching the hinges between adjacent pairs of 3's.

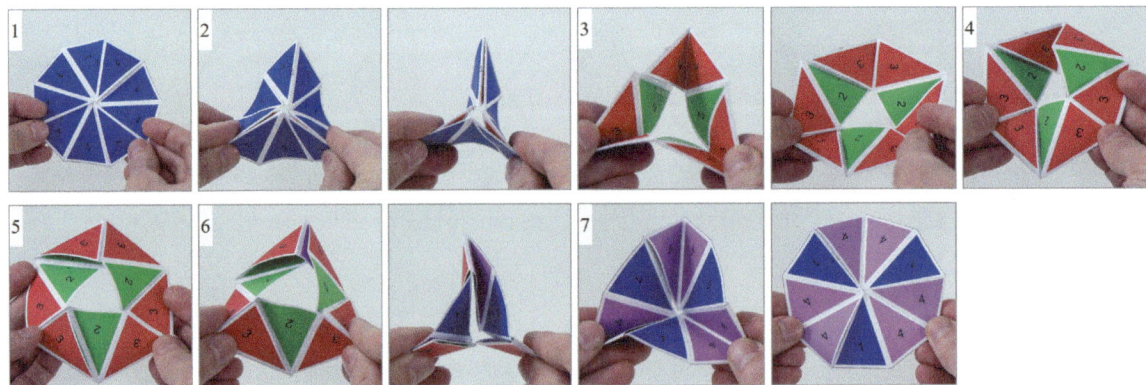

FIG. 6.3 Performing the P333 flex twice in a row on an enneaflexagon.

Note that this is very similar to doing two pinch flexes on a flexagon when it has leaves that aren't regular triangles. The first flex takes you to a state that isn't flat, and a second flex at the same hinges returns you to a flat state.

Next is an enneaflexagon that supports a larger number of P333 flexes. The template in figure 6.4 spirals on top of itself, so you need to assemble it from multiple pieces. After cutting out the three pieces, tape together the pair of edges labeled a, then tape together the pair of edges labeled b. After folding together adjacent 8's, 7's, 6's, 5's, 4's, and 3's, tape together the edges labeled c.

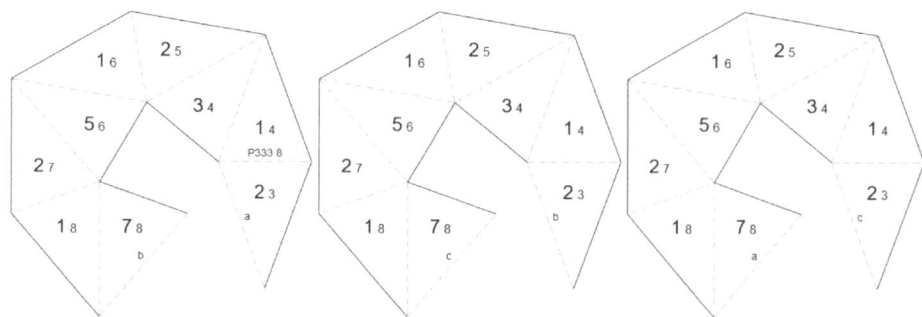

FIG. 6.4 Template for an enneaflexagon (leaves: 27) that supports multiple P333 flexes.

If you want a challenge, try exploring all the different arrangements of leaves reachable by applying P333 at various hinges from different faces. You may be able to find a state that has all odd numbers on one face (3, 5, and 7) and all even numbers on the opposite face (4, 6, and 8). Note that every flat state is followed by a cupped state, and that you can get two cupped states in a row.

Something you may notice from these enneaflexagons is that the concept of a **unified face**, where every leaf on a face shows the same label, no longer applies. This means there's no obvious way to say how many faces it has, so we'll instead use the number of leaves to differentiate the flexagon in figure 6.2, *enneaflexagon (leaves: 15)*, from the one in figure 6.4, *enneaflexagon (leaves: 27)*.

Recall that, with the pinch flexes we previously looked at, if a flexagon had N pats, then every label appeared N times in the template. In contrast, though these enneaflexagons have nine pats with nine 1's and nine 2's, the other numbers only appear six times each. This is because leaves travel in groups of six or three when using P333 instead of groups of nine. We have chosen to label the leaves in such a way that P333 reveals interesting patterns, but keep in mind that we could just as easily pick a different labeling.

The concept of a sector is also a bit less clear. When we previously looked at the pinch flex, we saw that each pair of adjacent pats around the flexagon had the same structure, and we called each pair of pats a sector. We could count the leaves in a sector to figure out how many faces the flexagon had. But after a single P333 on the enneaflexagon (leaves: 27), three pairs of adjacent pats changed in the same way, but the other three pats didn't change at all. In figure 6.3, the six purple pats have changed, but the three blue pats have stayed the same. If you shift one hinge and do a second P333 on this enneaflexagon, you'll change a different set of six pats.

P3333, P444, and P66

Next is a **dodecaflexagon** (12 pats per face) made from isosceles triangles, which supports several pinch flex variants. See figure 6.5 for the template. After cutting out the four pieces, tape the edges labeled *a* together, then the *b*'s, and finally the *c*'s. After folding, tape together the edges labeled *d*.

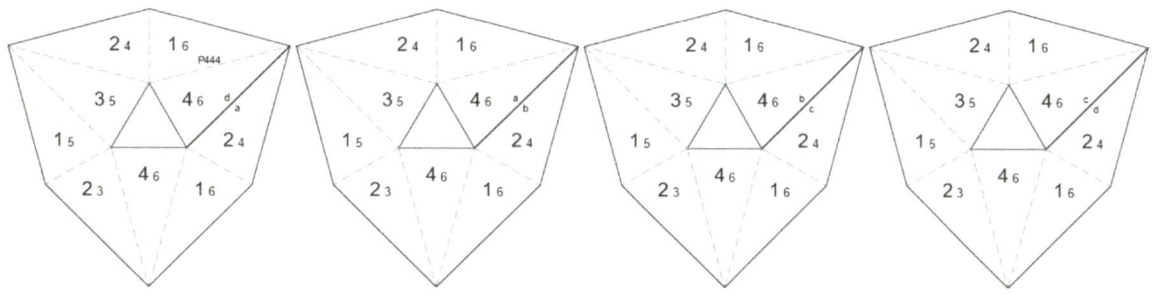

FIG. 6.5 Template for an isosceles dodecaflexagon that supports P3333, P444, and other pinch variants.

There are lots of possible pinch variants on this dodecaflexagon. For example, you could pinch every third hinge (P3333), every fourth hinge (P444), every sixth hinge (P66), or some other combination. You may need to be a bit more careful with these flexes, however, since this flexagon can get tangled.

P444

Next, we'll experiment on a bronze dodecaflexagon, where the leaves are 30°-60°-90° bronze triangles. With the pinch flex variations we've seen so far, after a single flex, the flexagon doesn't lie flat and there's a hole in the middle (as we saw in figure 6.3), then we flex a second time to return to a flat state. Due to the symmetries of the bronze dodecaflexagon, it lies flat after a single P444 flex.

However, while the result might be a flat hexagon in shape, the pats are arranged in a different configuration. From here, a second P444 takes you to a pinwheel shape. A third P444 brings you back to the original state. Figure 6.6 shows these arrangements.

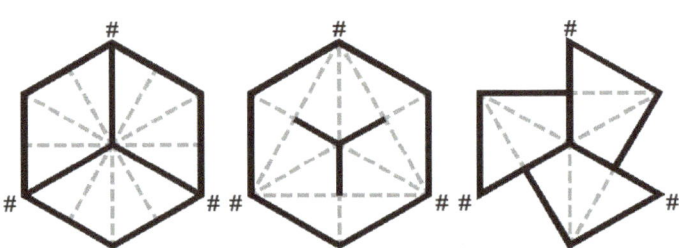

FIG. 6.6 Different arrangements of pats on a bronze dodecaflexagon using the P444 flex.

To try out these flexes, make a flexagon from the template in figure 6.7.

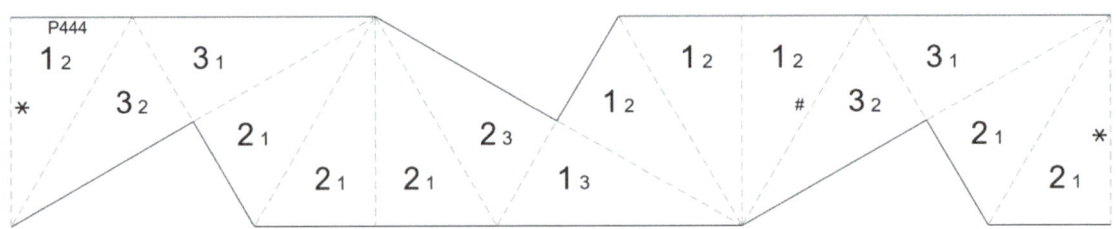

FIG. 6.7 Template for a bronze dodecaflexagon that supports the P444 flex.

Follow along with figure 6.8 to do a series of three P444 flexes. After folding the template from figure 6.7, start from flat state 1/2.

1. Place your left hand at the hinge marked with #. Place your right hand four hinges to the right.
2. With each hand, pinch the leaves together backward. The hinges will form a mountain pointed toward you. The hinge at the top should also form a mountain.
3. Open the flexagon from the center to reveal a face with a mix of 3's and 2's. This completes the first P444.

4. Shift your hands to the hinges between the adjacent pairs of 2's that form the corners of the triangle made of 2's, then pinch.
5. Open from the center, revealing face 2, which is shaped like a pinwheel. This completes the second P444.
6. Shift your hands to the pinwheel hinges and pinch.
7. Open from the center to reveal the original face.

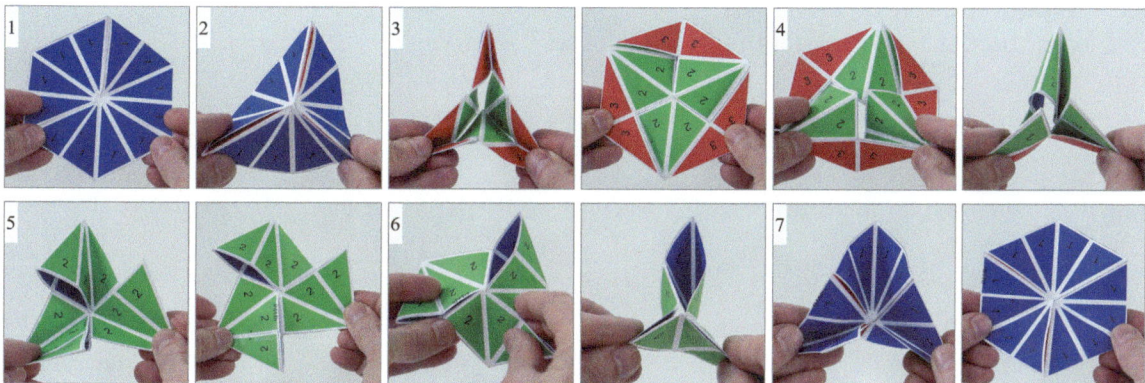

FIG. 6.8 Performing a series of P444 flexes on a bronze dodecaflexagon.

P223

The final pinch variant we'll look at is the P223 on a **heptaflexagon** (*hepta* means seven, so seven pats per face). This is a bit like a standard pinch flex on a hexaflexagon, except that two of the three states don't lie flat. The surface of the flexagon in figure 6.9 in one of its non-flat states is a Möbius strip, which means that it only has a single surface. You can see this by running your finger along the outside of the flexagon when you reach step 7.

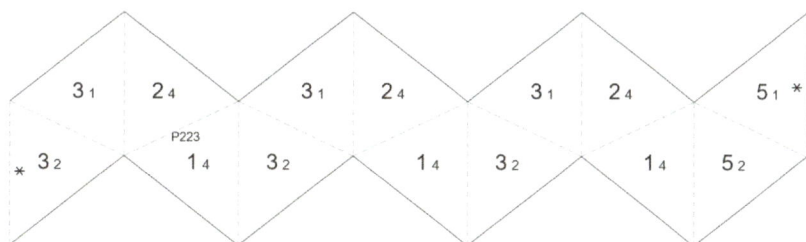

FIG. 6.9 Template for a heptaflexagon that supports the P223 flex.

Follow along with figure 6.10 to do a series of P223 flexes. This is a more difficult flex and requires some patience.

1. Position face 1 such that the hinge labeled P223 is at the bottom. Then place your hands at the two hinges on either side of the P223 hinge, as shown in the first photo.
2. With each hand, pinch the leaves together backward. The hinges will form a mountain pointed toward you, with a third mountain at the top.

3. Open the flexagon from the center to reveal a face with a mix of 3's and 4's. You will only see six leaves on this face, and there will be a tent sticking off the back of the flexagon.
4. Shift your hands one hinge to the right. Push up on the tent on the back face so that the tent shifts from the back face to the front face, revealing leaves with 2's on them.
5. With your left hand, pinch backward at the hinge between the adjacent pats marked 2 and 3, and with your right hand, pinch backward between the adjacent pats marked 3 and 4.
6. Open from the center, revealing face 2 with two pats sticking up off the face.
7. Shift your hands one hinge to the right.
8. Pinch the pats together.
9. Open from the center to reveal the original face.

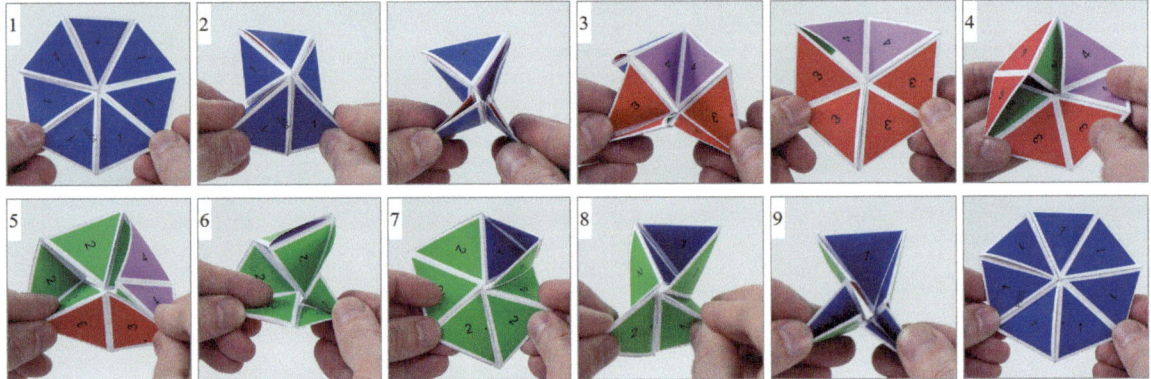

FIG. 6.10 Performing a series of P223 flexes on a heptaflexagon.

Note that you can position any of the seven hinges at the bottom when you start this flex. Each hinge leads to a different combination of numbers on the first face you reveal. See if you can find a solid face of 3's or 4's, or a face that's a mix of 3's, 4's, and 5's.

Summary

In this chapter, we generalized the idea of the pinch flex by skipping extra hinges. These pinch flex variations mix up the colors of the faces and can even change the shape of the flexagon. They also allow us to work with flexagons with an odd number of pats.

Chapter 7

The Book Flex

All the flexagons we've explored so far have been made from triangles, but other polygons can be used as well. To expand the possibilities, next we look at flexagons with faces made of four squares, so they're called **square tetraflexagons**. We'll manipulate them using the **book flex**, named because it feels like opening a book. The abbreviation for the book flex is B.

Square tetraflexagon

To make the minimal square tetraflexagon that supports a single book flex, use the template in figure 7.1. This folds into a square tri-tetraflexagon, where *square* refers to the shape of the leaves, *tri* means it has three faces, and *tetra* indicates it has four pats per face.

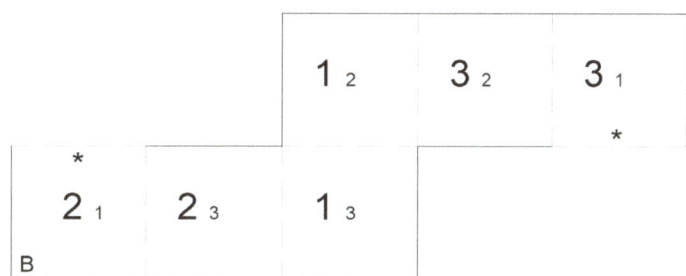

FIG. 7.1 Template for the square tri-tetraflexagon, the minimal square tetraflexagon that supports the book flex.

To perform a book flex on this flexagon, start with face 1 on the front and face 2 on the back. Make sure the 1's from the front of the template are right side up. Refer to the pictures in figure 7.2 while following these steps:

1. Fold the left and right edges backward till they meet.
2. Open up the center, revealing face 3.

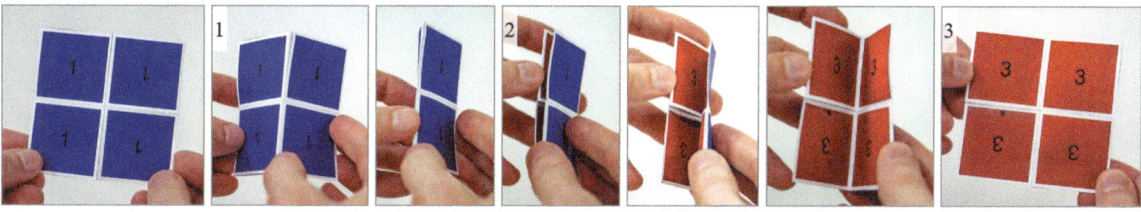

FIG. 7.2 The steps in performing a book flex.

DOI: 10.1201/9781003433538-8

The book flex is basically the pinch flex done on a square flexagon, so we will sometimes use *pinch flex* to cover this generalization. However, something you might notice as you try out the tri-tetraflexagon is that it doesn't cycle like the pinch flex on a tri-hexaflexagon. Once you book-flex to face 3, you can't do another book flex from that face. You instead have to turn the flexagon over and do a book flex on the opposite face in order to book-flex back to the original configuration.

We can describe this sequence of flexes as B^B^ using **flex notation**, a shorthand for describing multiple flexes. We'll learn more about flex notation in later chapters, so here we'll cover just enough so we can use it to understand the flexagons in this chapter. Here's what each of those symbols mean:

- **B:** Do a book flex. For the flexagon from figure 7.1, from the initial state, with face 1 on the front and face 2 on the back, do a book flex to a state with face 3 on the front and face 1 on the back.
- **^:** Turn over the flexagon. For the flexagon in this example, turn the flexagon over from left to right so face 1 is on the front and face 3 on the back.
- **B:** Then we do a book flex again. For the flexagon in this example, after the second book flex, you should have face 2 in front and face 1 on the back.
- **^:** Turn over the flexagon. For the flexagon in this example, turn over the flexagon so you're back at the initial state, with face 1 on the front and face 2 on the back.

Flexes are performed relative to a **reference hinge**, which helps us keep track of how to do multiple flexes. For the book flex, the reference hinge is the vertical hinge between the two pats at the top of the flexagon of the flexagon. When we turn it over, we turn it over left to right so that the reference hinge is still between the top two pats. In the sequence B^B^, this means that the flexagon is oriented properly after the first book flex so we can perform the second book flex. This notation lets us describe the behavior of a flexagon.

While the B^B^ flexagon requires us to turn the flexagon over in order to return to the original state, we can also create square flexagons that can cycle, returning to the starting point without having to turn it over. Use the template in figure 7.3 to make a square tetra-tetraflexagon that supports a book-flex cycle. After each book flex, shift the reference hinge one to the right before doing the next book flex. Doing this four times in a row brings you back to the initial state. This is represented by the sequence B>B>B>B>, where > indicates to shift the reference hinge one hinge to the right. Similarly, < indicates to shift one hinge to the left. You can shorten this representation by putting the repeated sequence in parentheses followed by the number of times to repeat: (B>)4.

- **B:** Do a book flex.
- **>:** Shift the reference hinge one to the right.
- **(B>)4:** Repeat B> four times.

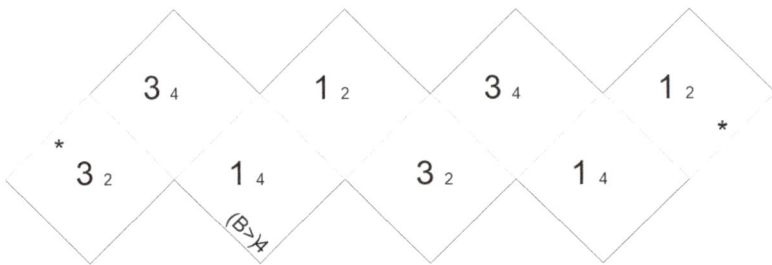

FIG. 7.3 Template for a square tetra-tetraflexagon supporting (B>)4.

If you make the square tetra-tetraflexagon from figure 7.4, you'll find that it supports two subtly different book flexes. When you're looking at face 1 with face 2 on the back, the book flex opens up hinges from the upper right and lower left to reveal face 4. If you now shift one hinge and do another book flex, the book flex instead opens hinges from the upper left and lower right. If you want to differentiate between these two variants, the one that opens from the upper right is called the **book-right flex**, Br, and the one that opens from the upper left is called the **book-left flex**, Bl. When the difference doesn't matter, we'll simply call them the book flex.

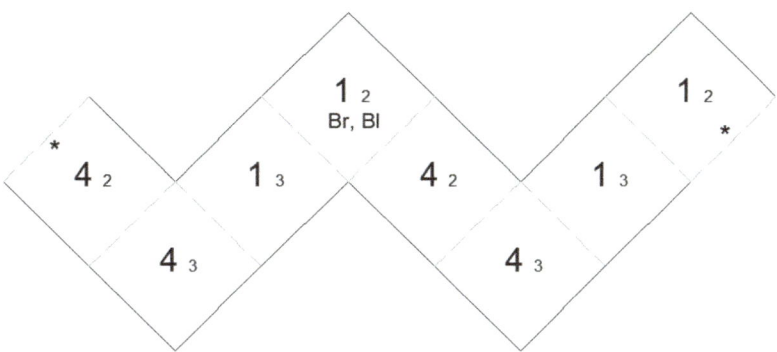

FIG. 7.4 Template for a square tetra-tetraflexagon that supports the book-right and book-left flexes.

Regardless of the flexagon, if you continue to do a book flex at the same hinge rather than shifting hinges, you eventually run into a dead end. Figure 7.5 contains a template for a square tetra-tetraflexagon that supports BB. Figure 7.6 contains a template for a square hexa-tetraflexagon that supports BBBB. For both of these flexagons, start flexing from face 2.

3_4	2_4	1_3	1_2
1_2	1_3	2_4	3_4

BB

FIG. 7.5 Template for a square tetra-tetraflexagon supporting BB.

5_6	4_6	3_5	2_4	1_3	1_2
1_2	1_3	2_4	3_5	4_6	5_6

BBBB

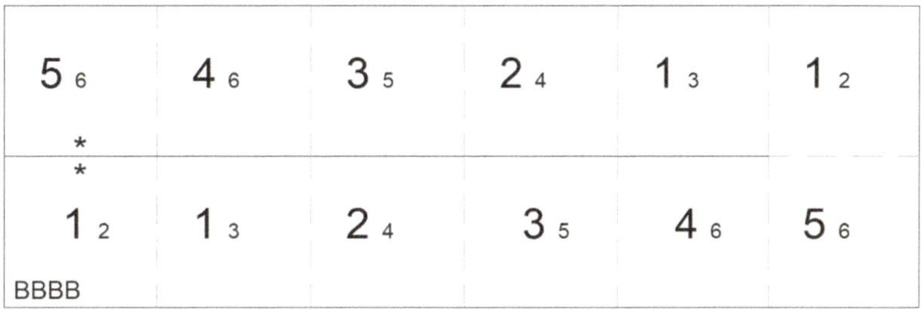

FIG. 7.6 Template for a square hexa-tetraflexagon supporting BBBB.

You can extend this pattern further, allowing you to keep opening up new faces at the same place. However, since this pattern folds the paper into a spiral, you quickly run into a practical limit due to the thickness of the paper.

Figure 7.7 contains the template for a square flexagon that supports the sequence (B^>)4. Start in state 1/2. Orienting the flexagon so that the numerals are sideways, do a book flex to get to state 3/1. Turn it over to get back to face 1, and rotate the flexagon 90° so you shift your hands one hinge to the right; this is the sequence B^>. Repeating B^> takes you through the series 4/1, 1/4, 5/1, 1/5, 6/1, and 1/6. To return to 1/2 from 1/6, position the numerals upright, and repeat the sequence (B^>)4.

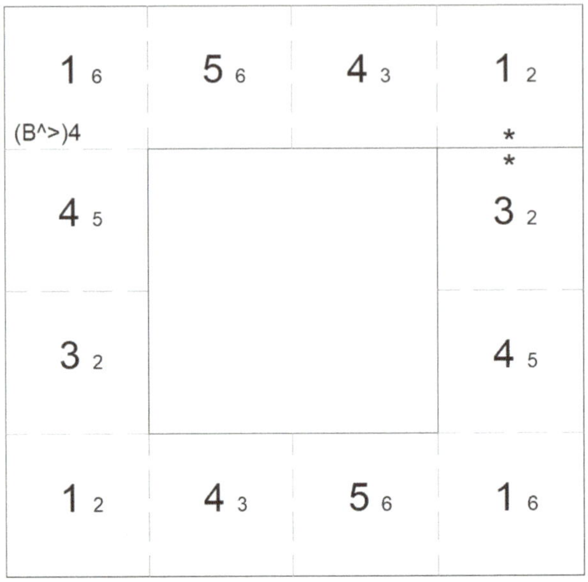

1 ₆	5 ₆	4 ₃	1 ₂
(B^>)4			*
4 ₅			3 ₂
3 ₂			4 ₅
1 ₂	4 ₃	5 ₆	1 ₆

FIG. 7.7 Template for a square hexa-tetraflexagon that supports (B^>)4. To remove the center, either cut along the solid line between the stars, or fold in half before cutting.

Pentagon tetraflexagon

Next, let's look at the book flex on a **pentagon flexagon**, a flexagon with leaves that are pentagons. Figure 7.8 contains the template for a regular pentagon tri-tetraflexagon – it has leaves that are regular pentagons, three faces (*tri*), and four leaves per face (*tetra*). Note that the regular pentagon tetraflexagon doesn't lie flat, since four 72° angles meet in the middle.

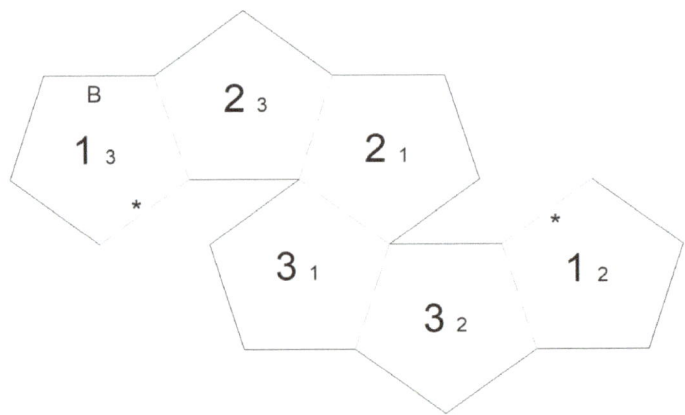

FIG. 7.8 Template for a pentagon tri-tetraflexagon. After folding it up, fold the flexagon in half to easily tape the starred hinge. This works for the following pentagon flexagons as well.

In order to do a book flex from face 1 in state 1/2, the hinge with the stars should make a valley shape, as should the hinge directly across from it. The other two hinges should make a mountain shape. Flatten the flexagon by bending along the mountain hinges, bringing the back leaf-faces together. Like the square tri-tetraflexagon, you can book-flex between two different states (1/2 to 3/1), but you can't do a cycle.

The shortest cycle of pinch flexes in a *triangle* flexagon is *three*, and the shortest cycle of book flexes in a *square* flexagon is *four*, so it should come as no surprise that the shortest cycle of book flexes in a *pentagon* flexagon is *five*. Figures 7.9 and 7.10 contain two different templates for pentagon penta-tetraflexagons that support the cycle (B>)5. Note that the second flexagon is less stable than the first one.

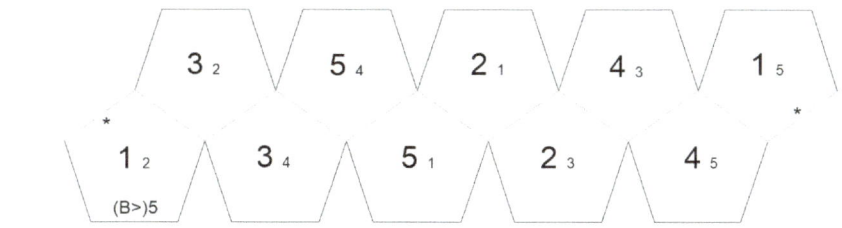

FIG. 7.9 Template for a pentagon penta-tetraflexagon.

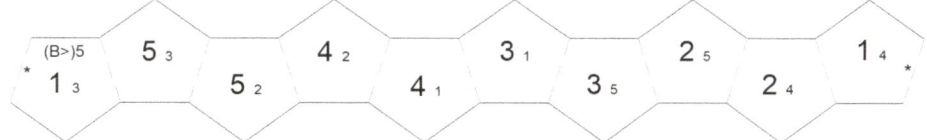

FIG. 7.10 Template for another pentagon penta-tetraflexagon.

With both these pentagon penta-tetraflexagons, start with the hinge marked with stars as a valley, with each of the four hinges alternating between mountains and valleys. From face 2 of state 2/1, do a book flex to reveal face 3. As you open the face, keep going until you've folded the flexagon in half the opposite direction. Then you can open the center to reveal face 4. Again, keep opening, allowing the leaves to pivot smoothly until it's folded in half again. Opening it reveals face 5. After a final pivot, opening it up shows face 1. And a final book flex takes you back to state 2/1.

When you make the next pentagon tetra-tetraflexagon in figure 7.11, there are two possible book flexes you can do from face 1 of state 1/2. One will take you to state 3/1 and the other will take you to state 4/1. To switch between the two options, you need to "snap" the flexagon, turning the mountain folds into valley folds and the valley folds into mountain folds.

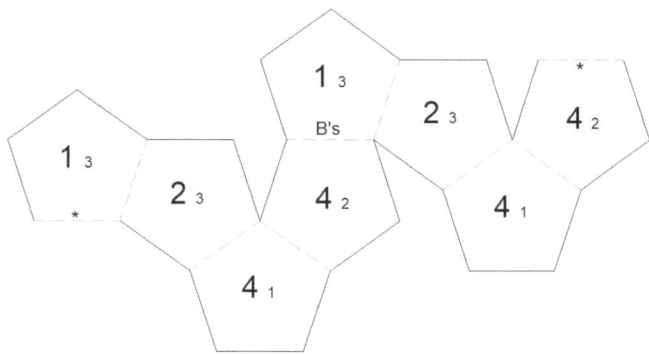

FIG. 7.11 Template for a pentagon tetra-tetraflexagon that supports two book flexes from state 1/2.

Hexagon tetraflexagon

Using patterns similar to the square and pentagon flexagons, we can make flexagons from polygons with even more sides, such as a **hexagon flexagon**, which has leaves that are hexagons. For example, figure 7.12 contains the template for a hexagon hexa-tetraflexagon – leaves that are hexagons, six faces (*hexa*), and four pats per face (*tetra*).

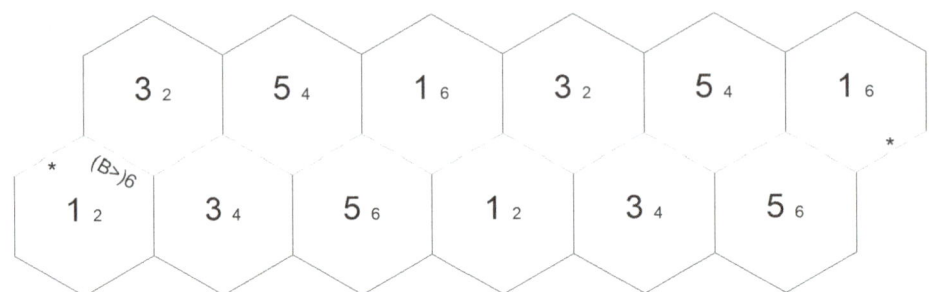

FIG. 7.12 Template for a hexagon hexa-tetraflexagon.

Pentagon hexaflexagon

As with triangle flexagons, you can create new templates by varying the angles of the leaves or by repeating a pattern a different number of times. Figure 7.13 contains the template for a pentagon penta-hexaflexagon where the angles of the leaves are 60°, 120°, 120°, 120°, and 120°. This flexagon has three elegant flat arrangements reachable with the book flex (aka pinch flex), as seen in figure 7.14.

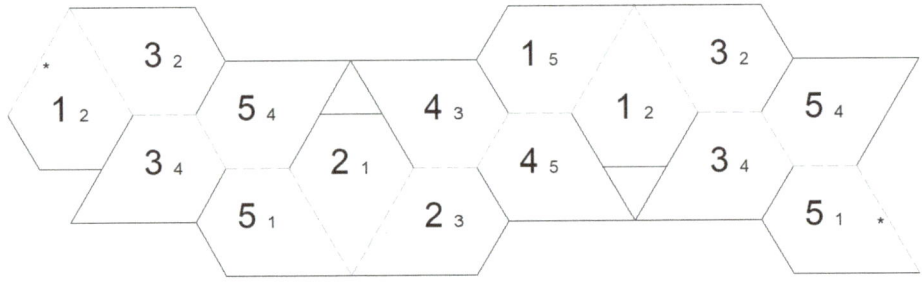

FIG. 7.13 Template for a pentagon penta-hexaflexagon.

FIG. 7.14 The three flat positions of the pentagon penta-hexaflexagon. The leftmost image shows 1/2, the middle shows 2/4, and the rightmost image is 3/5.

Summary

The book flex is the equivalent of the pinch flex when the leaves have more than three sides. We used it on flexagons made from squares, pentagons, and hexagons. Then we varied the angles of the leaves to make an interesting pentagon hexaflexagon that changes shape as you flex it.

Chapter 8

The Box Flex

The **box flex** is another easy flex on flexagons where the leaves are squares, pentagons, hexagons, etc. – in other words, polygons other than triangles. It's called a box flex because you open up the flexagon into a box shape as one of the intermediate steps.

Square

As we will see, there are two ways to open up the flexagon after you make a box shape, from the top or the bottom. These two variants require different internal structures, hence they have different minimal flexagons. We use Bt to indicate the **box-top flex** and Bb the **box-bottom flex**.

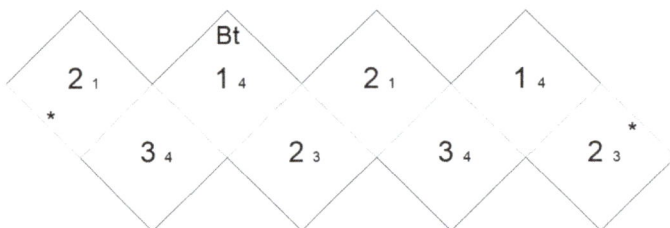

FIG. 8.1 Template for the minimal square tetra-tetraflexagon that supports Bt, the box-top flex.

Figure 8.1 contains a template for the minimal square tetraflexagon for the box-top flex. To perform the box-top flex on it, start with face 1 in the front, with face 2 on the back, and with Bt in the top right square. Refer to the pictures in figure 8.2 while following these steps:

1. With single-leaf pats in the upper left and lower right corners, fold backward along the vertical axis until the left and right halves meet in the back.
2. Open up the center into a box shape.
3. Continue pulling the left and right sides outward, flattening the box from the top and bottom.
4. Open the flexagon from the front by lifting the top leaves upward and pulling the lower leaves downward to reveal face 4.
5. From state 4/1, you can do a book flex to state 3/4, a box-top flex to 2/3, and finally a book flex to return to 1/2. You need to rotate the flexagon 90° after each flex.

DOI: 10.1201/9781003433538-9

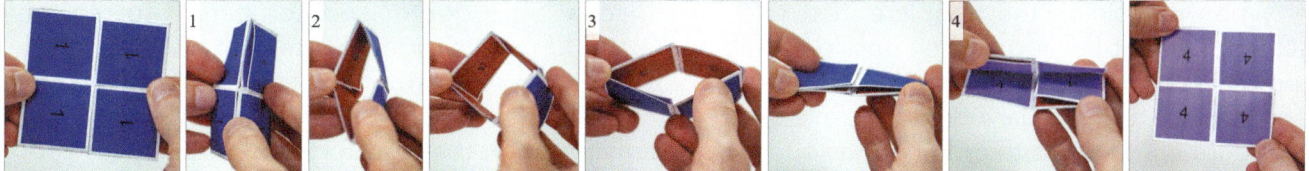

FIG. 8.2 The steps in performing the box-top flex.

Figure 8.3 has a template for the minimal square tetraflexagon for the box-bottom flex. It's almost the same as the box-top flex, except that you open from the back instead of the front in step 4. Refer to the pictures in figure 8.4. Note that you start in state 1/2 and end in state 1/4. Can you figure out how to use a combination of book flexes and box-bottom flexes to return to state 1/2?

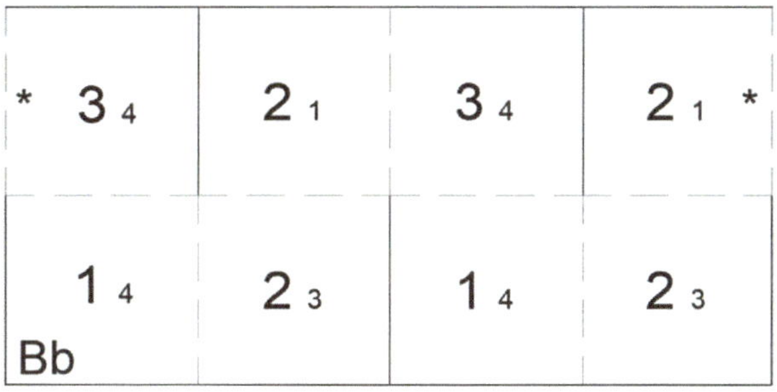

* 3 4	2 1	3 4	2 1 *
1 4 Bb	2 3	1 4	2 3

FIG. 8.3 Template for a square tetra-tetraflexagon that supports Bb, the box-bottom flex.

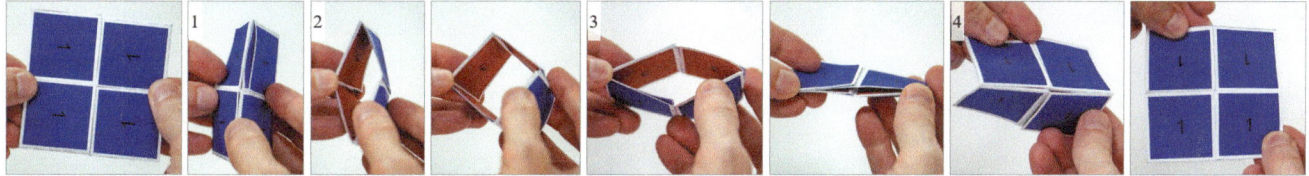

FIG. 8.4 The steps in performing the box-bottom flex.

Figure 8.5 contains a template for a square penta-tetraflexagon that supports both the box-top and box-bottom flexes from the same hinge. Thus, after opening and flattening the box in step 3 of the box-top flex, you can open the flexagon from either the top or the bottom. The box-top flex will lead you to state 4/1, while the box-bottom flex will reveal state 1/5.

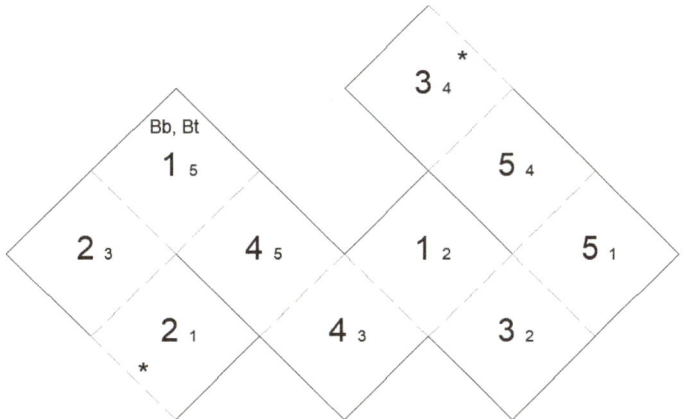

FIG. 8.5 Template for a penta-tetraflexagon that supports both Bt and Bb as well as the book flex.

When examining the template, first fold the adjacent 5's together and compare it to the template for the Bt flexagon. Then unfold the 5's and just fold the 4's together and compare that to the template for the Bb flexagon. You should be able to see how the structures for both box flexes are embedded in this square penta-tetraflexagon.

Figure 8.6 shows a template for a square deca-tetraflexagon that supports the book flex and box flexes in many places. Can you find all ten faces? *Hint:* From face 1 of state 1/2, you can do a book flex in two different places, a box-top in two different places, and a box-bottom in two different places, where each flex leads to a new face. Since this flexagon has lots of leaves, make sure that when you open up a box for a box flex, all the numbers on the inside are the same.

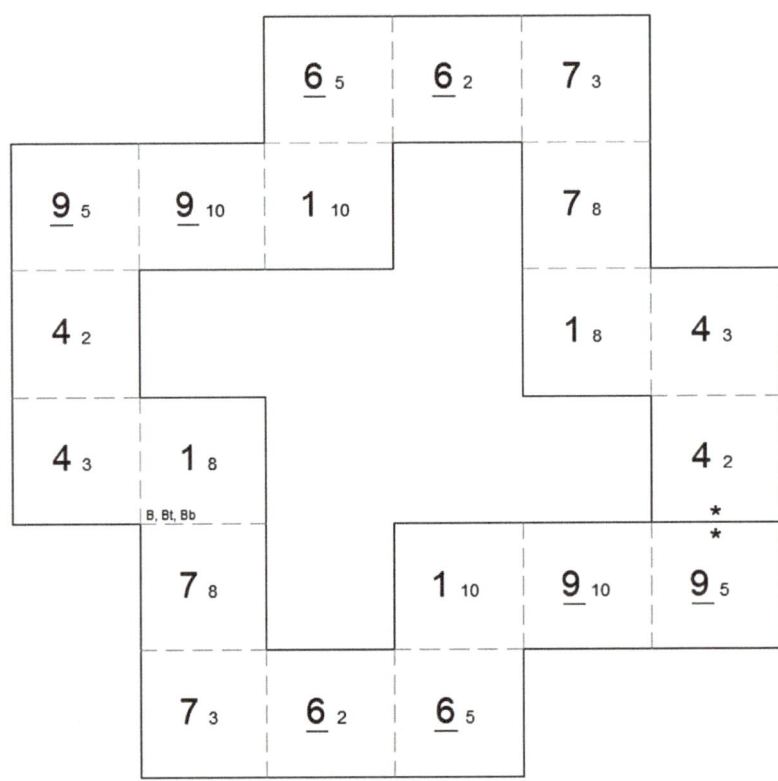

FIG. 8.6 Template for a square deca-tetraflexagon that supports B, Bt, and Bb. Note that the 6's and 9's are underlined to help you keep track as the leaves are rotated and rearranged.

Pentagon

The two box flexes also work on pentagon flexagons. Figure 8.7 shows the template for the minimal pentagon tetraflexagon for the box-top flex, which can take you from state 1/2 to state 4/1. As you did in figure 8.2, make sure you have the pats with the single leaves in the upper left and lower right, and fold along the vertical axis. Lift up the pentagons from the front.

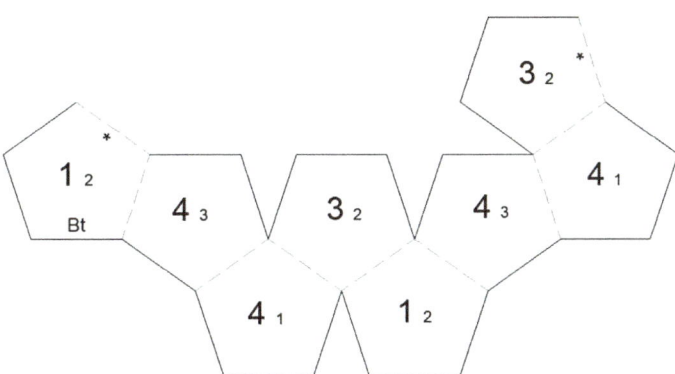

FIG. 8.7 Template for the minimal pentagon tetraflexagon that supports the box-top flex.

Figure 8.8 shows the template for the minimal pentagon tetraflexagon for the box-bottom flex, which can take you from state 1/2 to state 1/4.

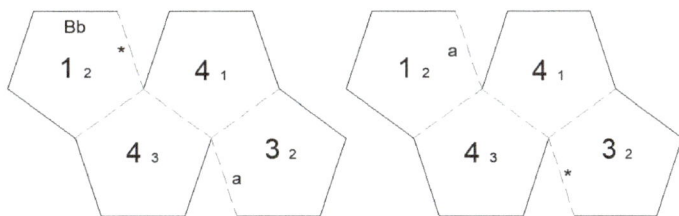

FIG. 8.8 Template for the minimal pentagon tetraflexagon that supports the box-bottom flex.

Summary

We tried out the box flex on square and pentagon flexagons. We also made a more complex square flexagon that supports both the book flex and box flex, which makes it trickier to find all the hidden faces.

PART TWO

Asymmetric Flexes

In Part One, we experimented with symmetric flexes, flexes that swap out entire faces or symmetrically make changes around the entire flexagon. In Part Two, we'll see a variety of rotationally **asymmetric flexes**, flexes that change faces in irregular ways and hence can mix up the faces.

DOI: 10.1201/9781003433538-10

Chapter 9

Flex Diagrams

Flex diagrams are a useful way of visually understanding how to perform a flex. We will use this visual language throughout the rest of this book to explain flexes, sometimes combined with corresponding photos. These symbols build on the language of origami, supplemented with flexagon-specific concepts.

Symbols

Symbols used to describe the static state of a flexagon (see also figure 9.1):

- An **edge** is the cut edge of a leaf or pat and is indicated by a solid black line.
- A **hinge** occurs between adjacent pats and is where you can fold back and forth. A flat hinge is indicated by a dotted line. A hinge in a valley fold uses a blue dashed line, while a hinge in a mountain fold (the opposite of a valley) uses a red dashed and dotted line.
- To help keep track of how the leaves get rearranged as you fold and flex, the leaves in flex diagrams follow a common coloring scheme based on face number: 1-blue for the face that's initially in front, 2-green for the face that's initially in back, and 3-red for the first new face you reveal when you unfold the flexagon. The next four faces that appear when unfolding a pat are 4-purple, 5-yellow, 6-brown, and 7-gray.

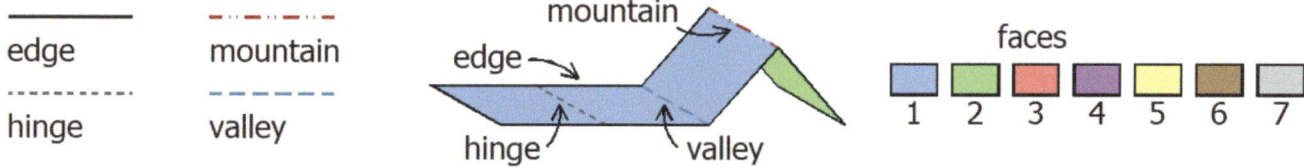

FIG. 9.1 Visual notation for edges, hinges, and faces.

Figure 9.2 shows common symbols for describing how to change a flexagon by folding pats together, unfolding a pat, and turning over the flexagon. Note that the turn-over-side-to-side arrow is the same as ∧ in flex notation, which we first saw in chapter 7. The arrow is designed to be understood visually, while the ∧ symbol works well with the text used for flex notation. If the front leaf-face is labeled as *a*, the back leaf-face is −*a*.

DOI: 10.1201/9781003433538-11

valley fold
mountain fold
open
turn over side to side
a b
-b -a
turn over top to bottom
a b
-a -b

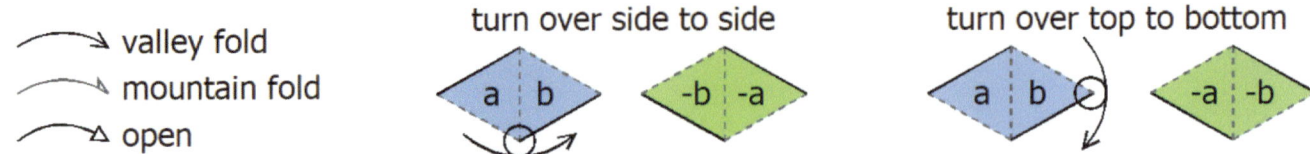

FIG. 9.2 Visual notation for folding pats and turning over the flexagon.

Different ways to fold two pats together are represented by the symbols in Figure 9.3. Note that the face colors in each sequence are based on the initial view, starting with blue in the front and green on the back.

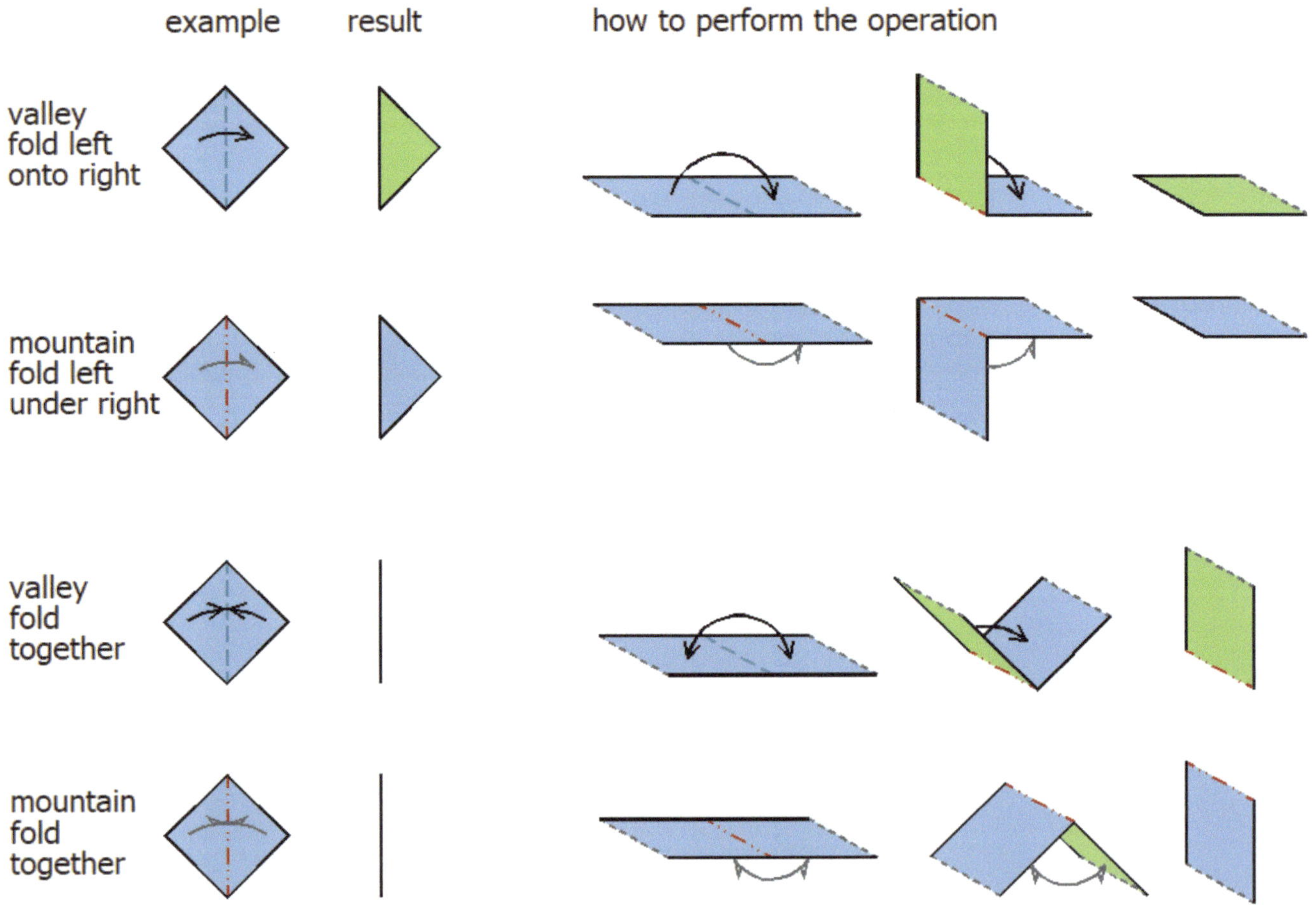

example
result
how to perform the operation
valley fold left onto right
mountain fold left under right
valley fold together
mountain fold together

FIG. 9.3 Visual notation for folding two pats together.

Figure 9.4 presents various ways to unfold a pat. Red represents the leaf-face that is hidden before unfolding.

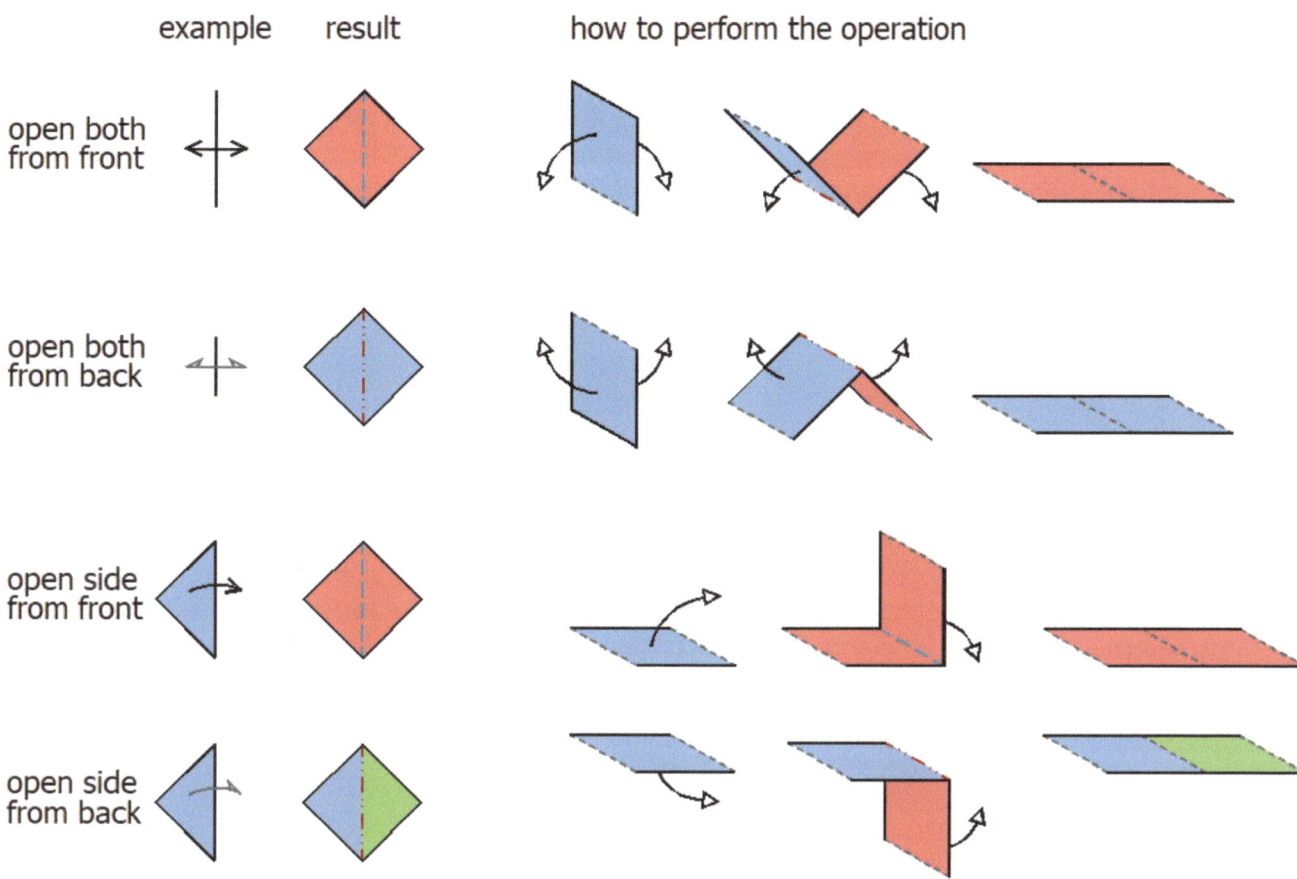

FIG. 9.4 Visual notation for unfolding a pat.

Something we observed as we used the pinch flex was that it made a difference where we did the flex: sometimes we could pinch-flex at a hinge and sometimes we couldn't. This is because the pinch flex has specific requirements for how each leaf is folded together into a pat. In general, each flex has its own requirements for the internal structure of the pats.

Therefore, we add two additional pieces of information to a flex diagram:

- Reference hinge: The steps for performing a given flex are defined relative to a reference hinge, marked in flex diagrams with #. Defining the reference hinge before and after a flex allows us to keep track of how to perform a whole sequence of flexes, even as they rearrange the internals of each pat.
- Required internal structure for each pat: This more complex requirement is described in the next section.

Pat structure

To describe the required internal pat structure of triangle flexagons, we use a dash (or other symbol) to represent a leaf. Two symbols are surrounded by square brackets when two leaves or pats are folded together. For example, a single leaf is simply a dash, -, while two leaves folded together are [- -], with each dash representing a leaf and the square brackets telling you that they're folded together. Folding a third leaf on top of that pair is represented as

[- [- -]]

while folding a third leaf underneath the pair is [[- -] -]. Each set of square brackets represents leaves or pats that are folded together.

All the possible ways to fold pats with 1, 2, 3, or 4 leaves (if you also include 4a and 4c turned upside down) appear in figure 9.5, with each image representing a stack of leaves in a pat. The figure uses the numbering and coloring scheme mentioned at the beginning of this chapter: The top of a pat is 1-blue. The bottom is 2-green. The first time you unfold the pat, the two leaf-faces you see are 3-red. In the representation of the folded pat, you can see how these two red leaf-faces are face to face in the stack of leaves. The next two unfoldings give you 4-purple and 5-yellow. The arcs in the diagrams indicate that two leaves are connected by a hinge, while the arrows show which leaves are connected to adjacent pats. Note that triangular leaves have three edges, but since they're only connected to two other leaves, the leaves in these diagrams are simplified into rectangles that only connect on either side. This means that the arcs and arrows show when two leaves are connected but not how those connections are actually oriented. Figure 9.6 contains sample templates for each of the arrangements in figure 9.5 (excluding the single-leaf base case), which you can fold up and compare to the diagrams. Note that each template folds up to a pat with an extra leaf on either side so you can see how the pat is connected to its two neighbors.

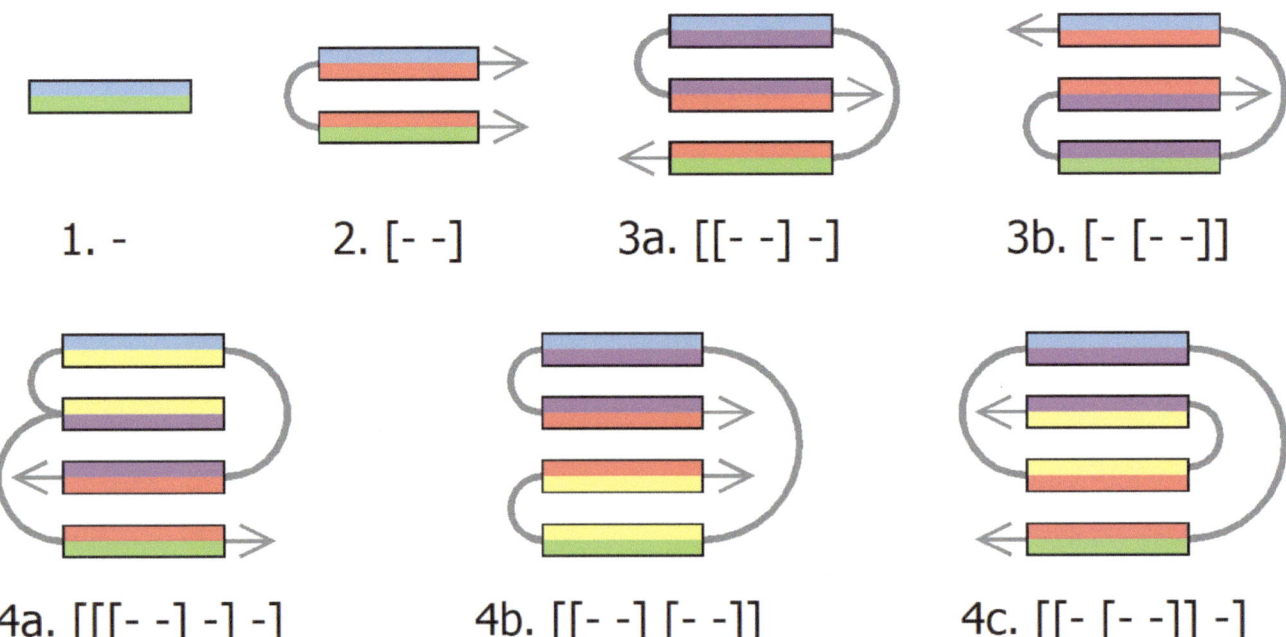

FIG. 9.5 Notation for pat structure and what the notation represents.

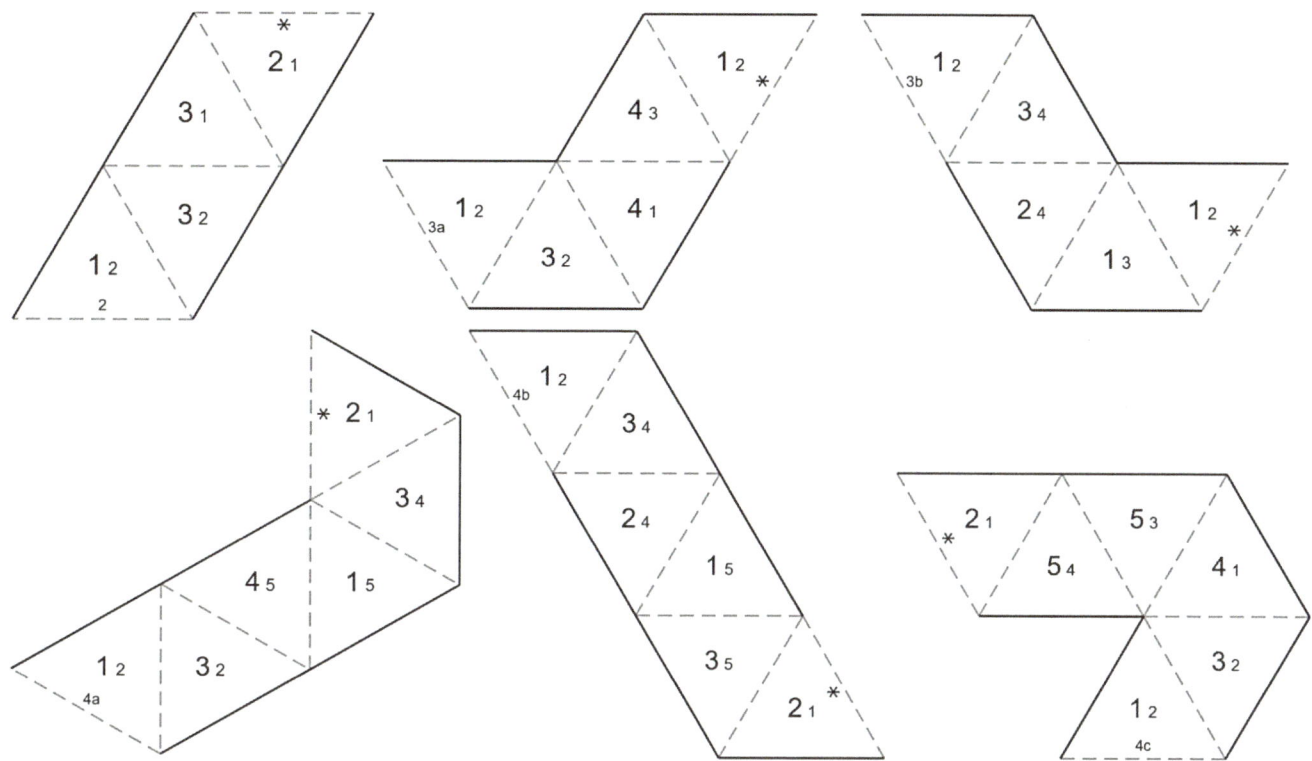

FIG. 9.6 Sample templates for pats with 2, 3, and 4 leaves. These correspond to figure 9.5.

In a flex diagram, this notation appears next to a pat if more than a single leaf is required. Note that this is for the *minimum* required structure; the flex still works if the pat contains additional leaves. For example, if the flex diagram says the minimum structure of a pat is [- -], the flex still works if the corresponding pat is [[- -] -] or [[- [- -]] -], because [- -] is a subset of the more complex pat.

Flex diagrams for the pinch flex

We put this all together for the pinch flex in figure 9.7. Step 1 shows the required pat structure, the location of the reference hinge, and the initial folds to perform. The second image is what it looks like when you're partway through folding together the pats. Step 2 shows the result of those initial folds plus directions on what to do next. It looks like a propeller because the adjacent pats have been folded completely together. The next image shows what it looks like when you're partway through opening it back up. Step 3 shows the end result, including where the final location of the reference hinge is, which is useful for understanding how to string together a series of flexes. This is a visual way of describing how to pinch together adjacent pats and open them back up to see a new face.

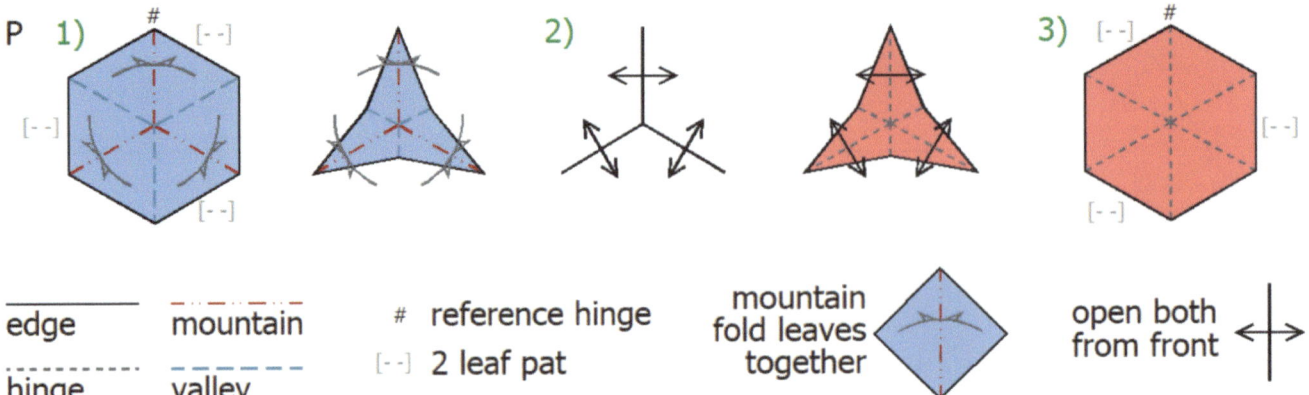

FIG. 9.7 Visual notation describing the pinch flex on a hexaflexagon, including a key with the symbols it uses.

If we compare this to the flex diagram for the book flex in figure 9.8, it becomes more obvious how it's analogous to the pinch flex: its pat structure requirement follows the same pattern and the folding instructions are the same. However, the fact that a square has one more edge than a triangle gives us the two slightly different versions we previously saw, Bl (book-left) versus Br (book-right), which have different requirements for their internal pat structure. The flex diagrams indicate this by placing [- -] next to the corresponding pats. Note that we've left out the intermediate states that we showed in figure 9.7, reducing it to just the basic folding and unfolding steps.

FIG. 9.8 Visual notation for the book-right and book-left flexes on a square tetraflexagon.

Figure 9.9 shows the reverse pass-through flex on a silver octaflexagon, which was described in chapter 5. The second step shows the symbol for reversing valley folds and mountain folds.

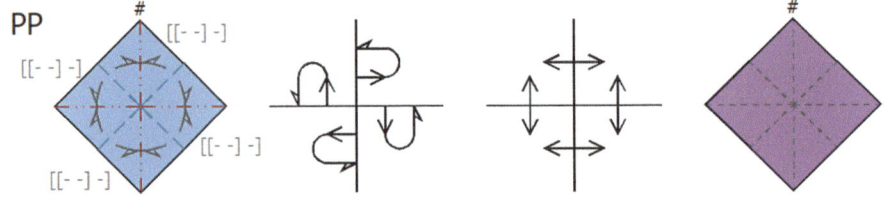

FIG. 9.9 Visual notation describing the reverse pass-through flex on a silver octaflexagon.

Flex notation

We use flex notation as shorthand for describing a series of flexes. We first used it in chapter 7, and we'll expand on it later in chapter 18. Every flex has an abbreviation, such as P for the pinch flex and Bt for the box-top flex, which we'll note when we introduce a flex. Here are a few other symbols used in flex notation:

^	turn the flexagon over left to right
<	shift the reference hinge one hinge counterclockwise (to the left if the reference hinge is oriented at the top)
>	shift the reference hinge one hinge clockwise (to the right if the reference hinge is oriented at the top)
A'	the inverse of flex A, which undoes the effects of A
(A)n	repeat flex sequence A n times

Flex summary

Each chapter describing a new flex concludes with a summary of the flex, showing its name, the shorthand to use, how many pats it modifies, the simplest flexagon that supports the flex, and a picture showing the effect of the flex on the simplest flexagon. Figure 9.10 shows how the pinch flex changes a regular hexaflexagon, including the minimal pat structure and how the leaves rotate. Note that you'll find a similar pattern when looking at other flexagons that support the pinch, such as the tetraflexagon (the minimal flexagon for the pinch) or the octaflexagon (with eight pats).

full name:	**pinch flex**	shorthand:	**P**
modified pats:	**all**	minimal flexagon:	**tetraflexagon**

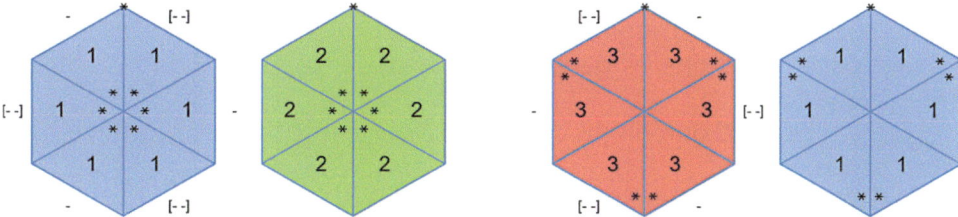

FIG. 9.10 Front and back of the starting state (left) and the state after applying the pinch flex (right). The star markers in the corners of the leaves track the angles that start in the center of the flexagon, showing how they rotate.

Summary

For a summary of the visual language for describing flexes used in flex diagrams, review figure 9.11.

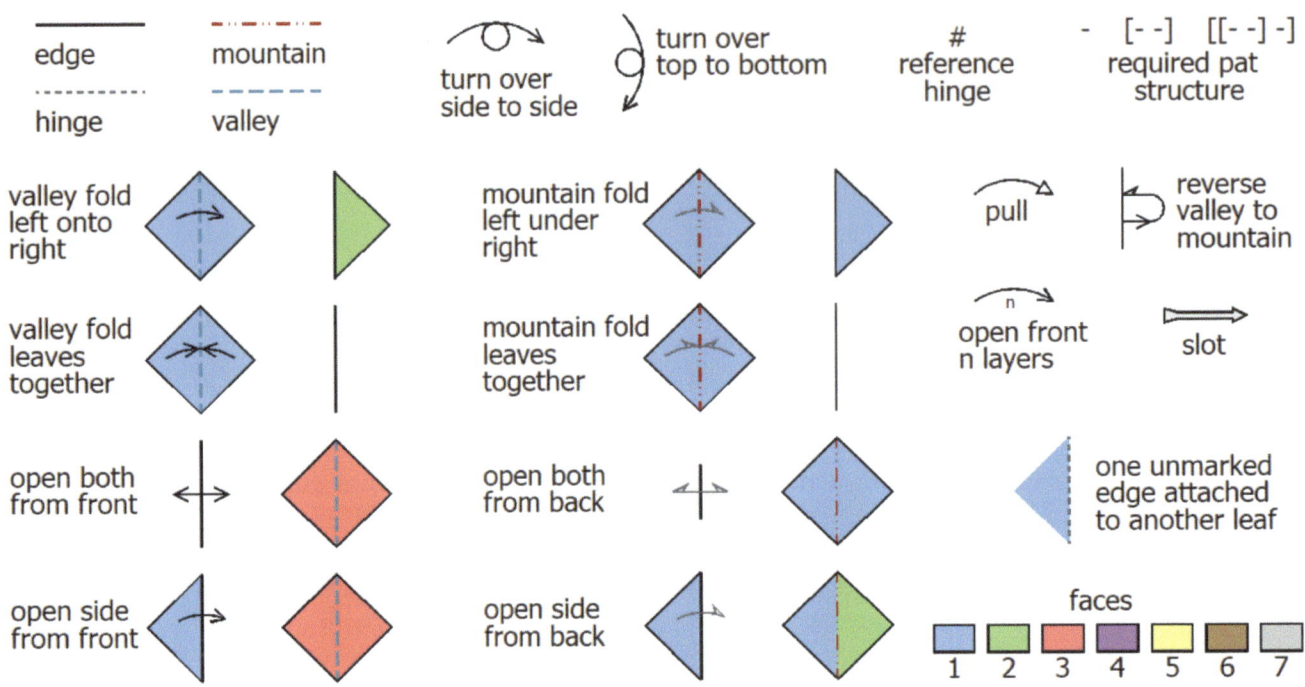

FIG. 9.11 The visual language of flex diagrams.

Also recall the symbols used in flex notation:

∧   turn the flexagon over left to right

<   shift the reference hinge one hinge counterclockwise (to the left if the reference hinge is oriented at the top)

>   shift the reference hinge one hinge clockwise (to the right if the reference hinge is oriented at the top)

A′   the inverse of flex A, which undoes the effects of A

(A)n   repeat flex sequence A n times

Chapter 10

The V-Flex

Our tour of asymmetric flexes starts with the **v-flex**. T. Bruce McLean introduced the v-flex in his 1979 paper "V-Flexing the Hexahexaflexagon". The "v" was in honor of Bob Verrey and Alan Moluf (imagine the A as an upside-down V) who had experimented with the flex years earlier. Before that, people generally explored hexaflexagons using only the pinch flex. On the hexa-hexaflexagon made from a straight template (variation 6a), you can pinch-flex between nine different states. McLean showed that using both the pinch flex and v-flex allows you to get to 3420 different states! This opened up a whole new world to explore.

Like the pinch flex, the v-flex rotates and modifies every pat. But unlike the pinch flex, it doesn't keep all the leaves on a face together, instead mixing up the faces. The v-flex works best on the hexaflexagon. We abbreviate the v-flex as V.

A good way to learn a new flex is on its minimal flexagon, a flexagon that contains only the structure needed for the flex and no more. See figure 10.1 for the minimal hexaflexagon for the v-flex. As with other templates, you may find it useful to copy the stars at the two ends onto the back face along the same edge to help as you tape together the proper edges. When learning the flex, you may wish to color the leaves as described in chapter 9 so that the colors correspond to the illustrations used to explain the flex. The relevant colors on this flexagon are blue for 1, green for 2, red for 3, and purple for 4. Note that this template has six 1's and 2's but only two 3's and four 4's.

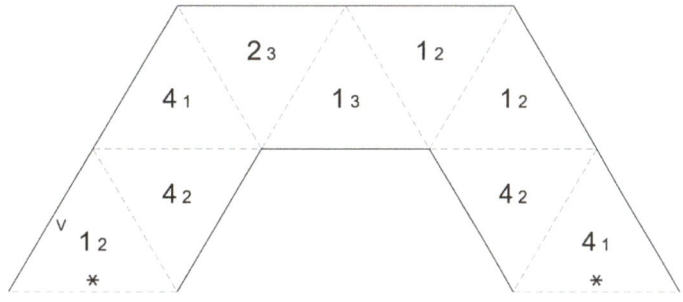

FIG. 10.1 Minimal hexaflexagon for the v-flex.

Once you've folded and connected the template, find the reference hinge on face 1 marked with the star. Orient the hexagon so the reference hinge is pointed up. Then refer to figures 10.2 and 10.3 along with the following directions to perform the v-flex:

DOI: 10.1201/9781003433538-12

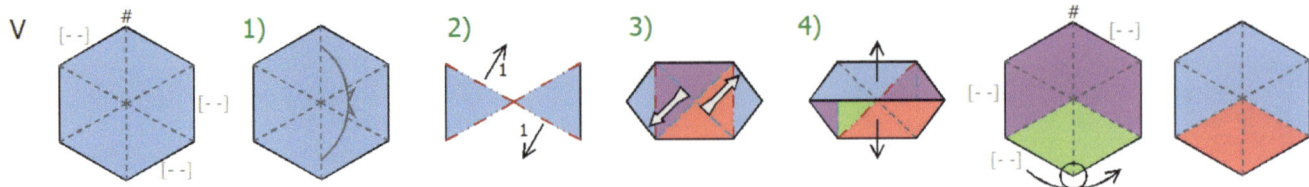

FIG. 10.2 Flex diagram for the v-flex on a hexaflexagon.

1. Fold the top and bottom corners backward until they meet each other in the back to make a bow tie shape.
2. Open the center of the flexagon from the front, exposing the inside of a pyramid.
3. Slide the left half of the pyramid (the two 4's) down and to the left, while sliding the right half (the two 3's) up and to the right.
4. Open from the center of the flexagon. The front should have four 4's and two 2's, while the back should have four 1's and two 3's.
5. To return to the original state, turn the flexagon over from left to right and repeat steps 1–4.

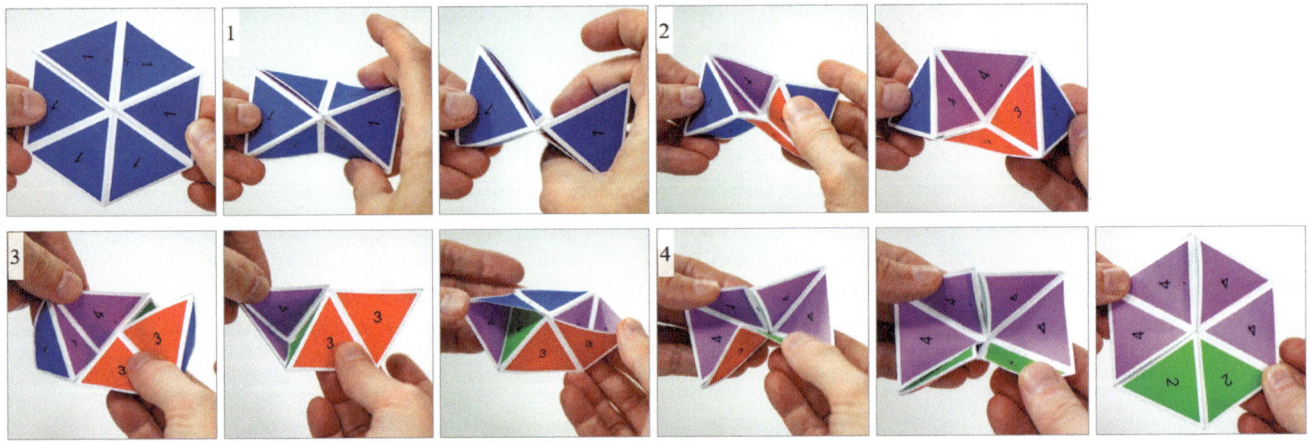

FIG. 10.3 Pictures of the v-flex on a hexaflexagon.

Examine the faces before and after the v-flex to get a feel for how it changed the flexagon. Before the flex, every pat had 1 on the front face and 2 on the back face. After the flex, four pats have 4 on the front and 1 on the back, while the other two pats have 2 on the front and 3 on the back. If you put marks on the corners of the triangles in the center of face 1 before flexing, you'll find that the marks end up along the edge of the hexagon after the v-flex. The v-flex has changed and rotated every pat and also mixed up the faces. The changes have **bilateral symmetry**, meaning that it looks the same if you mirror it across a line that goes between the two leaves labeled with 2's, but not **rotational symmetry**, because there's no way to rotate the faces such that two rotations look the same. We've included the v-flex in this section because it's **rotationally asymmetric**.

While the minimal flexagon is useful for learning how to do the v-flex, it gets much more interesting when you can do multiple v-flexes on a flexagon. The template in figure 10.4 folds into a hexaflexagon that supports a cycle of six v-flexes, meaning that it returns to the original state. Start the first v-flex from face 1 with the hinge marked with stars at the top. After each v-flex, rotate the flexagon 60° counterclockwise before doing the next v-flex, which shifts the reference hinge one to the right (to the hinge that was 60° clockwise from the top). It's helpful to make sure the flexagon matches the steps in figure 10.5 before doing the next v-flex so you know you've done each step correctly. This cycle is represented in flex notation as (V>)6.

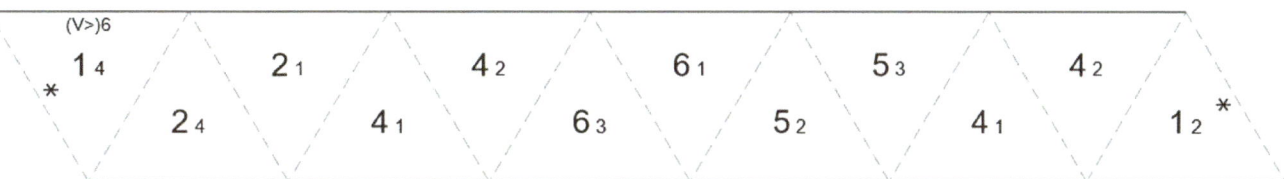

FIG. 10.4 Regular hexaflexagon that supports a cycle of six v-flexes: (V>)6.

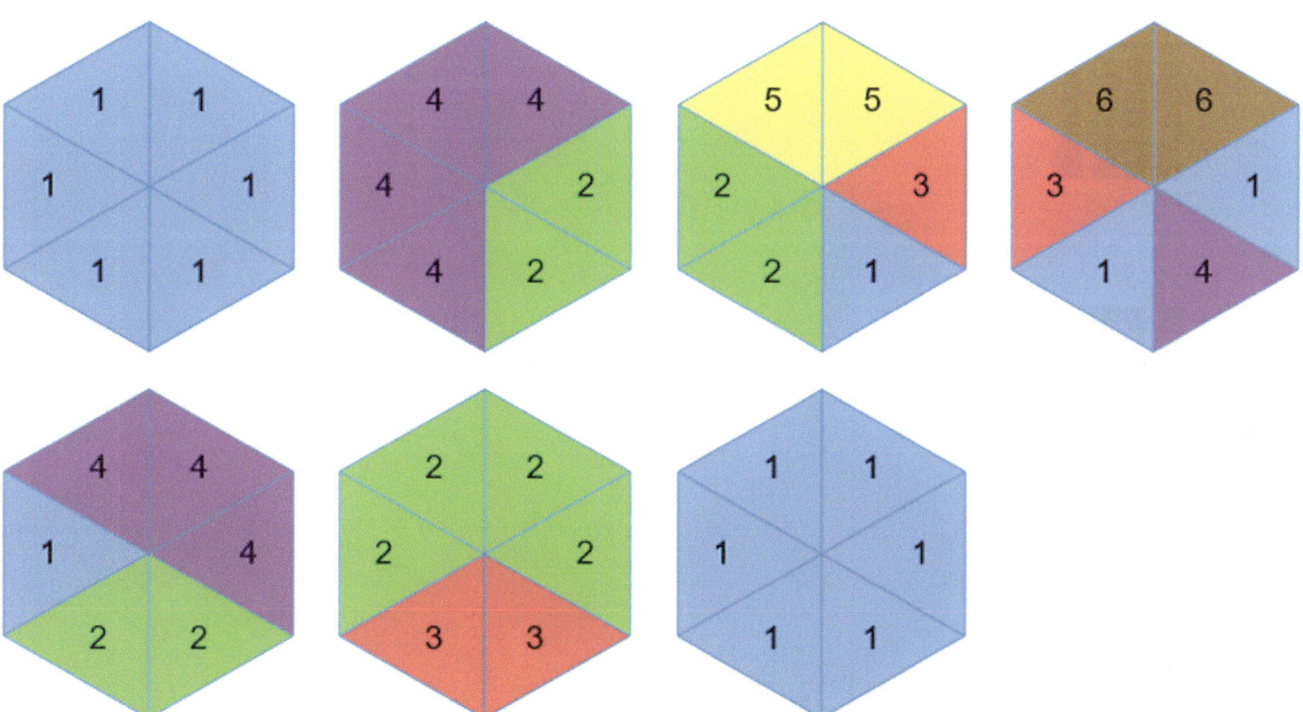

FIG. 10.5 The states that appear during the (V>)6 cycle, with the reference hinge for the next v-flex at the top.

If you've mastered doing the cycle of six v-flexes, you can try doing an even longer cycle. The template in figure 10.6 folds into a hexaflexagon that lets you do 15 v-flexes in a row before returning to the original state. Start with the hinge marked (V<)15 at the top. After each v-flex, rotate the flexagon 60° clockwise (so the reference hinge shifts one to the left) before doing the next v-flex. See figure 10.7 for what each state looks like immediately before doing the next v-flex. This can be a tricky sequence, so following along with these pictures can be helpful. This cycle is the flex sequence (V<)15.

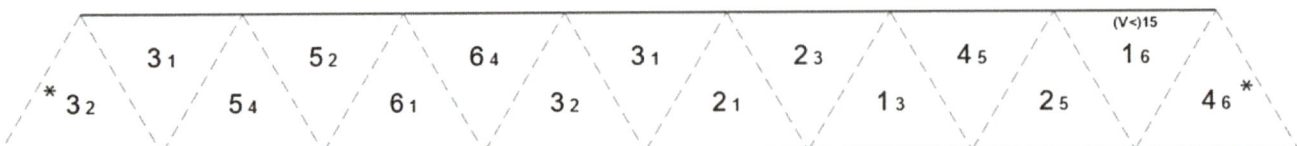

FIG. 10.6 Regular hexaflexagon that supports a cycle of 15 v-flexes.

FIG. 10.7 The states that appear during the (V<)15 cycle. Note that the reference hinge for the next v-flex is at the top.

While it's possible to use the v-flex on flexagons with an even number of pats greater than six, it's tricky to perform. Since the v-flex rotates the leaves in a similar fashion to the pinch flex, often this means that the flexagon won't lie flat after the flex, depending on the angles of the leaves. Therefore, we'll generally restrict our usage of the v-flex to the regular hexaflexagon.

However, there are some flexagons where the v-flex can travel between flat states, such as the silver octaflexagon. This can work when the leaves are isosceles triangles where the flexagon lies flat when either of the two equal angles of the triangle is in the middle of the flexagon. If you v-flex from the correct hinge, then when the leaves rotate during the flex, the other equal angle will end up in the center, resulting in another flat state. See figure 10.8 for the template for the minimal silver octaflexagon for the v-flex.

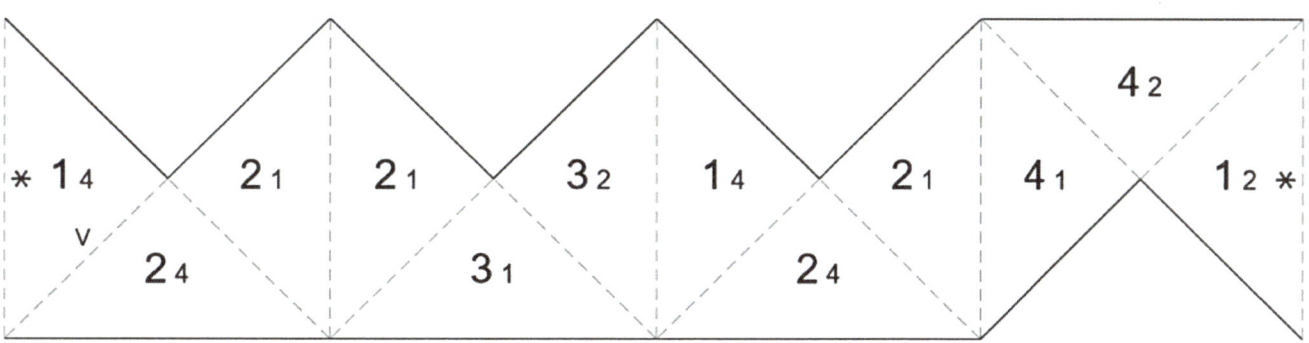

FIG. 10.8 Template for the minimal silver octaflexagon supporting the v-flex.

Position the hinge with the stars at the top, then refer to figure 10.9 or figure 10.10 along with the following directions:

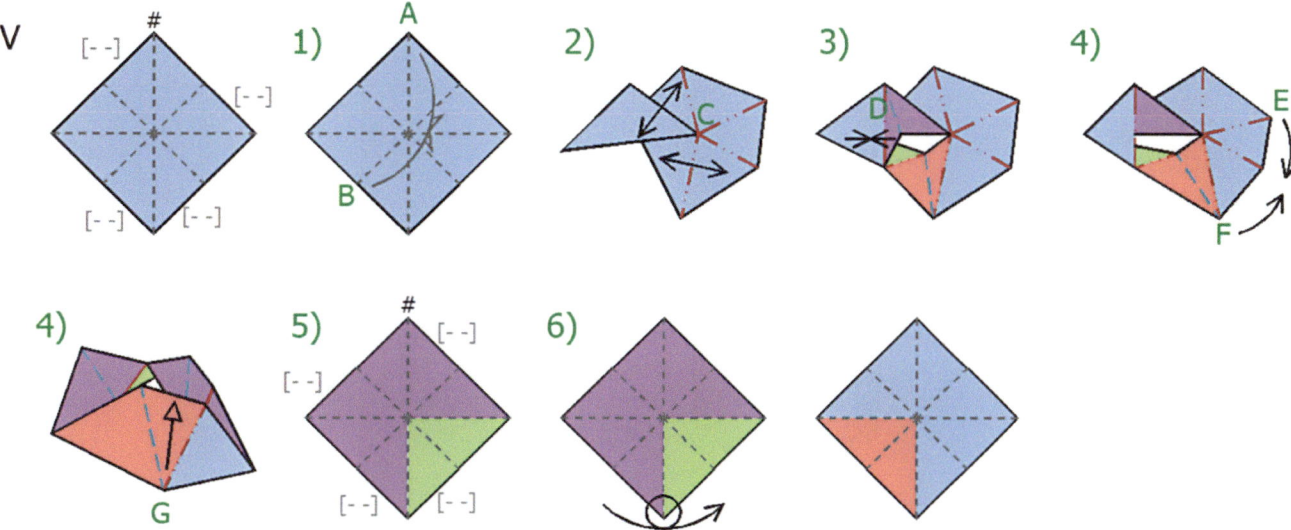

FIG. 10.9 Flex diagram for the v-flex on the silver octaflexagon.

1. Fold hinges A and B backward until they meet.
2. Open the flexagon from the center C along the pair of hinges marked with arrows.
3. Bring the two pats that are adjacent to hinge D together.
4. Bring corners E and F together in back.
5. Push corner G into the middle, and open from the center. When using the minimal flexagon, the final state will be mostly 4's with a couple 2's in the lower right corner.
6. To return to the original state, first turn the flexagon over, then repeat steps 1–4.

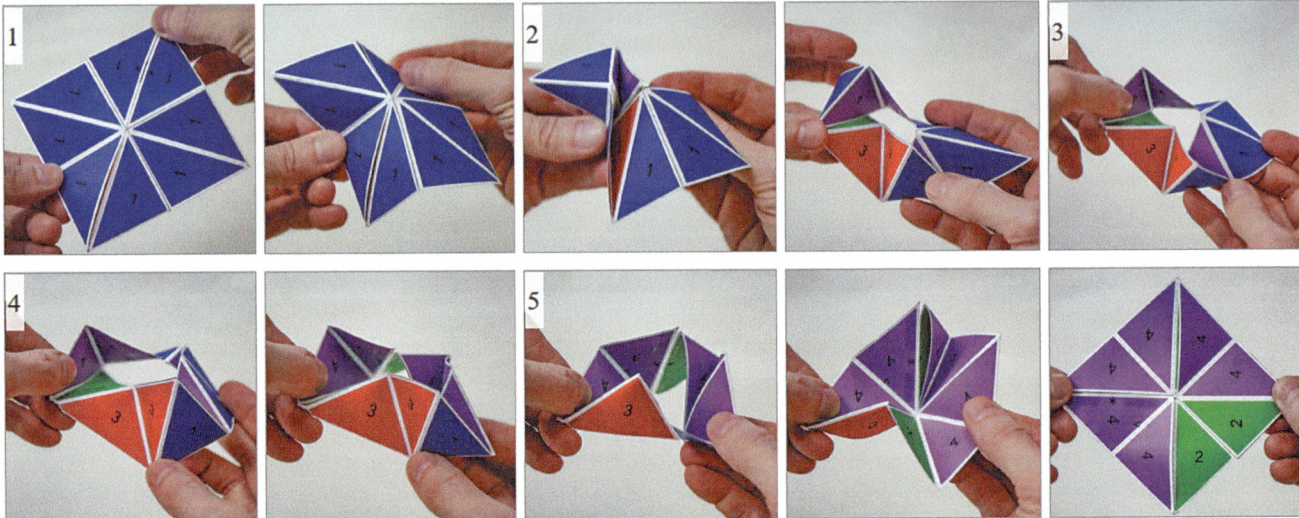

FIG. 10.10 Pictures of the v-flex on a silver octaflexagon.

Summary

The following table lists useful facts about the v-flex, and figure 10.11 shows how it changes the minimal flexagon.

full name: **v-flex** shorthand: **V**
modified pats: **all** minimal flexagon: **hexaflexagon**

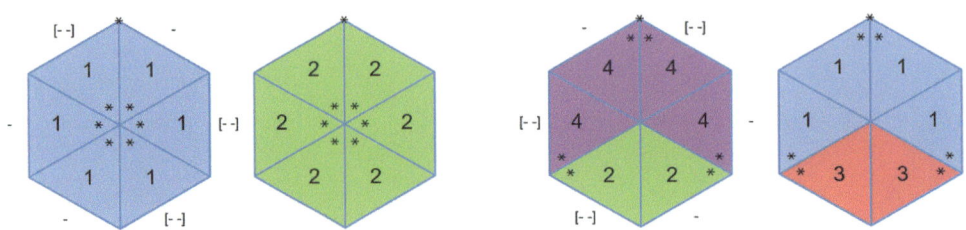

FIG. 10.11 Front and back of the starting state (left) and the state after applying the v-flex (right).

Chapter 11

The Tuck Flex

The **tuck flex** was first described by Conrad and Hartline in their 1962 technical report on flexagons, though they didn't give it a name. They dismissed it as a "distortion" and a "much dreaded aspect of the flexagon". One reason was because "it involves forcing of the supposedly rigid leaves". A second reason was because it mixes up the faces and "often results in a disassembly and consequent reassembly of the flexagon involved" (Conrad and Hartline, 1962, pp. 14–15).

They might not have complained about the "distortion" if they had noticed that *no bending is required* on many flexagons such as those made from right triangles, where it's a very elegant flex. In response to their complaint about mixing the faces, we merely observe that this was many years before the Rubik's Cube came out, which became popular precisely *because* mixing up faces and trying to put them back together again was a fun challenge. A missed opportunity perhaps?

To introduce the tuck flex, we first look at the minimal bronze hexaflexagon for the tuck, which requires no bending of the leaves. See figure 11.1 for the template.

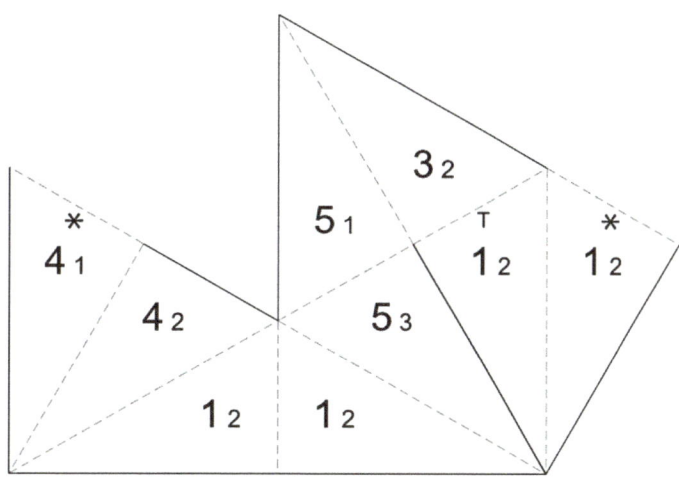

FIG. 11.1 Minimal bronze hexaflexagon for the tuck flex.

Once you've folded and connected the template, find the reference hinge on face 1, marked with T. Orient the flexagon so the reference hinge is pointed up. Then refer to figures 11.2 and 11.3 along with the following directions to perform the tuck flex. Start with the hinge marked T at the top.

DOI: 10.1201/9781003343538-13

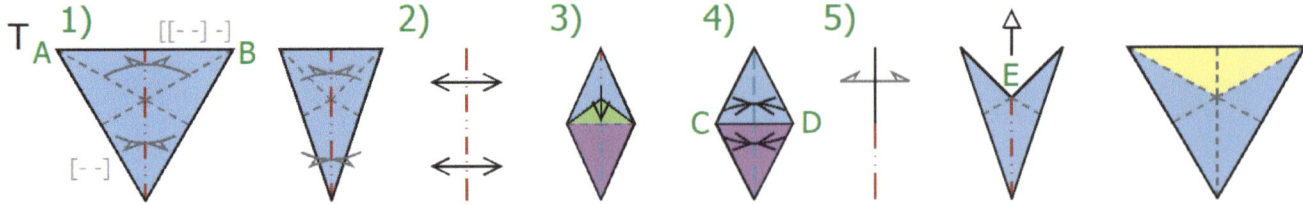

FIG. 11.2 Flex diagram for the tuck flex on the bronze hexaflexagon.

1. Fold the flexagon backward across the centerline, folding corners A and B together in the back.
2. Open the center of the flexagon from the front.
3. Pull the top hinge down, flattening into a diamond shape.
4. Fold opposite corners C and D forward together.
5. Open the center of the flexagon from the back. Corner E should naturally move up, giving you a triangular face.

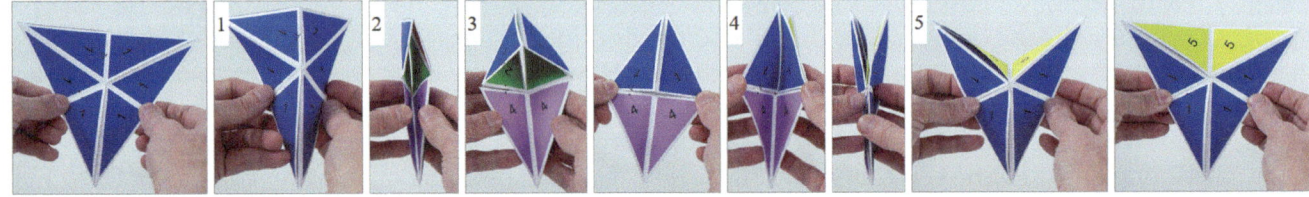

FIG. 11.3 Performing the tuck flex on the bronze hexaflexagon.

If you examine the flexagon after doing the tuck flex, you should see two of the pats have changed – 5-yellow is now on the front with 3-red on the back – and four of the pats have stayed the same – four 1-blue pats are still on the front and four 2-green pats are still on the back. Up until now, all the flexes we've tried have been **global flexes**, flexes that change and rotate pats across the entire flexagon. The tuck flex is our first example of a **local flex**, a flex that only changes a portion of the flexagon.

With many flexes, after you do the flex, you can simply turn the flexagon over and do the flex on the opposite side to restore it to its original position. Or, more precisely, if we have some flex A, we can describe this in flex notation as ^A^, which says turn the flexagon over, do flex A, then turn it over again so that you're looking at the original face with everything in its original state. This strategy works with the pinch flex, book flex, box flex, and v-flex.

With the tuck flex, however, this isn't always true. Depending on the internals of the pats, you may need to instead do all the steps of the tuck flex in reverse order. We use T as shorthand for the tuck flex and T' to indicate the **inverse tuck flex**, i.e., all the steps of the tuck flex are done in reverse order, from last to first. We add a prime symbol after the flex's shorthand symbol to indicate its inverse. See figure 11.4 for the flex diagram for the inverse tuck on the bronze hexaflexagon.

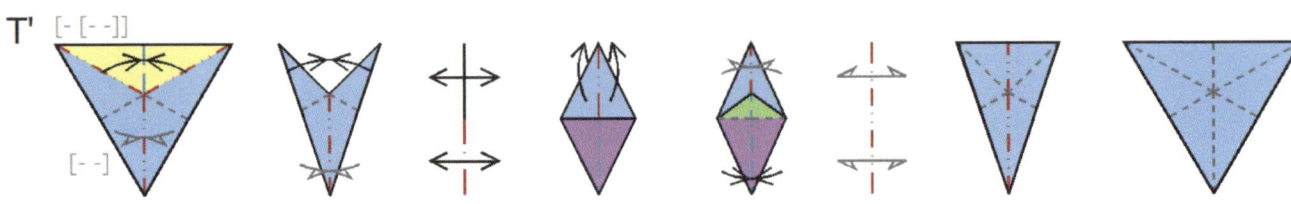

FIG. 11.4 Flex diagram for the inverse tuck flex on the bronze hexaflexagon.

Notice when you first do the tuck flex that the flap with 4's in it (shown in step 3 in figure 11.2) is required in order to provide enough freedom to open up the flexagon and perform the tuck, even though that flap doesn't change. If you pasted the 4's together, you'd have to damage the paper in order to perform the flex. For reference, this is called the **forced tuck flex**, with shorthand of Tf, and is of theoretical interest when trying to explore all the states of a hexaflexagon.

As Conrad and Hartline noted, doing the tuck flex on a regular hexaflexagon requires bending the leaves slightly in order to tuck one corner into the center during the flex. See figure 11.5 for the minimal regular hexaflexagon for the tuck flex and figure 11.6 for the flex diagram. Similar to the tuck flex on the bronze hexaflexagon, fold it in half backward, open from the center, tuck in the top corner, close it up, then open from the back.

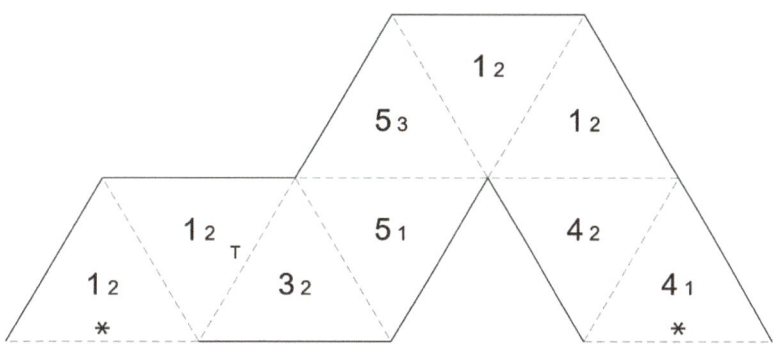

FIG. 11.5 Minimal regular hexaflexagon for the tuck flex.

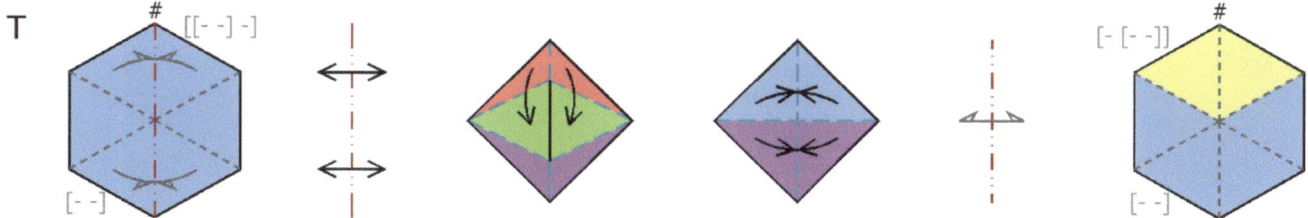

FIG. 11.6 Performing the tuck flex on the regular hexaflexagon.

But pulling that tucked piece back out when doing the inverse tuck can be a bit harder than it is on the bronze hexaflexagon, so we'll describe it in a bit more detail. After you finish the tuck flex from figure 11.6, follow along with the flex diagram in figure 11.7 while using the following steps:

1. Push the top corner A back and down while folding the flexagon in half by bringing together corners B and C at the top and corners D and E in the back.
2. Open up the flexagon from the middle so that you're looking at the inside of a pyramid.
3. This is the trickiest step. Reach a finger into the middle of the pyramid and get it under the two blue pats. Pull the blue pats up.
4. Close the flexagon by bringing the left and right halves together.
5. Open it up from the back.

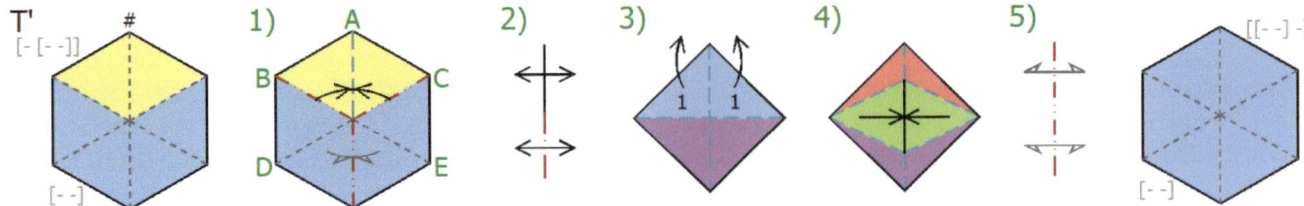

FIG. 11.7 Performing the inverse tuck flex on the regular hexaflexagon.

The silver octaflexagon in figure 11.8 supports doing the tuck flex at several different hinges from face 1. As we noted earlier, a tuck flex requires an extra flap in order to provide enough freedom to open up the flexagon, and this octaflexagon demonstrates that the flap can be at different hinges. Additionally, you can do a **double tuck flex**, where you do two tucks from opposite edges of the flexagon at the same time. See figure 11.9 for the flex diagram for the double tuck.

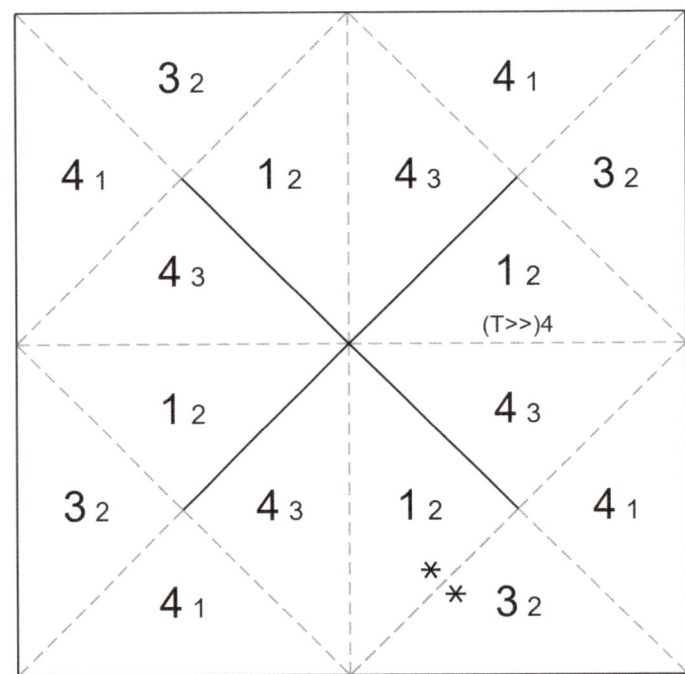

FIG. 11.8 Silver octaflexagon that supports the tuck flex in four places from the original face.

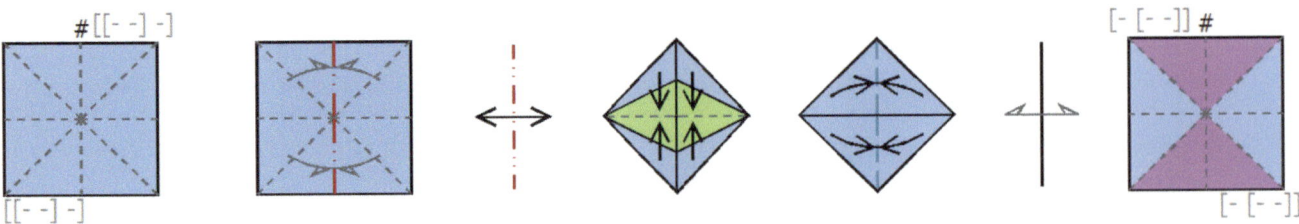

FIG. 11.9 Flex diagram for the double tuck flex.

On a hexaflexagon, there's only one possible location for the extra hinge that needs to open in order to do the tuck flex, which is directly across from the reference hinge. But for heptaflexagons (seven pats per face) and greater, there are multiple hinges that could open. If you need to distinguish which hinge is required, you can use T1 for the leftmost hinge, T2 for the next hinge to the right, etc. Figure 11.10 illustrates the tuck variants on the hexaflexagon and octaflexagon.

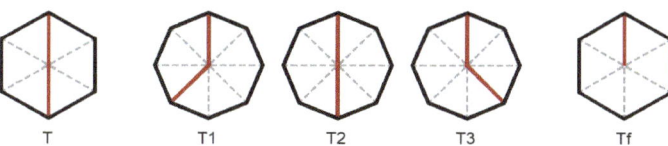

T T1 T2 T3 Tf

FIG. 11.10 The tuck flex on a hexaflexagon (left) requires being able to open top and bottom hinges. The tuck flex on an octaflexagon (center) requires opening the top hinge plus any one of the three hinges opposite from it. The forced tuck (right) only needs to open a single hinge, but requires severe bending in order to perform.

Note that we use Tf for the precise meaning of a tuck flex that doesn't require any extra flaps to perform. In later chapters, we will see flexagons with enough extra freedom of movement that their tuck flex is Tf.

Summary

The following table lists useful facts about the tuck flex, and figure 11.11 shows how it changes the minimal flexagon.

full name: **tuck flex**　　　　shorthand: **T**
modified pats: **2**　　　　minimal flexagon: **hexaflexagon**

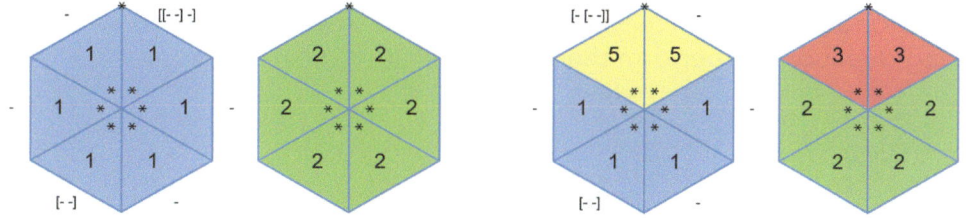

FIG. 11.11 Front and back of the starting state (left) and the state after applying the tuck flex (right).

Chapter 12

The Flip Flex

The **flip flex** is especially elegant on flexagons made from right triangles. It's called the flip flex because its primary motion is to flip pats from the front of the flexagon to the back. We'll try it out on several different types of flexagons to get a feel for how the flip flex behaves. Its shorthand is F.

Figure 12.1 shows the minimal silver octaflexagon for the flip flex.

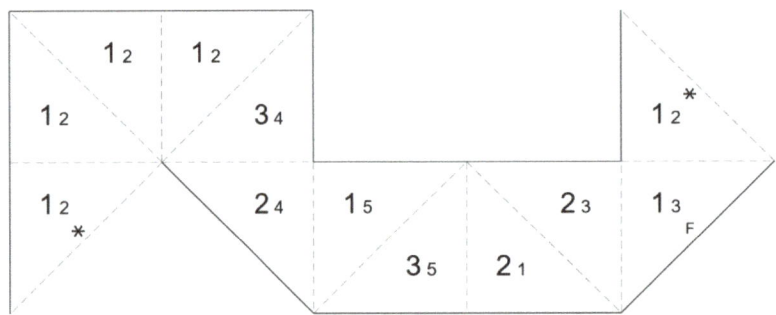

FIG. 12.1 Minimal silver octaflexagon for the flip flex.

Once you've folded and connected the template, find the reference hinge on face 1, marked with F. Orient the flexagon so the reference hinge is pointed up. Then refer to figures 12.2 and 12.3 along with the following directions to perform the flip flex.

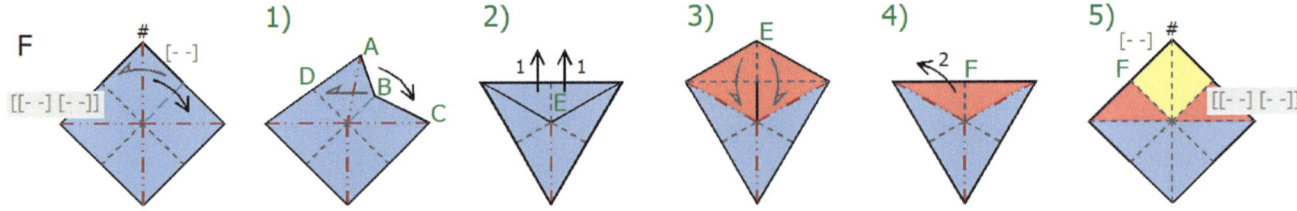

FIG. 12.2 Flex diagram for the flip flex on the silver octaflexagon.

1. Push corner B backward, bringing it against the back face of D. Fold A onto C.
2. Pull the center E upward, opening the top two pats from the front.
3. Continue by flipping E backward, eventually bringing it to the center on the back face.
4. Pull corner F to the left to open up the flexagon back to its original shape.
5. In the final state, note that the top four pats now show different leaves on the front and back, while the bottom four pats have stayed the same.
6. To return to the original state, turn the flexagon over from left to right and repeat steps 1–4.

DOI: 10.1201/9781003433538-14

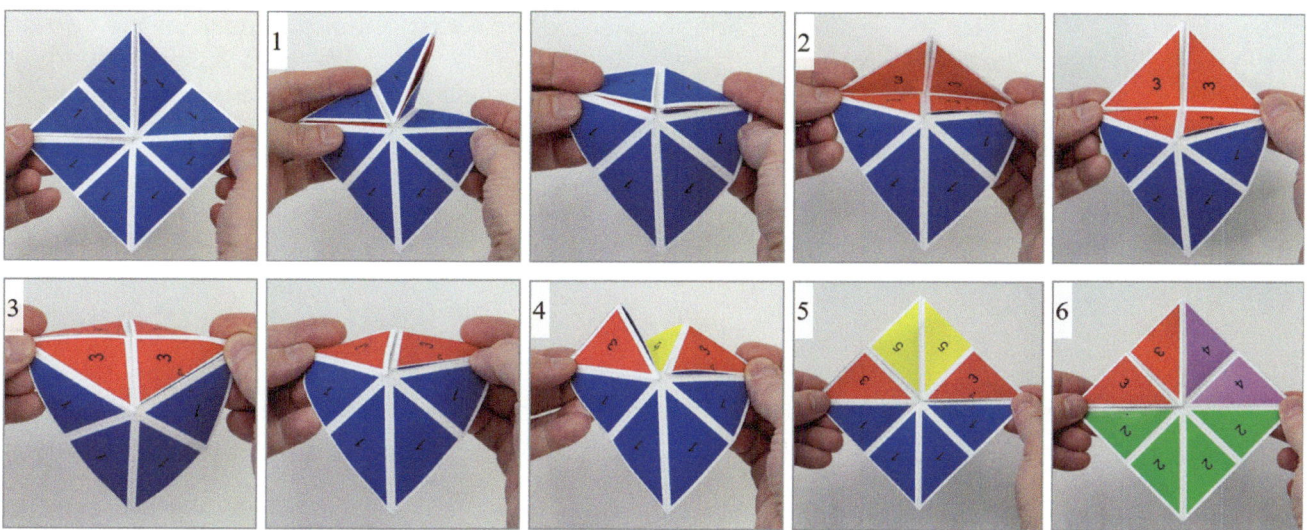

FIG. 12.3 Performing the flip flex on the silver octaflexagon.

Note that four of the eight pats have changed, while the others remain the same. Like the tuck flex, the flip flex is a local flex. But unlike the tuck, it changes four pats rather than just two.

While we saw the simplest *octaflexagon* that supports the flip flex, it's not the simplest possible flexagon. That honor goes to the bronze hexaflexagon, with six pats instead of eight. The minimal hexaflexagon only requires ten leaves to the octaflexagon's 12. Figure 12.4 shows the template for the minimal bronze hexaflexagon.

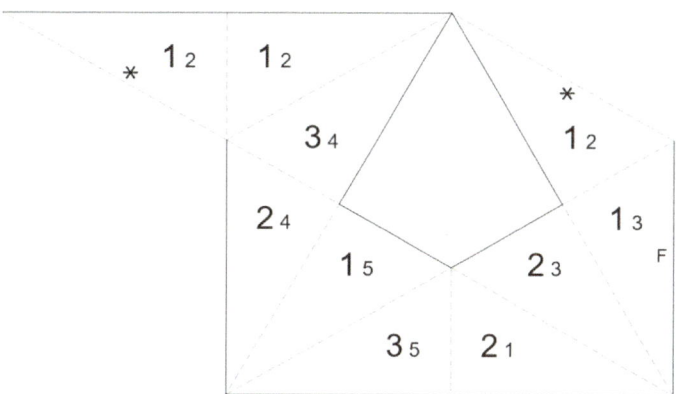

FIG. 12.4 Minimal bronze hexaflexagon for the flip flex.

You will need to be a little more careful doing the flip flex on the bronze hexaflexagon. During step 3, where you flip the pats back and tuck them into the middle, you may need to bend the pats a bit to get them into the center. This sort of thing happens with other flexes as well – a flex that's very elegant on one flexagon, such as the flip flex on the silver octaflexagon, may require a bit of bending on a more restrictive flexagon, such as the flip flex on the bronze hexaflexagon. Or it may work well at some hinges, but not others. In theory, you could even do the flip flex on a regular hexaflexagon made with regular triangles, but it would require so much bending that the paper might rip. We'll generally restrict ourselves to flexes that work within the tolerances of paper. They may sometimes require a bit of bending, but not too much.

Figure 12.5 is a **pentagonal decaflexagon** (ten pats per face, overall shape is pentagonal) that nicely supports the flip flex at five different hinges around the flexagon, one at each of the corners of the pentagon. Note that when you go to tape it, one of the leaves you need to tape will be in the middle of a pat. To make it easier to connect, put tape on the edge of the 3/4 leaf before you fold the flexagon so that it's easier to join the final two edges after folding.

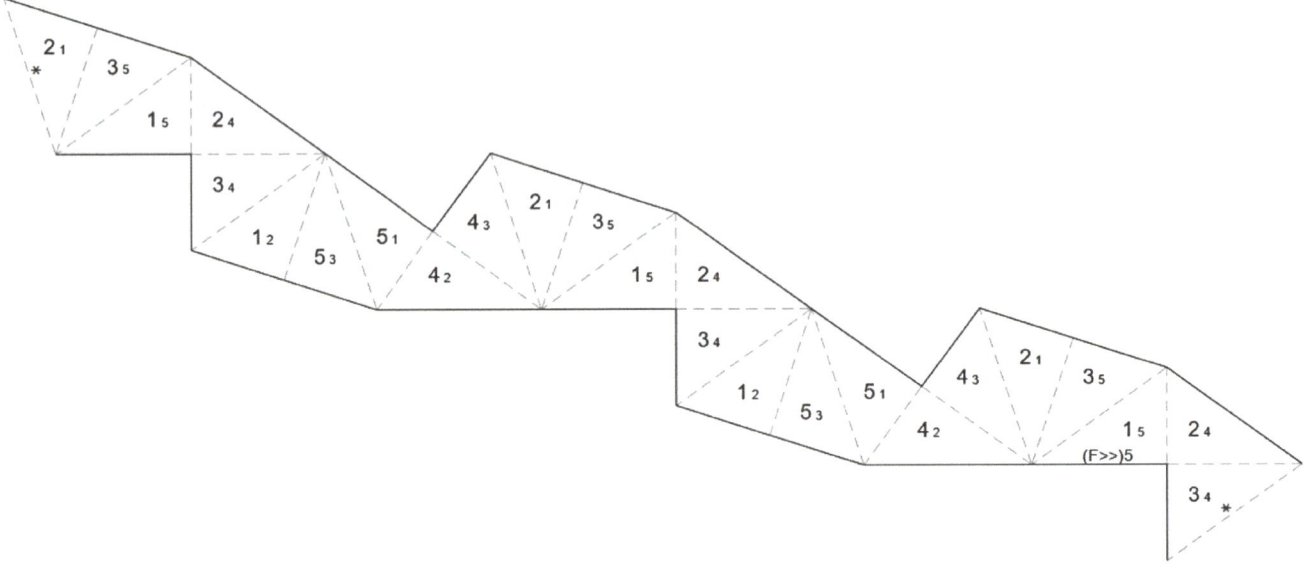

FIG. 12.5 Pentagonal decaflexagon that supports five flip flexes from its original state.

Figure 12.6 contains the template for a heptaflexagon that supports an elegant sequence of five flip flexes that brings you back to the original state. It's a little trickier to flex, however, since it has exactly enough freedom of motion to perform the flex without bending.

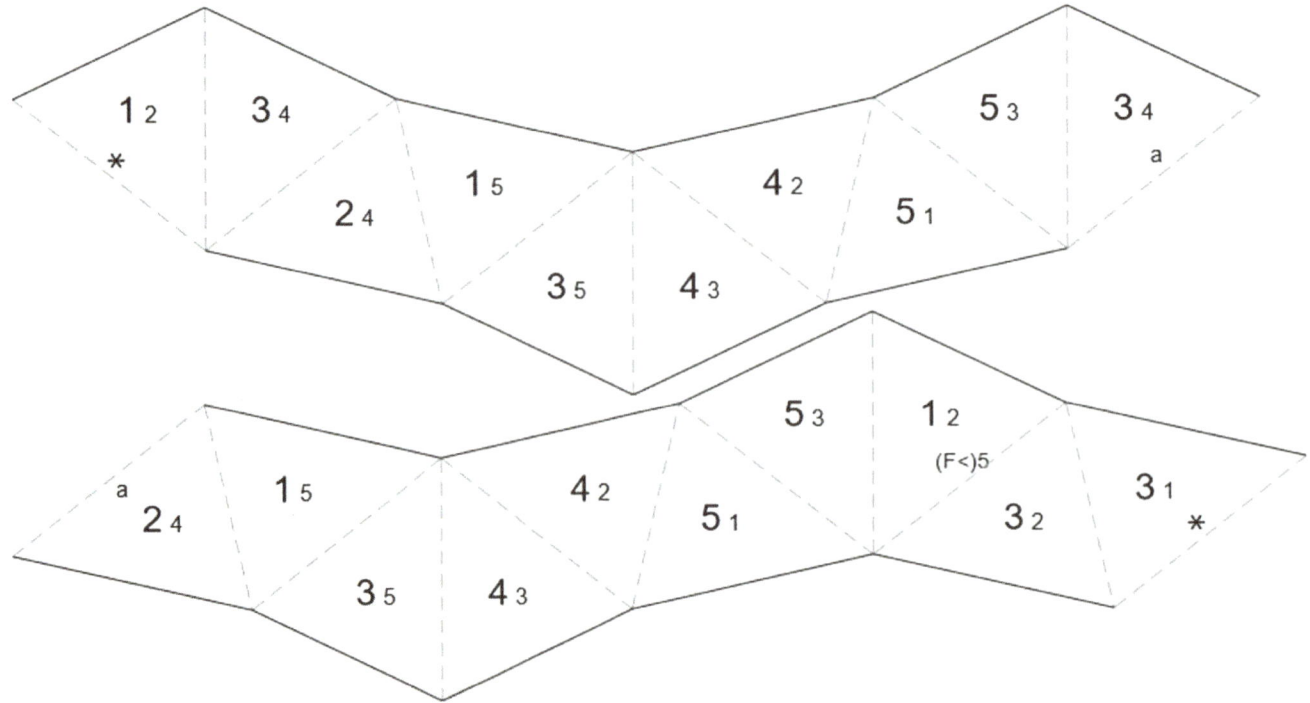

FIG. 12.6 Heptaflexagon that supports a series of flip flexes.

First, orient the flexagon so the reference hinge marked with (F<)5 is at the top. To perform a single flip flex on the heptaflexagon, refer to the flex diagram in figure 12.7 while following these steps:

1. Bring corners A and B together in the back.
2. Open a pocket from the center, pulling it up and to the left.
3. Push corner C up to meet the center of the pocket D in the back.
4. Open the pat at the bottom left. Note where corner E is to prepare for the next step.
5. This is the trickiest part, so look closely at the diagram. From behind, pull open the center toward the lower left. Push the top corner, E, backward while pulling corner F (in the center of the flexagon) down and to the left. Make sure that corner G in the back stays separate and moves to the right.
6. Push the top corner down, tucking it into the center of the flexagon in the back.
7. Open the first layer of the top left pat.
8. Continue opening the top left pat to flatten the flexagon back into a heptagon.
9. From here, you can turn it over and repeat steps 1–8 to return to the original position.

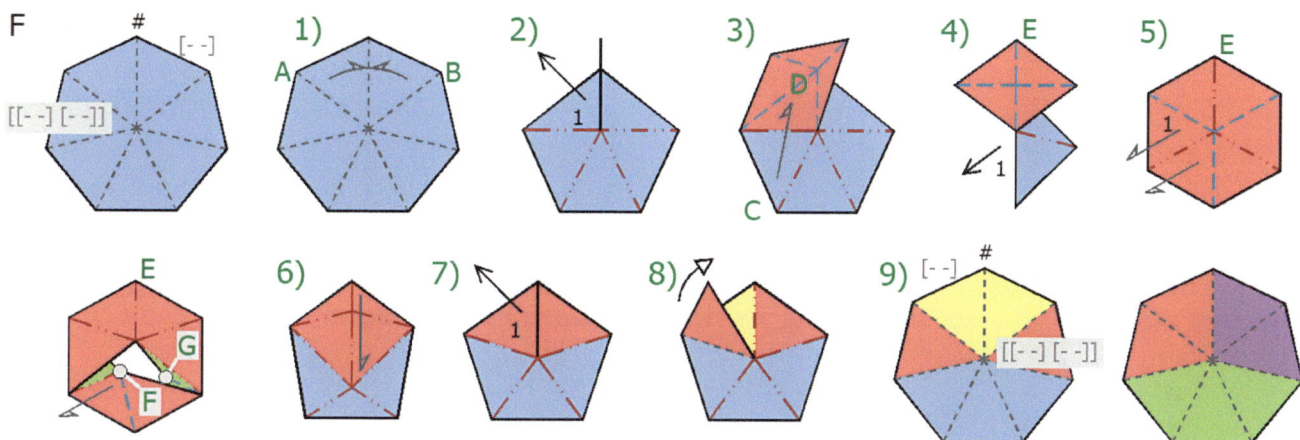

FIG. 12.7 Flex diagram for doing a flip flex on a heptaflexagon.

To do the full sequence of flip flexes, once you reach step 9, stay on the blue, red, and yellow face, shift one hinge counterclockwise and do a second flip flex. Repeat three more times, shifting to the counterclockwise hinge after each flip before doing the next flip flex. Figure 12.8 shows what the intermediate states look like immediately before you do the next flip flex, with the reference hinge oriented at the top. The result is that the flex sequence (F<)5 will cycle back to the original state with blue on the front and green on the back.

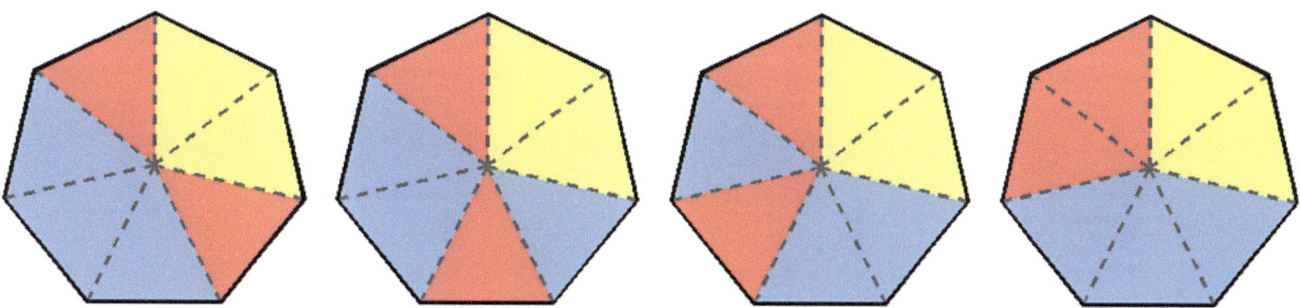

FIG. 12.8 Intermediate states for a sequence of flip flexes on the heptaflexagon.

Or, if you prefer, you can skip steps 7 and 8 where you open the flexagon to its flat state. From the non-flat pentagon state in step 7, you could instead rotate the pentagon clockwise one hinge and directly repeat steps 2–6. Repeat this process three more times before opening it back up to its original state. Figure 12.9 shows the intermediate pentagon positions you should see before doing the next flip flex.

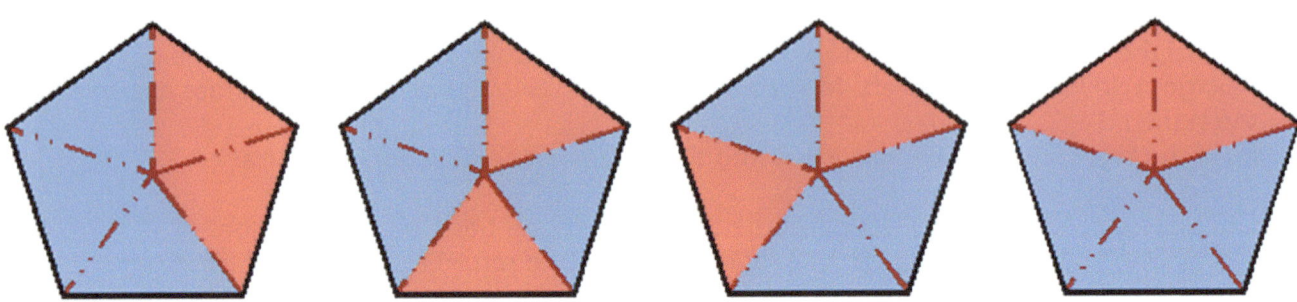

FIG. 12.9 Intermediate pentagonal states for a sequence of flip flexes on the heptaflexagon.

Summary

The following table lists useful facts about the flip flex, and figure 12.10 shows how it changes the minimal flexagon.

full name: **flip flex**   shorthand: **F**
modified pats: **4**   minimal flexagon: **bronze hexaflexagon**

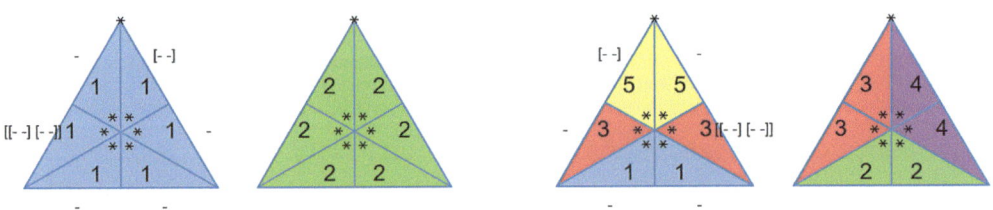

FIG. 12.10 Front and back of the starting state (left) and the state after applying the flip flex (right).

Chapter 13

The Pyramid Shuffle Flex

The **pyramid shuffle flex** gets its name from how the flex creates a pyramid and shuffles it to the opposite face before restoring the original shape. Figure 13.1 contains the template for the minimal hexaflexagon for the pyramid shuffle.

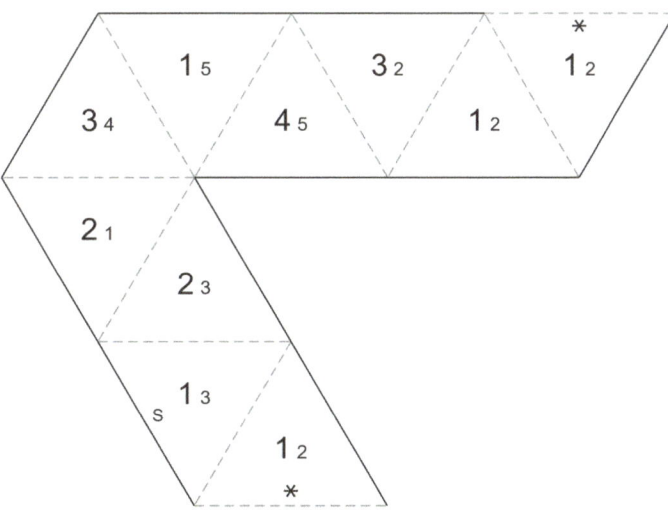

FIG. 13.1 Minimal hexaflexagon for the pyramid shuffle.

Refer to figure 13.2 or 13.3 while stepping through the following directions to perform the pyramid shuffle. Position the hinge labeled S at the top.

DOI: 10.1201/9781003433538-15

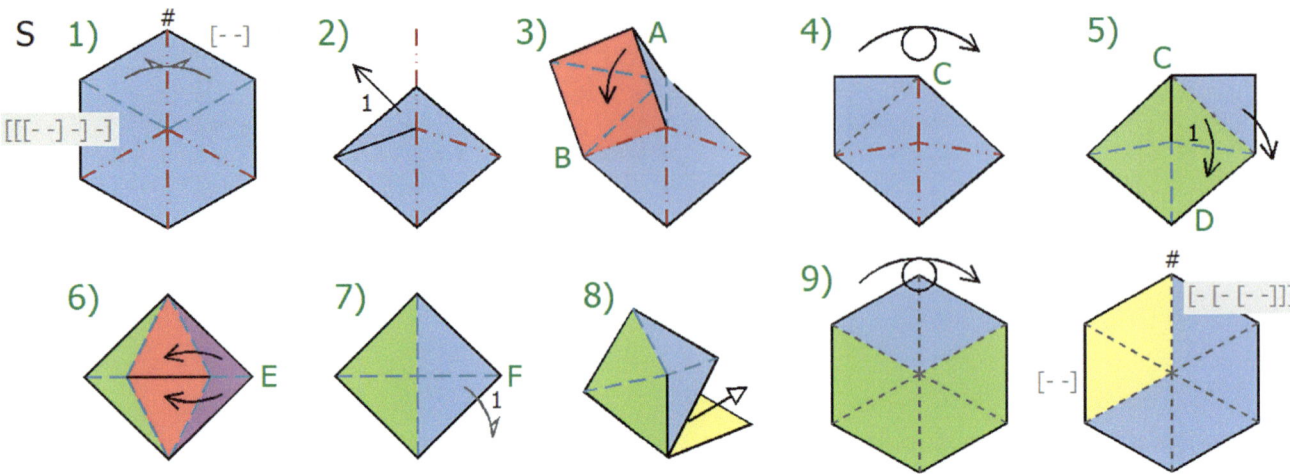

FIG. 13.2 Flex diagram for the pyramid shuffle on a regular hexaflexagon.

1. Fold together the back faces of the top two pats.
2. Open up the two upper left pats from the center of the flexagon.
3. Close the pocket by folding corner A against corner B so you're looking at the top of a pyramid.
4. Turn the flexagon over from left to right so you're looking at the inside of the pyramid.
5. Open the first layer of pats from corner C, folding it over to corner D.
6. Take corner E and tuck it into the center of the flexagon.
7. Unfold one layer of the pat from the back face of corner F.
8. Open the flexagon back out to a hexagon.
9. Turn it over from left to right to reach the final state.
10. To return to the original position, turn the flexagon over, rotate one hinge to the right (so 1 is to the left of the reference hinge at the top and 2 is to the right), and repeat steps 1–9.

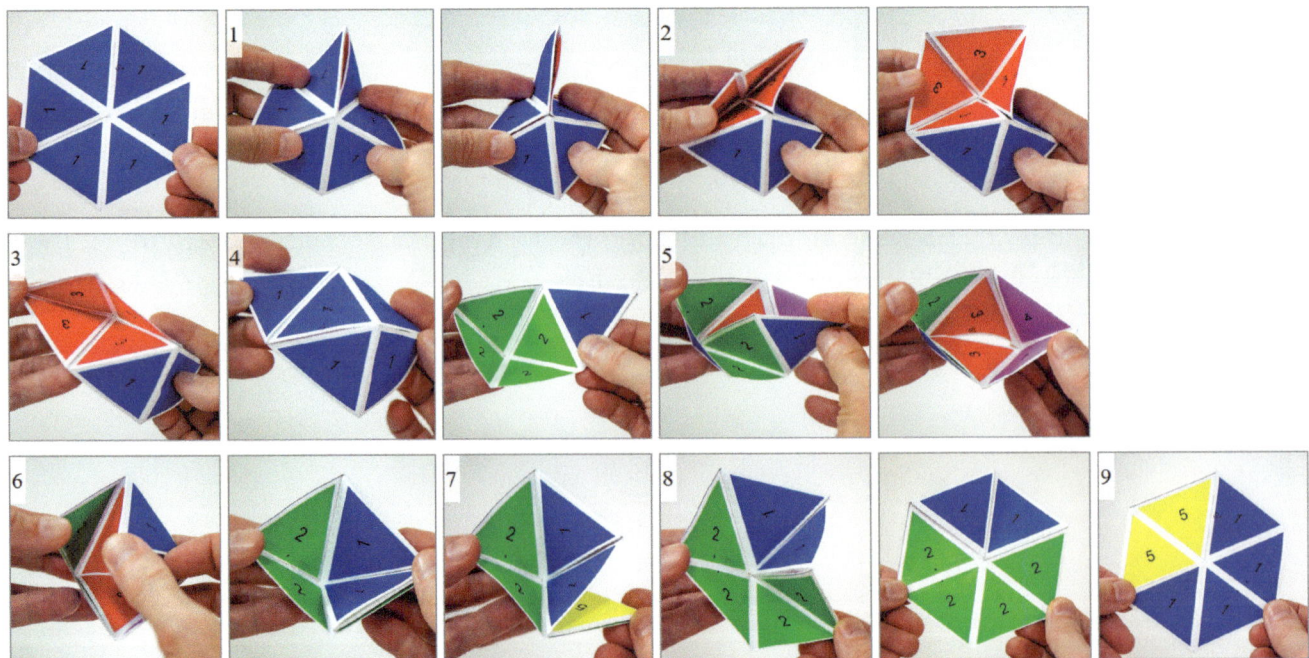

FIG. 13.3 Performing the pyramid shuffle on the hexaflexagon.

These directions have you turn over the flexagon for steps 4–8 to make it easier to see what you're doing. But you can flex more smoothly if you do the entire pyramid shuffle without turning it over, which also helps as you start combining multiple flexes together.

To do the pyramid shuffle entirely from one face, start with the same first three steps. But instead of turning over the flexagon at the beginning of step 4, use your right hand to open up one layer of leaves from the back side of corner C. Push them down so that you've opened a cup in the back (as shown in step 6). Take the corner sticking off to the left and tuck it into the center of the flexagon and open it back up to reveal the face with 1's and 5's. Once you get into the rhythm, you can think: open the pyramid, shuffle across the back face, tuck into the center, and open up.

Silver octaflexagon

Figure 13.4 contains the template for the minimal silver octaflexagon for the pyramid shuffle. Do the pyramid shuffle from face 1, with the reference hinge at the corner of the square marked with S. The numbers should match the above diagram, though the leaves are a slightly different shape. Note that, unlike the hexaflexagon, you don't need to bend the leaves in order to do the pyramid shuffle on the silver octaflexagon.

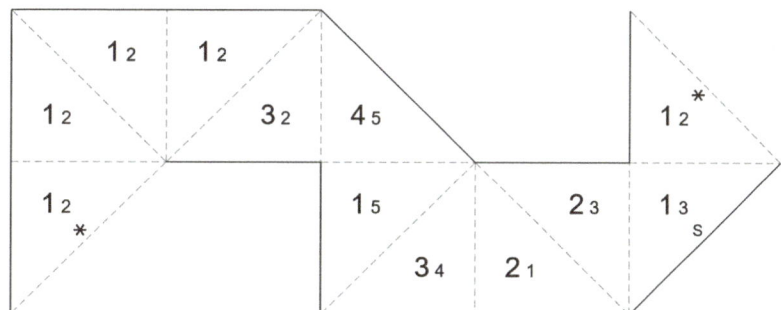

FIG. 13.4 Minimal silver octaflexagon for the pyramid shuffle.

As we saw with the hexaflexagon, you can return to the original state by doing a second pyramid shuffle, but this time, the reference hinge will be at an edge of the square rather than a corner. After the first pyramid shuffle, turn it over to the face that mostly has 2's on it. When you orient it so the two 1's are in the upper left corner, the reference hinge will be in the middle of the top edge, with 1's to the left and 2's to the right. Make a mountain fold at the reference hinge and lift up the left side from the center of the flexagon, revealing leaves with 3's and 4's. As in step #3 listed above, close the pocket by pressing its edges together. Turn it over left to right, taking you to a face with all 1's. With a single triangular pat sticking up, open the first layer in that pat and the one below it to reveal four 3's. Take the top corner and fold it into the center of the flexagon. You should now be able to open it up to face 1, returning you to the original state.

Pentaflexagon

The **pentaflexagon** (five pats on each face) in figure 13.5 is the simplest flexagon that supports the pyramid shuffle. It requires bending the leaves in order to do the flex.

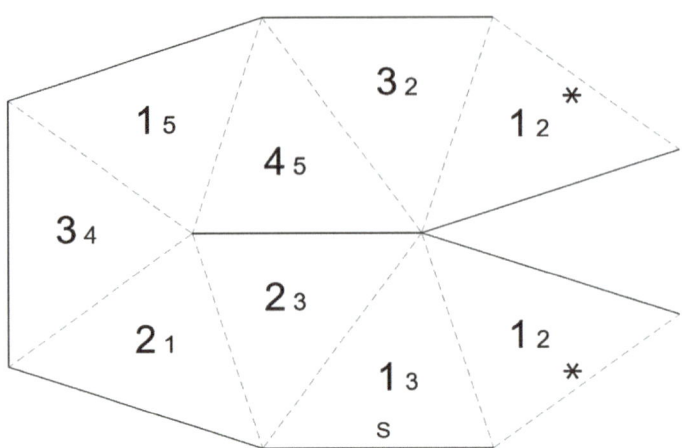

FIG. 13.5 Minimal pentaflexagon for the pyramid shuffle.

Try exploring the penta-hexaflexagon from figure 3.3 using only the pyramid shuffle starting from state 1/2. First off, you will notice that you can't do the pyramid shuffle from just any hinge. Examine the minimal pyramid shuffle hexaflexagon folded from figure 13.1 to see if you can recognize the conditions that support the pyramid shuffle. You should be able to reach eight different states with just this flex alone. Can you figure out how to do three pyramid shuffles to transform state 1/2 into state 5/1? This is an alternative to using three pinch flexes to make your way between those states.

Summary

The following table lists useful facts about the pyramid shuffle flex, and figure 13.6 shows how it changes the minimal flexagon.

full name: **pyramid shuffle**	shorthand: **S**
modified pats: **3**	minimal flexagon: **pentaflexagon**

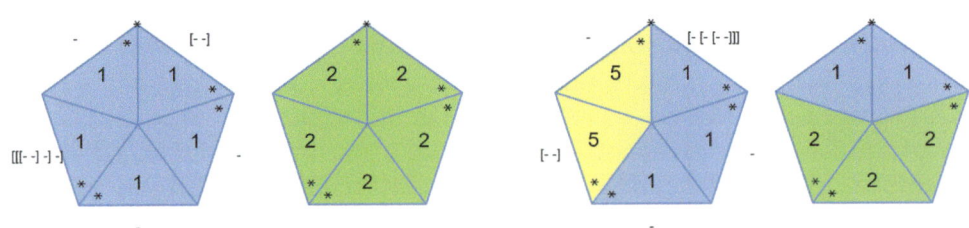

FIG. 13.6 Front and back of the starting state (left) and the state after applying the pyramid shuffle (right).

Chapter 14

Breaking Down Flexes

We've now seen a variety of different flexes across multiple types of flexagons. With the pinch flex, we discussed a technique that allowed us to come up with new variants: only pinch every few hinges to make the P333 or P345 or others. Is there also a way to generalize the local flexes we've seen – the tuck flex, flip flex, and pyramid shuffle – to create new flexes?

Splitting apart flexes

To answer that, let's start by taking a closer look at the flip flex on a silver octaflexagon as we saw at the beginning of chapter 12. Move slowly through the steps of the flip flex, as well as the inverse flip flex where you do everything in reverse. Notice that halfway through the flex (the beginning of step 3 in the flex diagram in figure 12.2), you reach a state that looks like a kite. From this position, you can either fold the top two pats backward and open up the flexagon or fold them forward and open up the flexagon.

The two halves of the flip flex are symmetrical, with the first half done from the front and the second half from the back of the flexagon. A flip flex on the silver octaflexagon turns the square into a kite by opening from the *front*, then restores the square by closing it from the *back*.

The **morphing flex** that changes the shape of the flexagon into a kite by unfolding from the front is called the **morph-kite front flex**, or Mkf. The flex that unfolds from the back is called the **morph-kite back flex**, or Mkb. We can state that the flip flex is the same as doing the *morph-kite front* followed by the *inverse of the morph-kite back*. We can express this as F = Mkf Mkb'.

See figures 14.1 and 14.2 for the flex diagrams for the Mkf and Mkb' flexes. Unsurprisingly, they should look just like the two halves of the flip flex. You can either practice them on any of the flexagons that support the flip flex or use the template in figure 14.3 to make the minimal flexagon for the morph-kite front. After you do the morph-kite front on it, you can turn it over to do the inverse of the morph-kite back.

To do Mkf, the morph-kite front flex, refer to figure 14.1 while following these steps:

1. Push corner B backward, bringing it against the back face of D. Fold A onto C.
2. Pull the center E upward, opening the top two pats from the front.
3. If you had a flat square in step 1, you will now have a kite that doesn't lie flat.

DOI: 10.1201/9781003443538-16

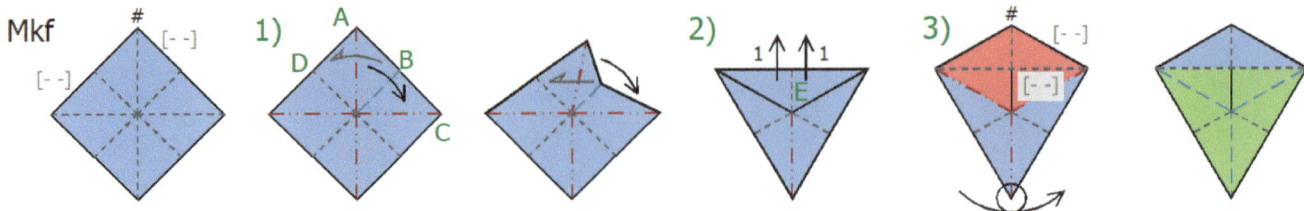

FIG. 14.1 Flex diagram for the morph-kite front flex (Mkf).

To do an inverse morph-kite back flex, or Mkb', refer to figure 14.2 while following these steps:

1. Fold the top corner A backward, eventually bringing it to the center on the back face.
2. Open from the top edge hinge B.
3. If you had a non-flat kite in step 1, you will now have a flat square.

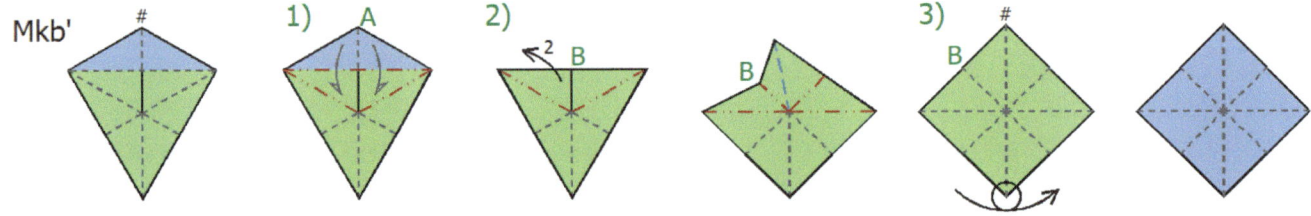

FIG. 14.2 Flex diagram for the inverse morph-kite back flex (Mkb'). This is colored assuming you're starting from the final step in figure 14.1.

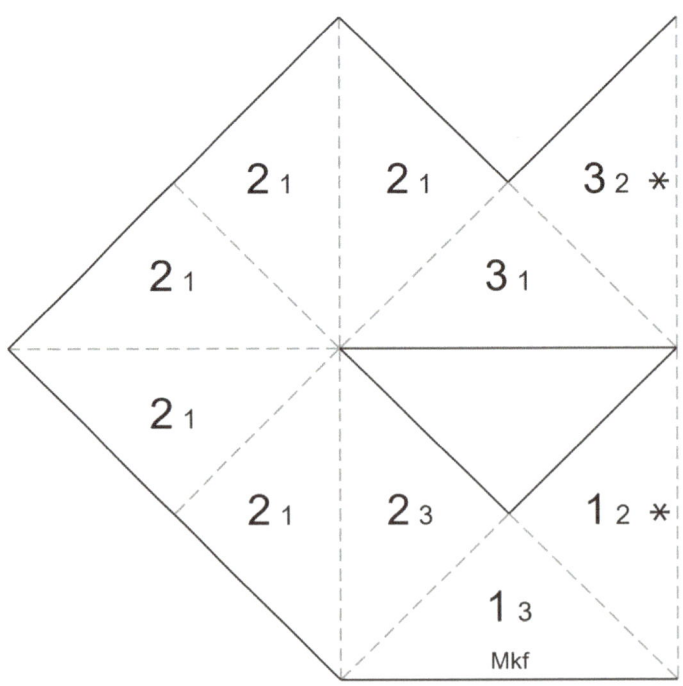

FIG. 14.3 Template for the minimal silver octaflexagon for the morph-kite front flex.

Next, let's try breaking the pyramid shuffle into two halves. You may wish to try this on the minimal silver octaflexagon from figure 13.4 to make it easier to compare to the morphing flexes we just explored. You should find that the second half of the pyramid shuffle is exactly the same as the second half of the flip flex, while the first half looks different. We call the first half the **morph-kite front shuffle flex**, or Mkfs. Therefore, the pyramid shuffle is the same as the *morph-kite front shuffle* followed by the *inverse of the morph-kite back*. To be precise, we take into account the reference hinge, giving us S = < Mkfs Mkb' >.

Figure 14.4 contains the flex diagram for the morph-kite front shuffle. The template for the corresponding minimal silver octaflexagon is in figure 14.5.

1. Fold together the back faces of the two pats in the upper right.
2. Open up the two upper right pats from the center of the flexagon.
3. Close the pocket by folding corner A against corner B.
4. From corner C, grab the back layer of pats and swing them across the centerline of the flexagon.
5. The resulting kite should have the same shape as the results of Mkf.

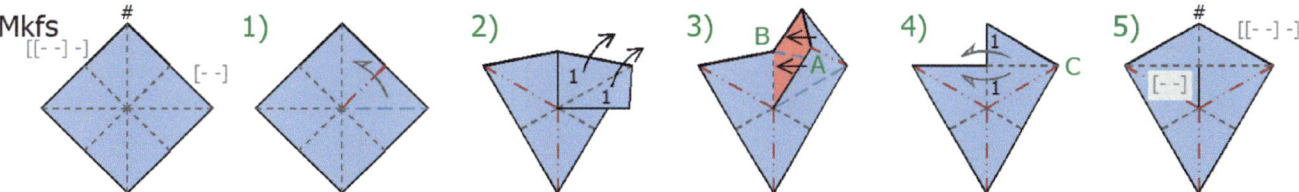

FIG. 14.4 Flex diagrams for the morph-kite front shuffle flex (Mkfs).

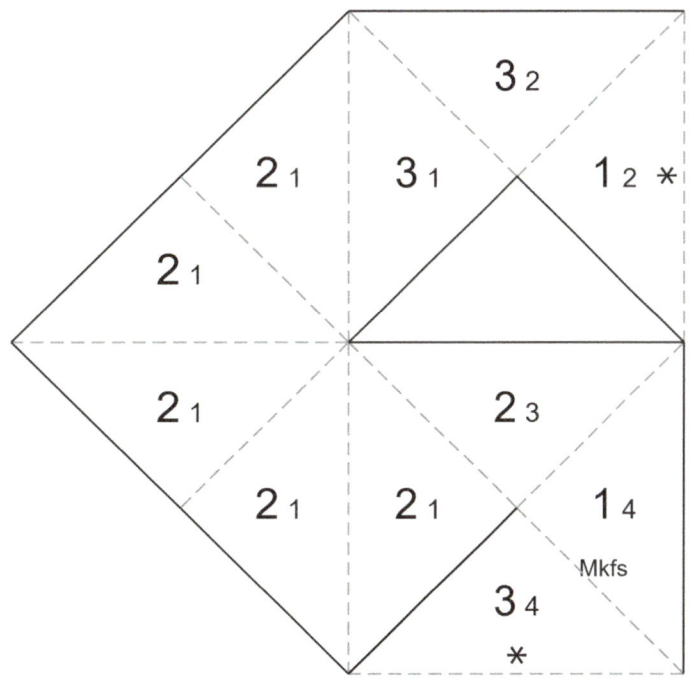

FIG. 14.5 Minimal silver octaflexagon for the morph-kite front-shuffle flex.

Finally, let's look at the tuck flex. Again, it may be useful to try it on a silver octaflexagon that supports the tuck flex, such as the one in figure 11.7. The second half should look familiar – it's the Mkf' we saw earlier. But the first half is new. We'll call it the **morph-kite right flex**, or Mkr. Therefore, the tuck is the same as the *morph-kite right* followed by the *inverse morph-kite front*. Factoring in the reference hinge gives us T = < Mkr Mkf' >.

See the flex diagram for the morph-kite right in figure 14.6 and the template for the corresponding minimal silver octaflexagon in figure 14.7.

1. Fold together the back faces of the two pats in the upper right.
2. Open up the three upper right pats from the center of the flexagon.
3. Close the pocket by folding corner A against corner B.
4. The resulting kite should have the same shape as the results of Mkf or Mkfs.

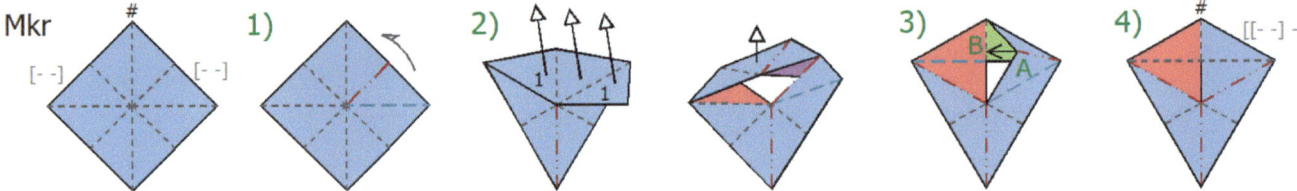

FIG. 14.6 Flex diagram for the morph-kite right flex (Mkr).

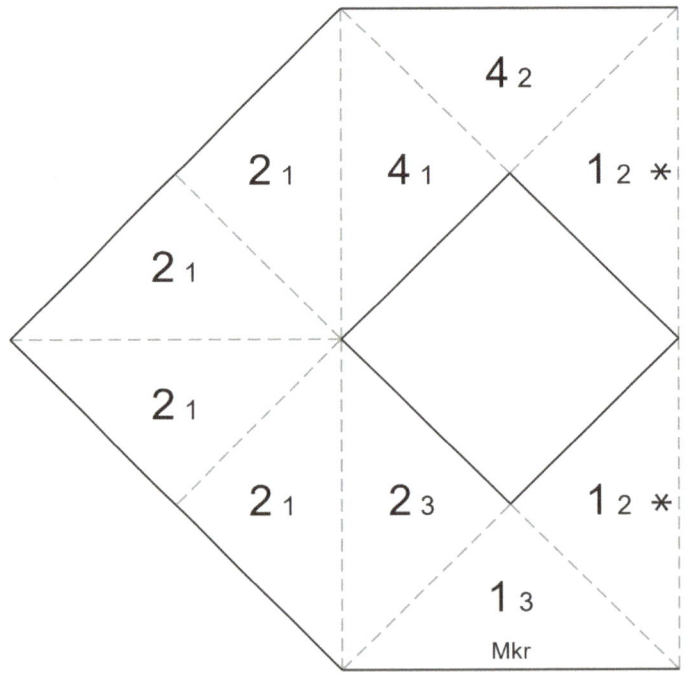

FIG. 14.7 Minimal silver octaflexagon for the morph-kite right flex.

Recombining the pieces

Now that we have several morph-kite flexes, we can try combining them in new ways. The template in figure 14.8 supports all the morph-kite flexes we've discussed, so it can be used to try out new combinations. Next, we'll walk through a few of the different combinations you can do on this flexagon.

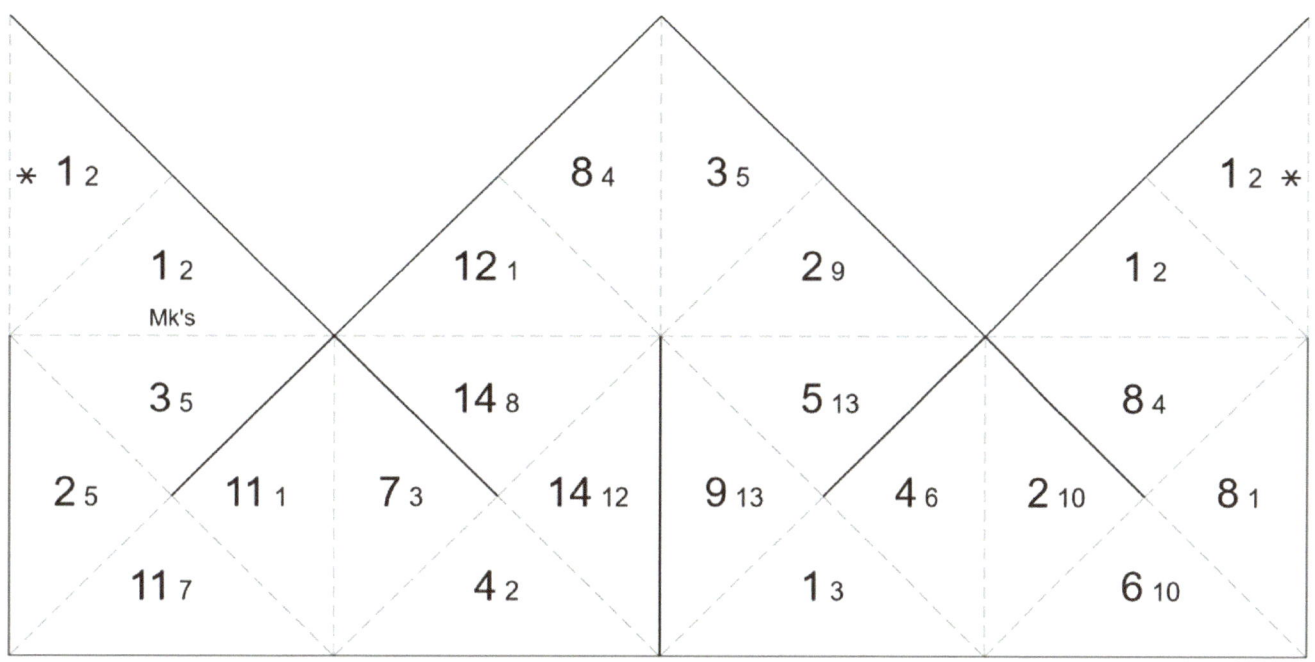

FIG. 14.8 Template for a silver octaflexagon that supports a wide variety of morphing flexes.

If we do Mkf followed by Mkfs', we get the **_silver tetra flex_** (St), so called because it's how you flex on the silver tetraflexagon. You can find the flex diagram for the silver tetra in figure 14.9. You can either use the general morphing flexagon from figure 14.8 or the minimal octaflexagon from figure 14.10. After you do the silver tetra, can you figure out how to do two tuck flexes to restore the flexagon to its original state? See figure 14.11 for the before and after states of the silver tetra.

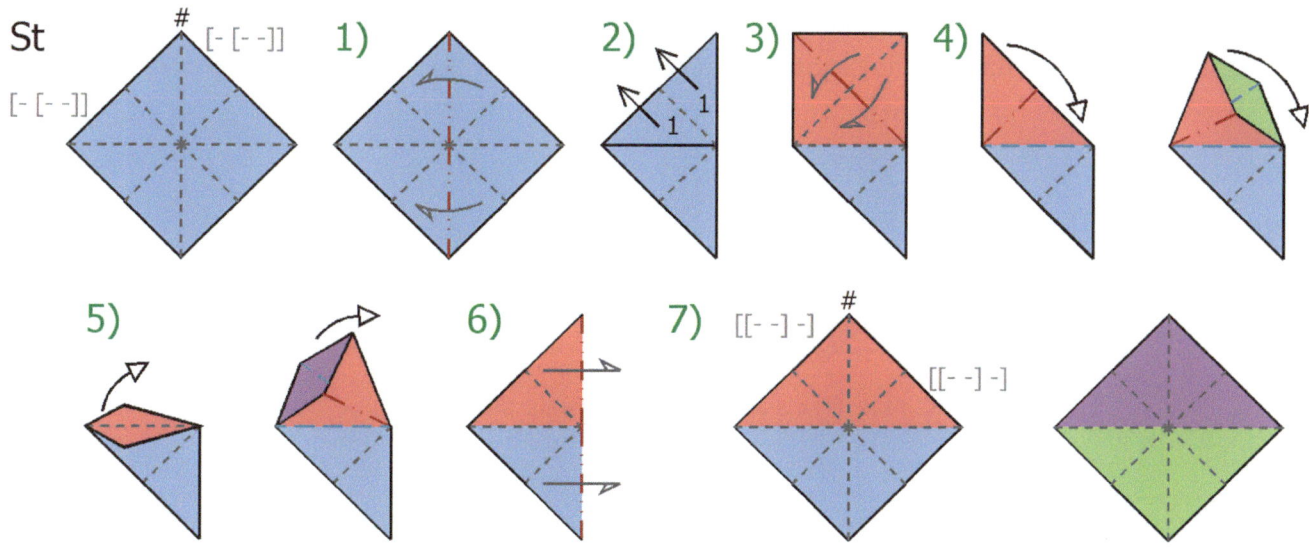

FIG. 14.9 Flex diagram for the silver tetra flex (St).

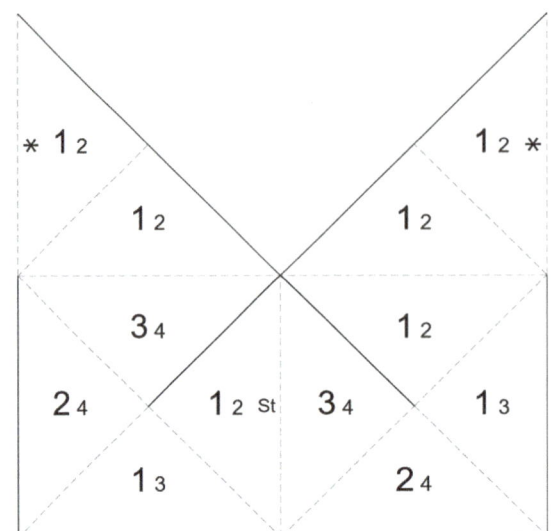

FIG. 14.10 Minimal silver octaflexagon for the silver tetra flex.

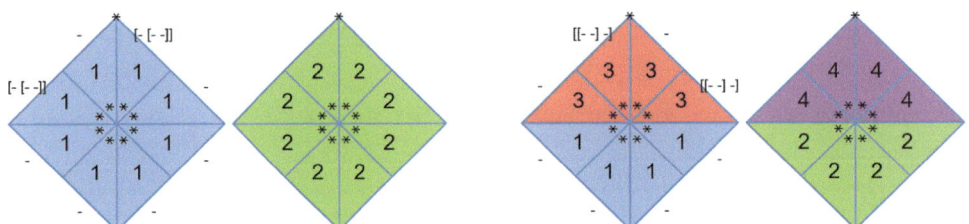

FIG. 14.11 Front and back of the starting state (left) and the state after applying the silver tetra flex (right).

If we try Mkr followed by Mkb', we get a flex called the **Möbius flip** (Fm), named because you first open up to a Möbius strip before doing a flip. Find the flex diagram in figure 14.12 and the template for the minimal octaflexagon in figure 14.13. Figure 14.14 shows the before and after states.

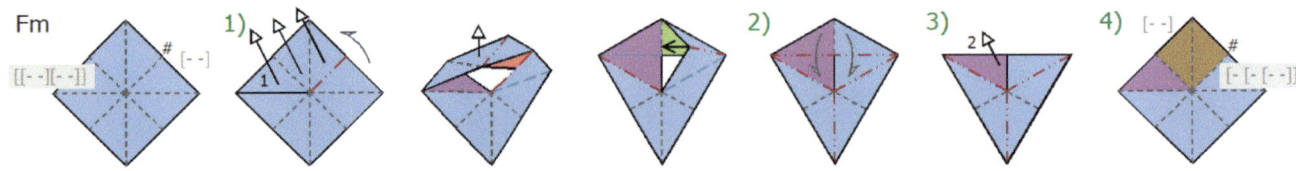

FIG. 14.12 Flex diagram for the Möbius flip flex (Fm).

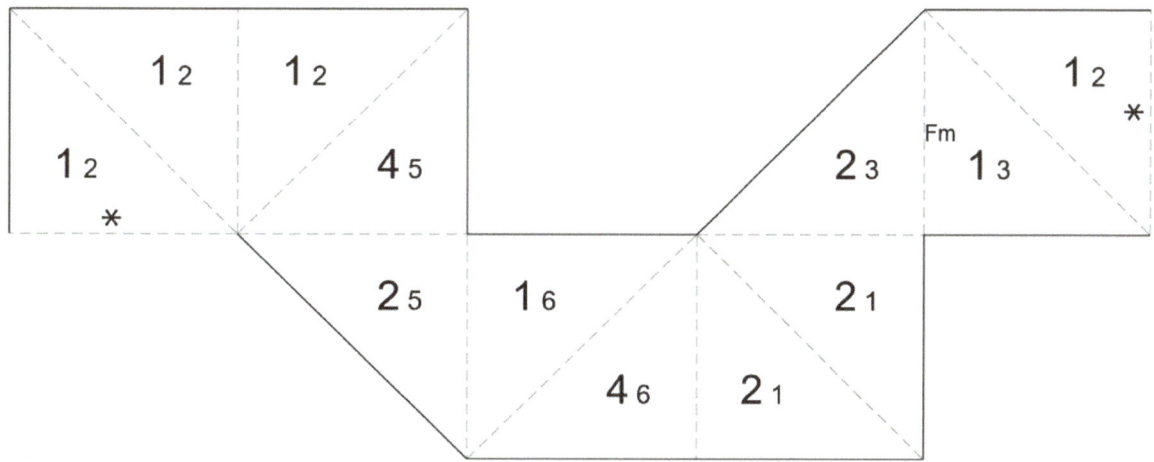

FIG. 14.13 Minimal silver octaflexagon for the Möbius flip flex.

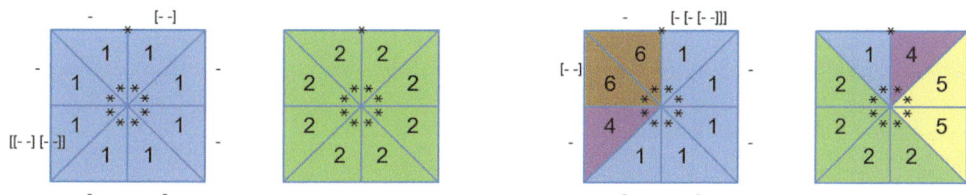

FIG. 14.14 Front and back of the starting state (left) and the state after applying the Möbius flip (right).

Next we define the ***morph-kite left flex***, or Mkl, as the mirror image of Mkr — same moves, but from the left instead of the right. If we try Mkr followed by Mkl', we get the ***pyramid shuffle 3 flex***, since it's very similar to the pyramid shuffle except that it works across three pats at a time rather than just two. See figure 14.15 for the flex diagram, figure 14.16 for the minimal octaflexagon, and figure 14.17 for the before and after states.

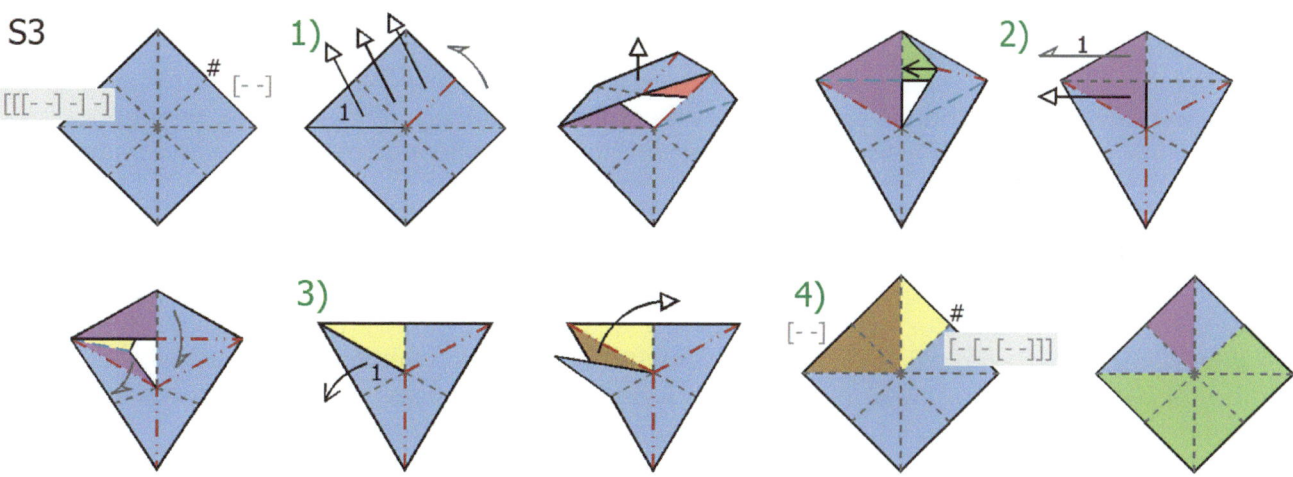

FIG. 14.15 Flex diagram for the pyramid shuffle 3 (S3).

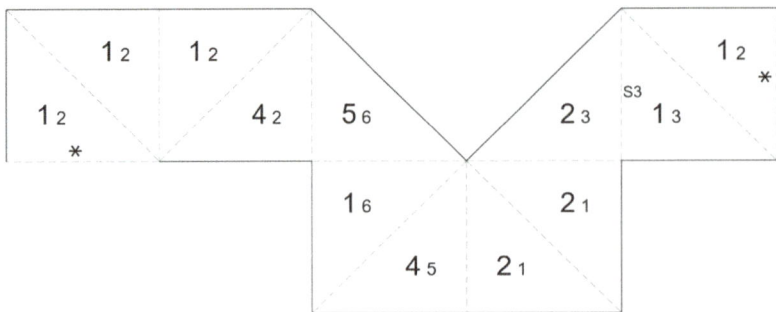

FIG. 14.16 Minimal silver octaflexagon for the pyramid shuffle 3 flex.

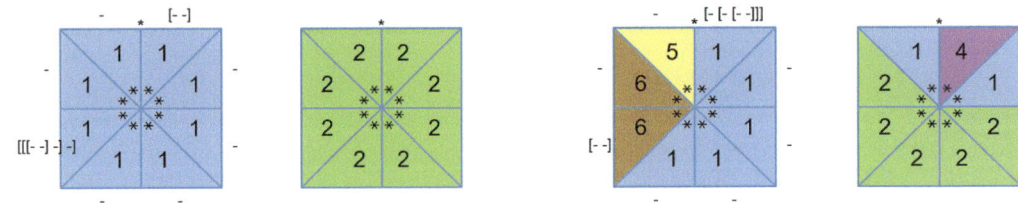

FIG. 14.17 Front and back of the starting state (left) and the state after applying the pyramid shuffle 3 (right).

We've now defined Mkf, Mkb, Mkr, Mkl, and Mkfs. This allows us to express several of the flexes we've seen in terms of these morph-kite flexes:

F	=	Mkf Mkb'	
S	=	< Mkfs Mkb' >	{when the structure supports the right side}
T	=	< Mkr Mkf >	{when the structure supports the right side}
St	=	Mkf Mkfs'	
Fm	=	< Mkr Mkb' >	
S3	=	< Mkr Mkl' >	

In the same way the morph-kite front has the similar morph-kite back and the morph-kite right has the morph-kite left, the morph-kite front shuffle has a similar morph-kite back shuffle. Can you figure out how to do the **morph-kite back shuffle flex** (Mkbs)? Can you create other new flexes by combining these six morph-kite flexes in new ways?

We used the square silver octaflexagon to demonstrate the morph-kite flexes, but they work on other flexagons as well. When we did the tuck flex on a hexaflexagon, the flip flex on a heptaflexagon, and the pyramid shuffle on a pentaflexagon, those flexes were made from two morph-kite flexes just as we saw on the octaflexagon. This means that our technique of breaking the flexes into two pieces works across a wide variety of other flexagons, which also means they can be combined on a wide variety of flexagons.

An interesting additional note about these morphing flexes is that the intermediate kite state they take you to is itself a valid flexagon, a non-flat kite-shaped flexagon with eight pats per face called the **kite silver octaflexagon**, as shown in figure 14.18. When you flex to a flexagon with the same number of pats that are arranged in a different overall shape, the other flexagon is called its **morph**. The square octaflexagon and kite octaflexagon are morphs of each other.

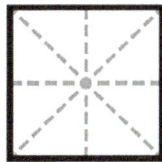 

FIG. 14.18 The square silver octaflexagon and one of its morphs, the kite silver octaflexagon. Note that the silver kite doesn't lie flat.

Reversing the morph-kite flexes

To recap, we just decomposed several flexes into two pieces and recombined them in different ways to create new flexes. These morph-kite flexes transform a set of pats that meet at a common point into a kite arrangement, with the inverse flexes taking us from the kite back to the original shape.

What if we reverse the order, instead starting and ending with the kite? We would then do an inverse morph-kite flex that takes us out of the kite shape, followed by a different morph-kite flex that returns the flexagon to the kite shape.

To investigate this idea, let's take another look at the kite octaflexagon. But this time, we'll make it out of bronze triangles rather than silver triangles so that it's flat in its main position. This gives us the **kite bronze octaflexagon** shown in figure 14.19.

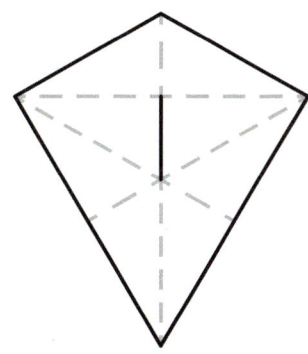

FIG. 14.19 The kite bronze octaflexagon.

As our first experiment, let's reverse the two halves of the flip flex, doing the inverse of the morph-kite front, Mkf', followed by the morph-kite back, Mkb. This means we'll do Mkf' Mkb rather than the flip flex's breakdown of Mkf Mkb'. Since this is the backward version of the flip flex, we call this the **backflip flex**, or Bf for short. Figure 14.20 provides the template for the minimal kite bronze octaflexagon for the backflip, and figure 14.21 shows the flex diagram.

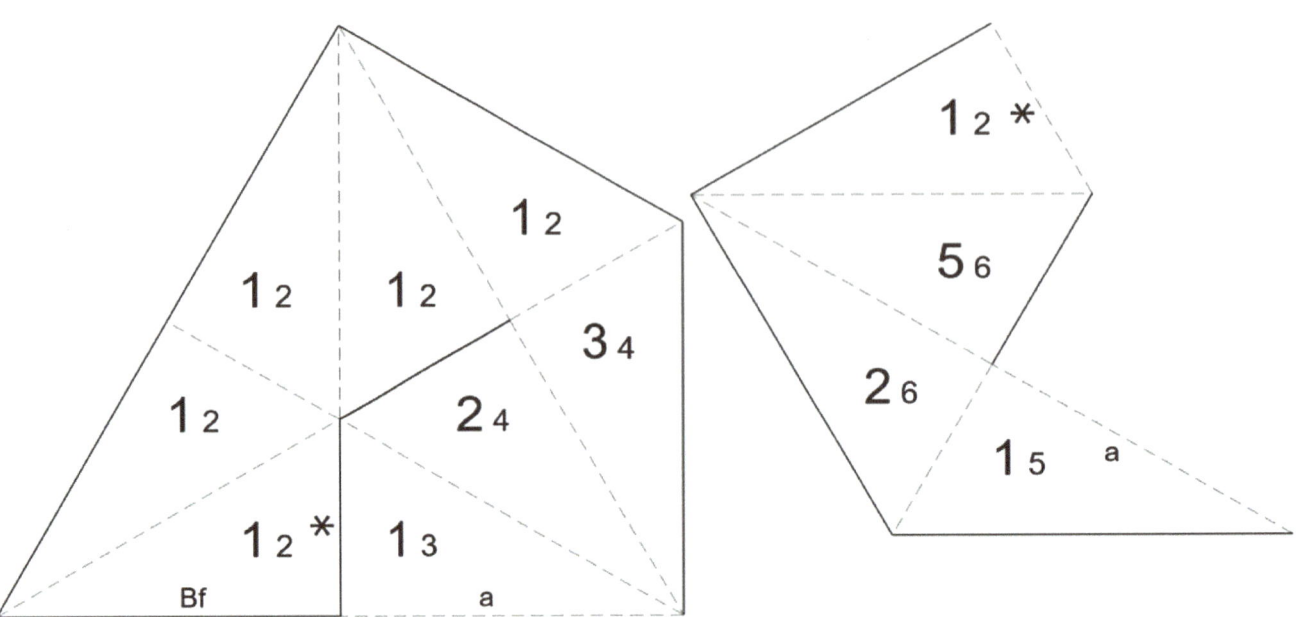

FIG. 14.20 Template for the minimal kite octaflexagon for the backflip flex.

1. Fold the top corner forward onto the center of the flexagon.
2. From corner A, grab the top two layers, pushing them to the left and opening them up. Push A over to opposite corner B while pushing corner C forward to the left.
3. Fold corner C onto D in the middle of the top edge.
4. Open the top layer of the back face from the center, folding it up from the top of the kite.

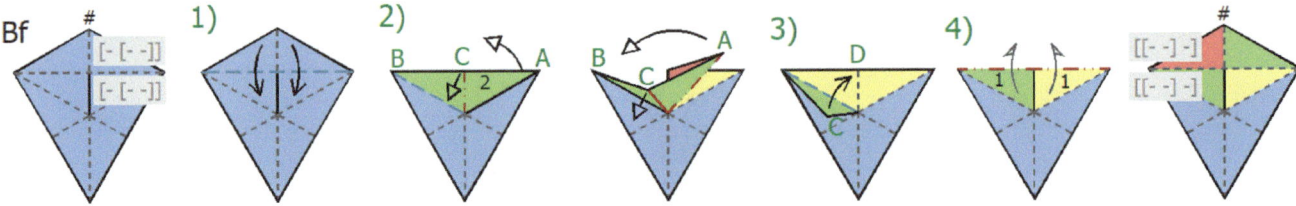

FIG. 14.21 Flex diagram for the backflip flex, Bf.

Just as we did with the square silver octaflexagon, we can make a kite bronze octaflexagon that supports all the morph-kite combos. See figure 14.22 for the template. With this flexagon, you can do any of the inverse morph-kite flexes from this chapter (Mkf', Mkb', Mkr', Mkl', Mkfs', and Mkbs'), followed by any other morph-kite to build a new flex that starts and ends with the kite shape. Besides the backflip, some other interesting combos to try are Mkr' Mkl, Mkf' Mkr, and Mkr' Mkfs.

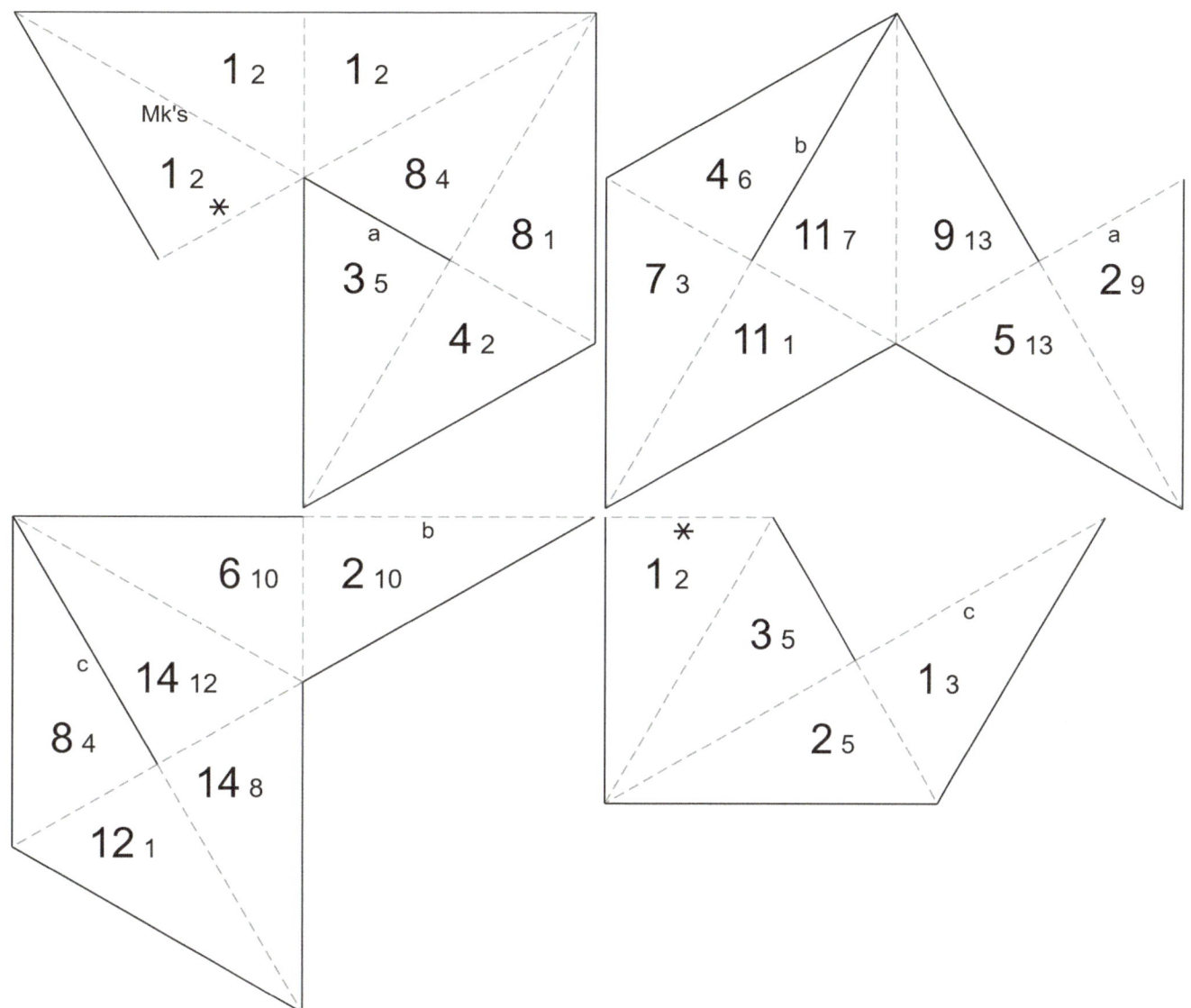

FIG. 14.22 Template for a kite bronze octaflexagon that supports a wide variety of morph-kite flexes. When you initially fold it, you may wish to use paper clips to keep the pats together.

Pocket flex

We first saw the pocket flex on the regular triangle tetraflexagon in chapter 4. As we noted then, it's the pinch flex when you only have four pats. Its shorthand is K.

But the pocket flex turns out to be especially interesting because pats on any triangle flexagon can possibly open into a pocket, sometimes even supporting a series of pocket flexes. The astute reader may notice that some of the flexes we've seen contain one or more pocket flexes as part of the overall flex movements. For example, the first step of the pyramid shuffle opens up a pocket, while the first step of the v-flex opens a pocket in the middle of the flexagon. This means that, like the morph-kite flexes, the pocket flex can be used as part of a more complex flex.

To get a better feel for the pocket flex, let's look at how we can do a series of pocket flexes to emulate the pinch flex. First, we'll do two pocket flexes on the tri-hexaflexagon we made in chapter 1. Position the tri-hexaflexagon so that one of the pats with two leaves is in the upper right, then walk through the following steps while referring to figure 14.23:

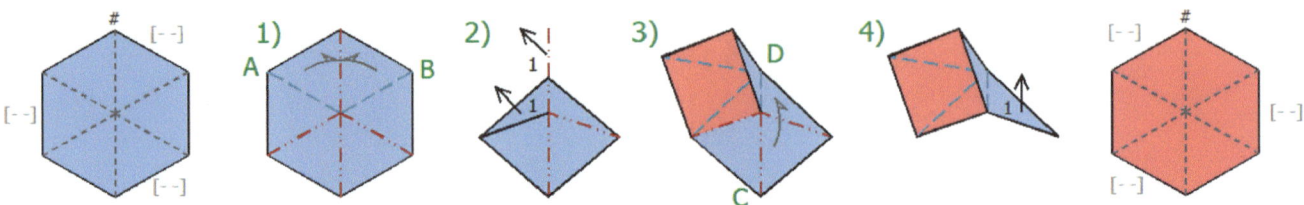

FIG. 14.23 Flex diagram showing two pocket flexes that have the same effect as a pinch flex.

1. Bring corners A and B together in the back of the flexagon.
2. Open a pocket from the center of the flexagon toward the upper left.
3. Push corner C up until it meets corner D.
4. Open the top layer of the rightmost pat to reveal face 3.

A more compelling example applies a series of six pocket flexes. Use the penta-hexaflexagon from figure 3.3 starting in state 1/2. Position one of the pats with multiple leaves so it's in the upper right, then walk through the following steps while referring to figure 14.24:

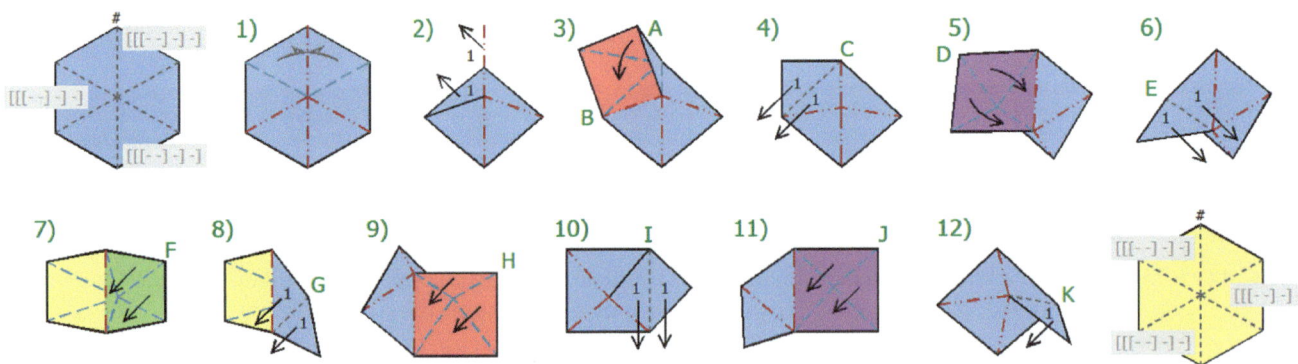

FIG. 14.24 Flex diagram showing six pocket flexes that have the same effect as three pinch flexes.

1. Bring the top left and right corners together in the back of the flexagon.
2. Open a pocket from the center of the flexagon toward the upper left by folding open the top layer of the two pats.
3. Close the pocket by folding corner A down to corner B.
4. Open a pocket from corner C by folding open the top layer of the two pats.
5. Close the pocket by folding corner D down to the center.

6. Open a pocket from corner E by folding open the top layer of the two pats.
7. You should now be looking at the inside of two adjacent pyramids, one with 5's and one with 2's. Close the pocket with 2's on the inside by folding corner F down to the left.
8. Open a pocket from corner G by folding open the top layer of the two pats.
9. Close the pocket by folding corner H down to the left.
10. Open a pocket from corner I by folding open the top layer of the two pats.
11. Close the pocket by folding corner J down to the left.
12. Open from corner K, turning over the flexagon to reveal face 5.

On the hexaflexagon as we just saw, you can perform two independent sets of pocket flexes from opposite ends of the flexagon. Additional pats mean additional possibilities for pocket flexes. To see an example, let's take a look at the **octagonal isosceles penta-octaflexagon** – octagonal in shape with isosceles triangles for leaves, five faces (*penta*), and eight pats (*octa*). First, fold the template in figure 14.25.

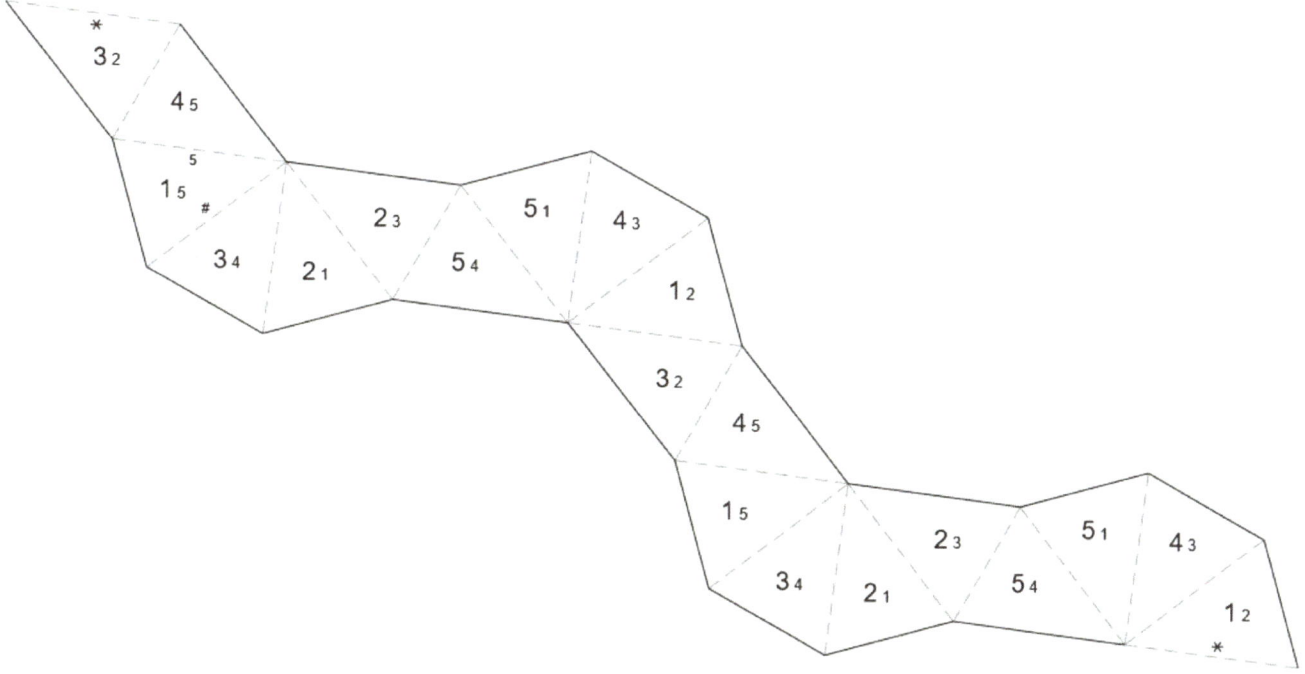

FIG. 14.25 Template for the octagonal isosceles penta-octaflexagon.

To do a series of three pocket flexes in the middle of this octaflexagon, refer to figure 14.26 while following along with these steps:

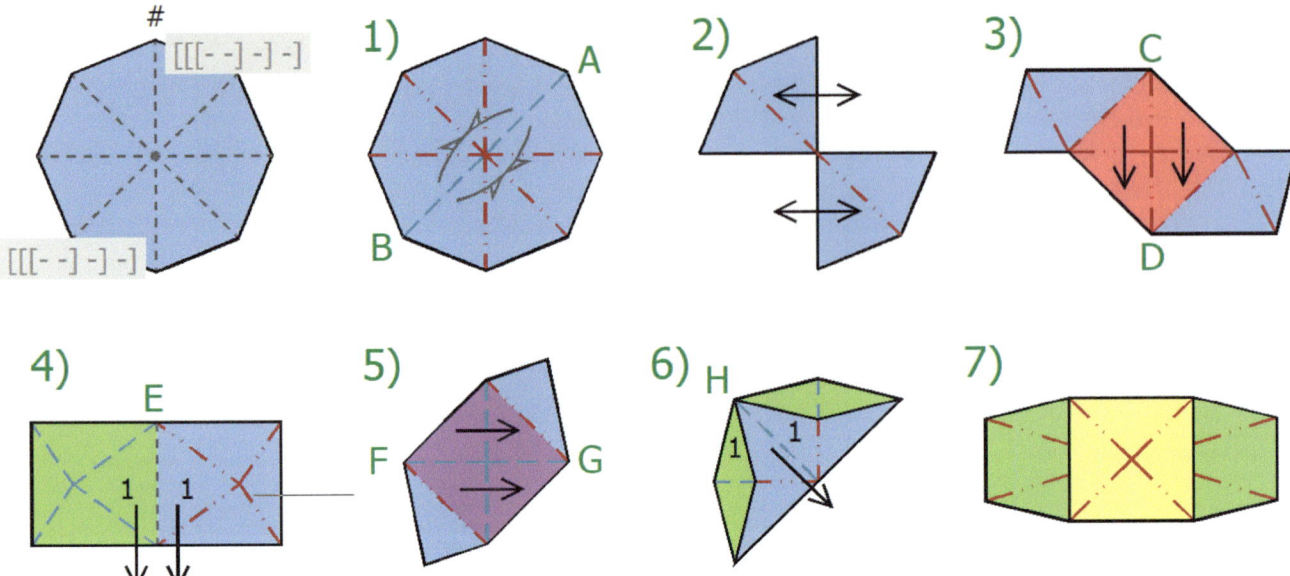

FIG. 14.26 Flex diagram for three pocket flexes in the middle of an octaflexagon.

1. Fold corners A and B backward until they meet.
2. Open a pocket in the middle by pulling back on the left and right halves of the flexagon.
3. Close the pocket by folding corner C down onto corner D.
4. Open a pocket by pulling the first layer of pats down from corner E.
5. Close the pocket by folding corner F across onto corner G.
6. Open a pocket by pulling the first layer of pats down and to the right from corner H.
7. You should now be looking at the insides of three adjacent pyramids. The middle pyramid has 5's on the inside while the two outside pyramids have 2's on the inside.

From here, you could do a series of three pocket flexes at each of the end pyramids that have 2's on the inside to take you to a flat position with 5's on one face and 1's on the other face.

Summary

The morph-kite and pocket flexes can be combined and varied in a lot of different ways to build up more complex flexes. We've seen how they make up many of the flexes we previously learned and how they can work across many different types of flexagons.

The following table summarizes the flexes introduced in this chapter:

full name	shorthand	modified pats
morph-kite front flex	Mkf	4
morph-kite back flex	Mkb	4
morph-kite front shuffle flex	Mkfs	4
morph-kite back shuffle flex	Mkbs	4
morph-kite right flex	Mkr	4
morph-kite left flex	Mkl	4
silver tetra flex	St	4
Möbius flip	Fm	4
pyramid shuffle 3 flex	S3	4
backflip flex	Bf	4

Chapter 15

Slot Flexes

The flexes we've explored so far consist of manipulating pats by folding them together, unfolding them, and sliding them along other pats. An additional technique to add to your tool kit is sliding pats through a narrow slot between other pats. The slot technique can be combined with other flexes we've seen to create new flexes called **slot flexes**.

Slots on the hexaflexagon

We start with a collection of six different slot flexes on a regular hexaflexagon. They all begin with the same set of moves, but finish in different ways. To make it easier to experiment with them, the template in figure 15.1 supports all six of these slot flexes. As with all the flex diagrams, each of the individual diagrams will show the pat structure of the minimal flexagon for the given flex, which will be a subset of what you see when using this slot flexagon.

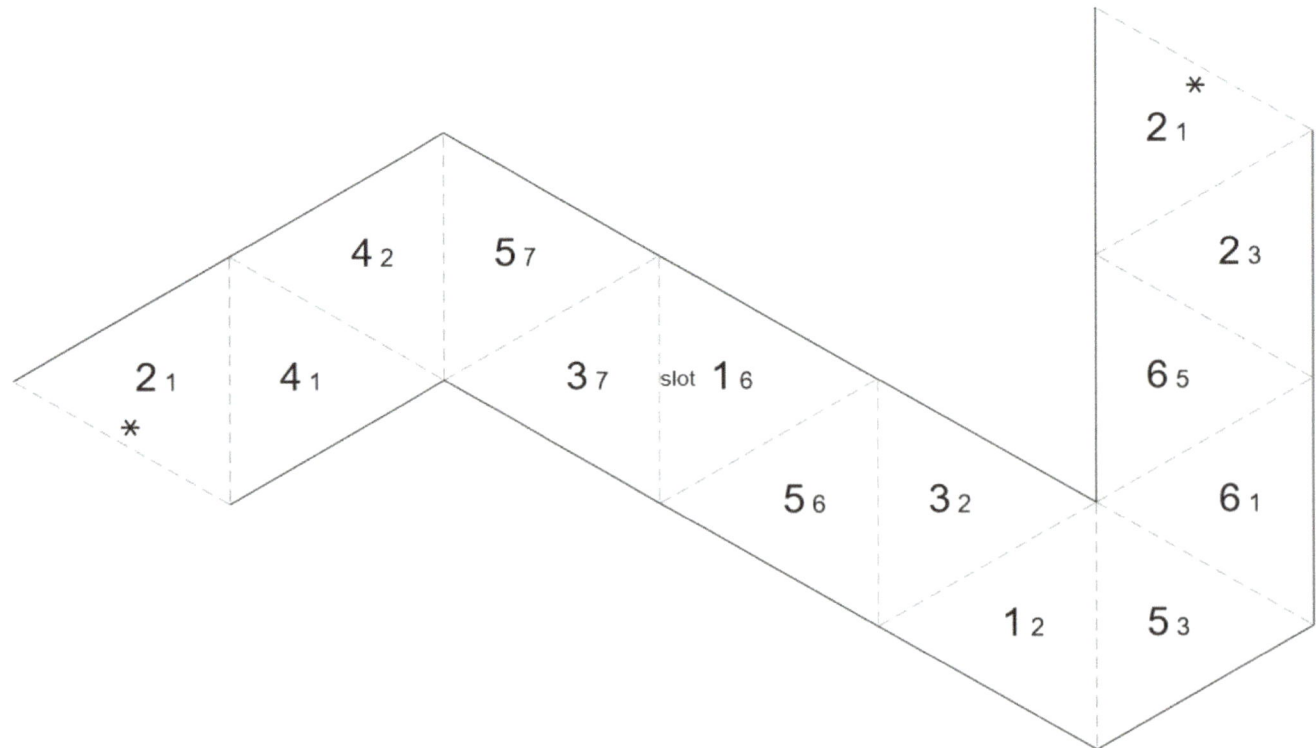

FIG. 15.1 Hexaflexagon template that supports a variety of slot flexes.

DOI: 10.1201/9781003433538-17

This set of slot flexes all begin by opening up a pocket and sliding pats through a slot, leading to an open cup position that looks like the midpoint of performing a tuck flex on the hexaflexagon. From this non-flat position, there are multiple ways to flex it before opening back up to a flat hexagon. To perform these common initial steps, position the hinge labeled "slot" at the top and refer to figure 15.2 or 15.3 while stepping through the following directions. Note that you will then need to use one of the slot flexes on the following pages to return to a flat arrangement.

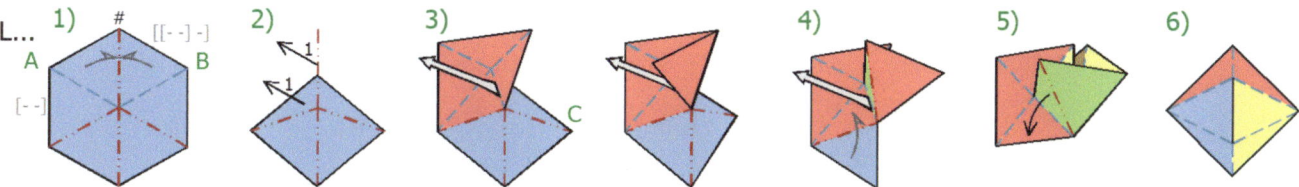

FIG. 15.2 Flex diagram for the first half of various slot flexes on the hexaflexagon.

1. Fold the top two pats backward along the top hinge, bringing A and B together in the back.
2. Open a pocket from the center of the flexagon, opening to the upper left.
3. Push corner C up to the left so that the pats slide through a slot. In step 3 in figure 15.2, the base of the arrow is positioned at the slot. Note that the pat going through the slot has multiple leaves in it, which could get caught on the slot, so you may need to be careful as you do this step.
4. After you've slid one pat through the slot, keep sliding so that a second pat goes through as well.
5. Fold the last pat that just emerged from the slot against the bottom inside of the pyramid.
6. Rotate the flexagon clockwise so you're looking at the inside of a cup or boat shape that consists of six pats, with a pair of 3-red at the top, a pair of 1-blue on the lower left, and a pair of 5-yellow on the lower right. This state is not flat, but you can use any of the flexes described in the rest of this chapter to return to a flat position.

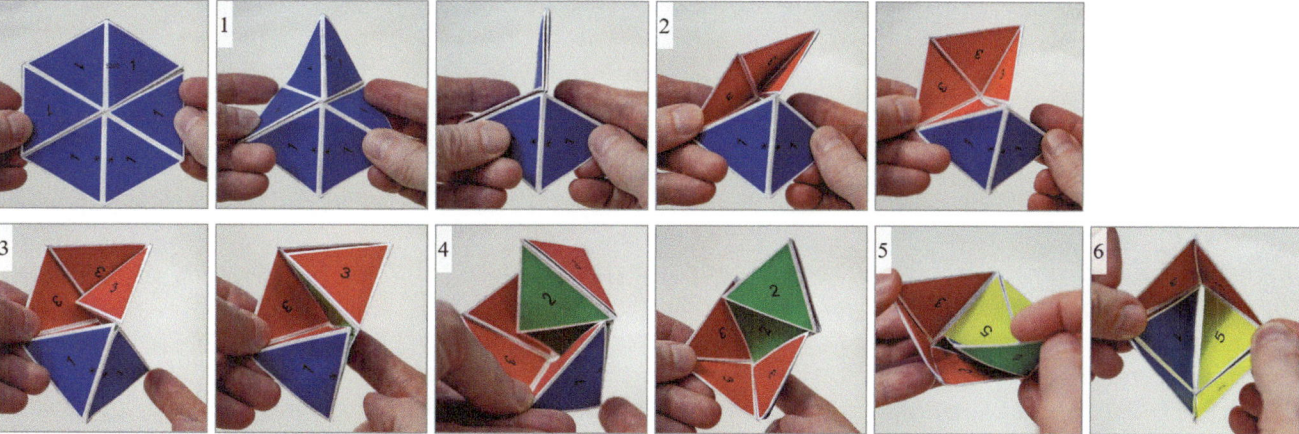

FIG. 15.3 Performing the first half of various slot flexes on the hexaflexagon.

It will make doing and undoing the following slot flexes easier if you practice these initial steps a few times, returning to the original position by sliding the pats back through the same slot before repeating the steps.

Note that the following flex diagrams all start with face 1 on the front so they can show the required pat structure. You then apply the common slot steps you just learned before continuing with the steps specific to a particular flex.

First up is the **slot-half flex** (Lh), so called because, after sliding pats through the slot, you fold the flexagon in half before opening it back up. Start with the common slot steps, then refer to figure 15.4 while stepping through these directions:

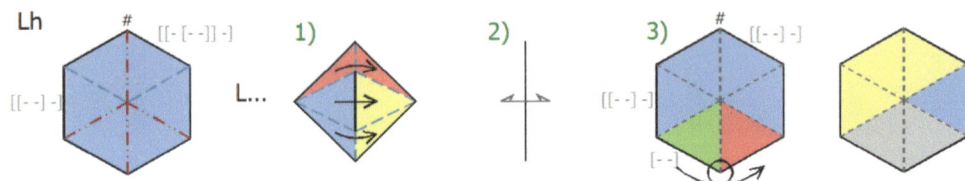

FIG. 15.4 Flex diagram for the second half of the slot-half flex (Lh) on the hexaflexagon.

1. From the cup position, close the cup by folding the left and right sides together.
2. Open it up from the back.
3. After finishing the flex, the new reference hinge will have two 1-blue pats on either side of it.

You can't perform a second Lh to return to the original state, so you instead need to perform these steps in reverse. Start by folding the left and right sides backward across the center hinges. Open up from the center to see the cup position before undoing the slot.

Next is the **slot-pocket flex** (Lk), named because you do a series of pocket flexes after the slot. After completing the common slot steps, follow these directions while referring to figure 15.5:

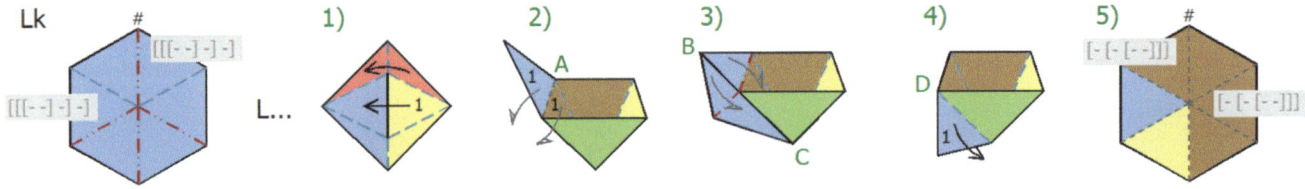

FIG. 15.5 Flex diagram for the second half of the slot-pocket flex (Lk) on the hexaflexagon.

1. From the cup position, fold the two upper right pats across to the left side of the cup. Note that you only fold across the top layer of pats from the middle pat.
2. From corner A, open a pocket from behind by opening the top layer backward.
3. Push corner B down to corner C at the back to close the pocket.
4. Open from corner D, rotating the flexagon so that the hinge being opened is now at the top.
5. To return to the original state, turn the flexagon over left to right and do another slot-pocket.

The next four slot flexes all include a tuck technique, but differ in whether you tuck from the top or the bottom and then whether you finish by opening from the front or back. Their names correspond to these top/bottom and front/back choices: the **slot-tuck-top-front flex** (Ltf), **slot-tuck-top-back flex** (Ltb), **slot-tuck-bottom-front flex** (Lbf), and **slot-tuck-bottom-back flex** (Lbb).

Use the following directions with figure 15.6 to do the slot-tuck-top-front flex:

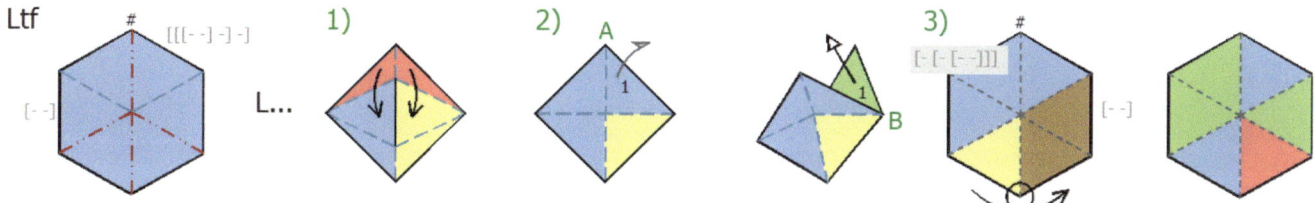

FIG. 15.6 Flex diagram for the second half of the slot-tuck-top-front flex (Ltf) on the hexaflexagon.

1. From the cup position, take the top corner and tuck it into the center of the cup.
2. Take the outside layer of pats from corner A and fold it backward. Then take the top layer of pats from corner B and open it up.
3. To return to the original state, turn it over left to right and do another slot-tuck-top-front.

Refer to figure 15.7 and follow these directions for the slot-tuck-top-back flex:

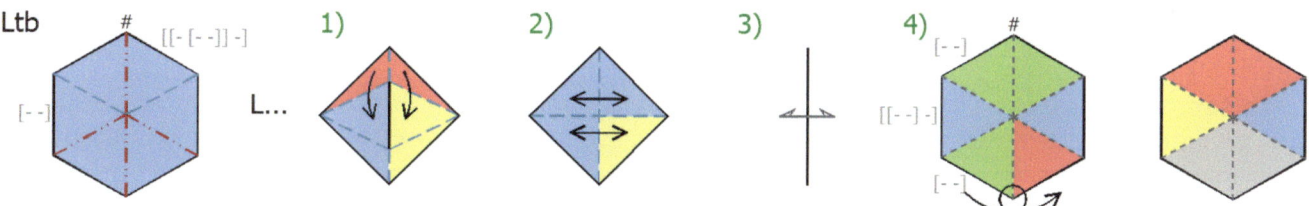

FIG. 15.7 Flex diagram for the second half of the slot-tuck-top-back flex (Ltb) on the hexaflexagon.

1. From the cup position, take the top corner and tuck it into the center of the cup.
2. Close the pyramid by folding the left and right sides together.
3. Open from the back.
4. To return to the original state, perform these steps in reverse.

For the slot-tuck-bottom-front flex, use the following directions with figure 15.8:

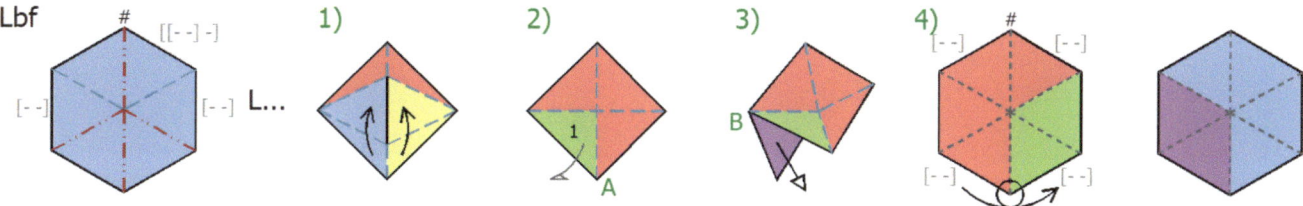

FIG. 15.8 Flex diagram for the second half of the slot-tuck-bottom-front flex (Lbf) on the hexaflexagon.

1. From the cup position, take the bottom corner and tuck it into the center of the cup.
2. Take the outside layer of pats from corner A and fold it backward. Then take the top layer of pats from corner B and open it up.
3. To return to the original state, perform these steps in reverse.

For the slot-tuck-bottom-back flex, refer to figure 15.9 while following these directions:

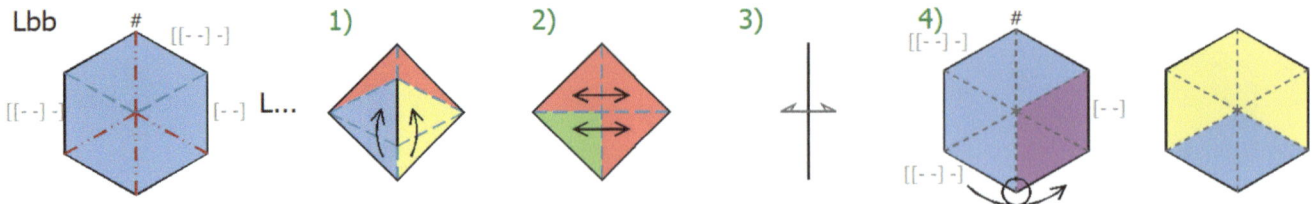

FIG. 15.9 Flex diagram for the second half of the slot-tuck-bottom-back flex (Lbb) on the hexaflexagon.

1. From the cup position, take the bottom corner and tuck it into the center of the cup.
2. Close the pyramid by folding the left and right sides together.
3. Open from the back.
4. To return to the original state, perform these steps in reverse.

Slots on other flexagons

So far, we've just looked at slot flexes on a hexaflexagon, starting by sliding two pats in a row through a slot. When we do the analogous flexes on flexagons with a different number of pats, we need to slide a different number of pats through the slot.

To see how this works, fold up the octaflexagon in figure 15.10. This flexagon supports both the slot-pocket and slot-tuck-top-front flexes. Start with the hinge marked "slot" at the top of the flexagon and follow the same common slot flex steps as we did for the hexaflexagon. When you do the slot step, keep sliding the pats through the slot until you can't slide them any further. You should end up sliding four pats through the slot instead of just two since the octaflexagon has two more pats than the hexaflexagon. To do the slot-tuck-top-front, finish by tucking the top of the pyramid with 3's on the inside into the center of the flexagon and open it up, similar to what figure 15.6 shows for the hexaflexagon.

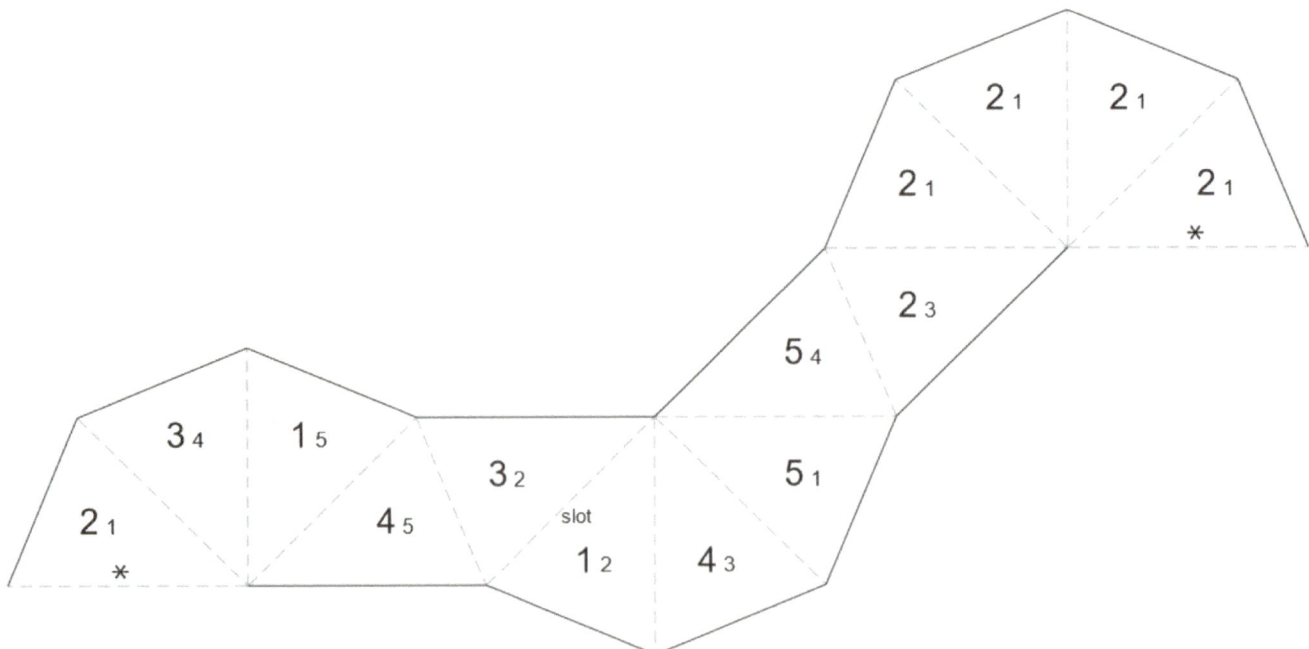

FIG. 15.10 Octaflexagon that supports the slot-pocket (Lk) and slot-tuck-top-front (Ltf) flexes.

Now use the template in figure 15.11 to try the same flexes on a pentaflexagon, with five pats per face. Start this one at the hinge marked L3. You should find that you can only slide a single pat through the slot. To make the slot flex a bit easier, you may wish to trim the corners of the pats that need to go through the slot.

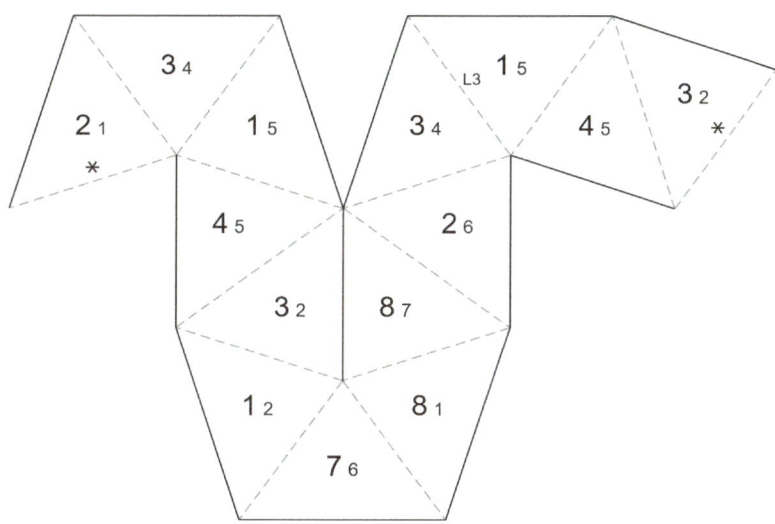

FIG. 15.11 Pentaflexagon that supports the slot-pocket (Lk), slot-tuck-top-front (Ltf), and slot-triple (L3) flexes.

At this point, given all the different ways we've combined flex techniques to build different flexes, you might start experimenting to see if there are other flex combinations that work on this pentafl-exagon. *Spoiler: the answer is yes.*

This pentaflexagon also supports the **slot-triple flex**, or L3. It starts at the same hinge as the other slots, but instead of doing the slot immediately after opening a pocket, you do a series of three pocket flexes before sliding a pat through the slot. Then you return it to a pentagonal shape by doing three more pocket flexes. See figure 15.12 for the flex diagram for the slot-triple flex. Note that you may need to occasionally rotate the flexagon slightly during the flex. After the L3, you can return to the original state by turning over the flexagon left to right and doing another slot-triple flex.

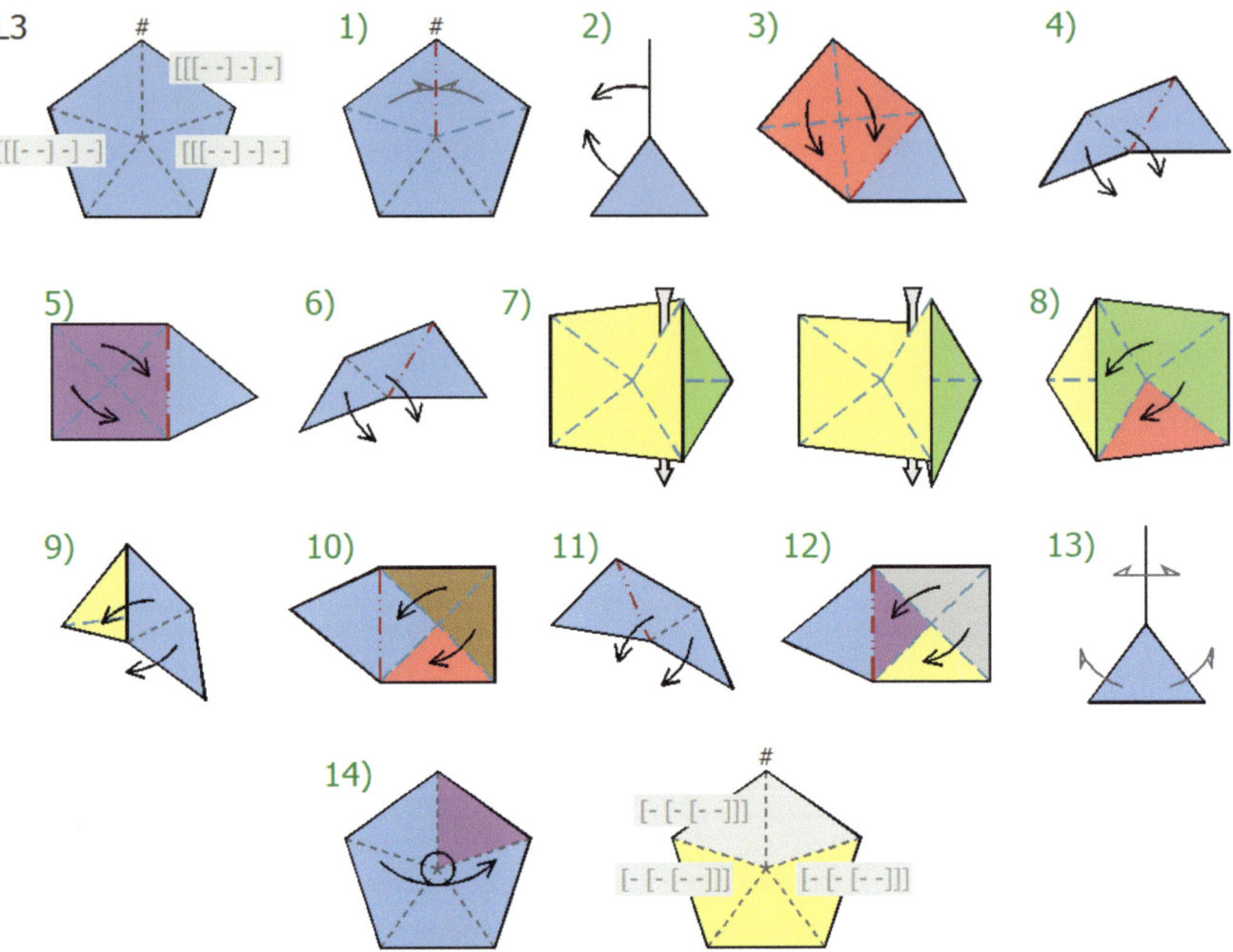

FIG. 15.12 Flex diagram for the slot-triple flex (L3) on the pentaflexagon.

This pentaflexagon is perhaps one of the more surprising flexagons we've seen because it seems like it should be completely inflexible. And yet, if you trim off the tips of the pats that slide through the slot, both the slot-pocket and slot-triple flexes can be done without bending the leaves at all.

Summary

In this chapter, we've seen how we can combine the technique of sliding pats through a slot with other techniques – such as the tuck and pocket flexes – to create a variety of new flexes. What happens if you try sliding pats a different number of times? Or combine the slot technique with the morphing flexes we saw in chapter 14? What do you see if you combine slots and pockets without worrying about returning to a flat position?

The following table summarizes the shorthand for the flexes introduced in this chapter:

full name	shorthand
slot-half flex	Lh
slot-pocket flex	Lk
slot-tuck-top-front flex	Ltf
slot-tuck-top-back flex	Ltb
slot-tuck-bottom-front flex	Lbf
slot-tuck-bottom-back flex	Lbb
slot-triple flex	L3

PART THREE

Learning about Flexagons

Now that we've seen a variety of flexagons and flexes, we'll dive into more technical details. In this part, we'll learn how to more precisely name and describe both flexagons and flexes. We'll discuss techniques for predicting and understanding how flexes change a flexagon's internal structure, which provides tools for designing your own flexagons with custom behavior. This theory leads up to definitions of the terms *flexagon* and *flex*.

The first three chapters on flexagon naming, state diagrams, and flex sequences are generally useful for exploring the flexagons in other parts of this book. The rest of the chapters in this part are optional. They provide a richer understanding of flexagons by discussing some of the mathematical concepts behind them.

DOI: 10.1201/9781003433538-18

Chapter 16

Flexagon Naming

Flexagons have a large variety of different attributes, each of which might be interesting in different contexts. Since it may only become apparent that a given attribute is interesting once you notice that it can vary, there isn't a single, consistent convention for assigning names to flexagon variants. Different books and papers on flexagons may use different names for the same concept.

We have adopted a naming convention that builds from the names of the original flexagons, like *hexaflexagon*, *trihexaflexagon*, *hexahexaflexagon*, and *tetraflexagon*. We've adjusted and augmented it to suit the wide range of flexagons and flexing behavior discussed in these pages. The three aspects of the outward appearance of a flexagon that we find most important to name are its *overall shape*, the *shape of the leaves*, and the *number of pats per face*. For some flexagons, it can also be useful to specify the *face count* when it's a whole number.

Throughout this book, we use the following naming convention, where each piece of the name is optional:

[overall shape] [leaf shape] [face count]-[pat count prefix]flexagon [(details)]

Overall shape

An adjective that describes the shape of the flexagon in the main position. See the examples in figure 16.1.
Examples: *triangular* flexagon, *square* flexagon, *pentagonal* flexagon, *hexagonal* flexagon.

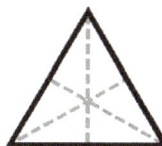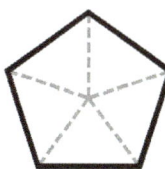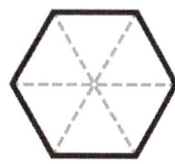

FIG. 16.1 Flexagons with different overall shapes: triangular, square, pentagonal, and hexagonal.

Leaf shape

The polygonal shape of the leaves. The name can be general, e.g., *triangle*, or more specific, e.g., *right triangle*. This includes geometric terms such as *regular* for leaves that are regular polygons (all sides and angles are equal) plus terms borrowed from origami: *silver triangle* (a triangle with angles of 45°, 45°, and 90°) and *bronze triangle* (a triangle with angles of 30°, 60°, and 90°). When unambiguous, you can simply use *silver* or *bronze*. See figure 16.2 for flexagons with the same overall shape but different leaf shapes.

DOI: 10.1201/9781003433538-19

Examples: *triangle* flexagon, *regular triangle* flexagon, *bronze* flexagon, *square* flexagon, *pentagon* flexagon.

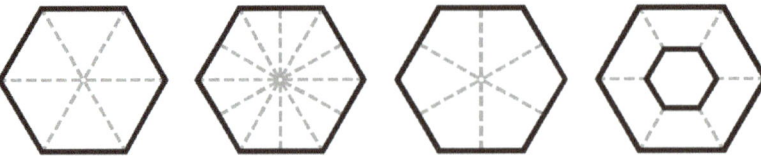

FIG. 16.2 Hexagonal flexagons with different leaf shapes: regular triangle, bronze triangle, kite, and trapezoid.

Face count

The theoretical number of independent faces, denoted by a Greek numeric prefix (e.g., *tri*, *tetra*, or *hexa*), typically used to indicate how many faces you can pinch-flex (or book-flex) between. This prefix ends with a hyphen to distinguish it from the pat count prefix that follows it. To avoid ambiguity when spoken, if you use this prefix, you should follow it with a prefix for the number of pats, e.g., hexa-hexaflexagon rather than hexa-flexagon.
Examples: *tri*-hexaflexagon (three faces), *hexa*-hexaflexagon (six faces).

Pat count

The number of pats in the flexagon's main position, indicated using a Greek numeric prefix. See the examples in figure 16.3.
Examples: *hexa*flexagon (six pats), *hepta*flexagon (seven pats), *dodeca*flexagon (12 pats).

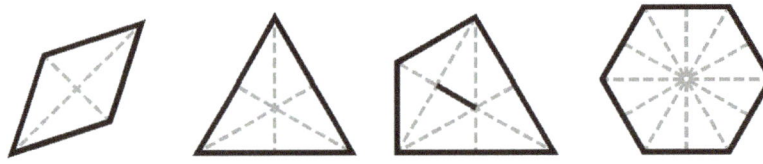

FIG. 16.3 Bronze flexagons with different numbers of pats: tetra (four), hexa (six), octa (eight), and dodeca (12).

Details

The flexagon name is optionally followed by additional details about the flexagon's internal structure enclosed in parentheses. Note that *generating sequence*, *pat directions*, and *pat structure* are described in later chapters.

- Leaves: Total number of leaves in the template, e.g., (leaves: 12), (leaves: 15).
- Generator: Flex generating sequence, e.g., (generator: F*>>S*).
- Directions: The directions from leaf to leaf in the template, e.g., (directions: //\\//).
- Pats: Pat structure in one state of the folded flexagon, e.g., (pats: [[1,2],3,4,[[5,6],7]]).

Here are some common Greek numeric prefixes used in flexagon names:

4	tetra	10	deca
5	penta	12	dodeca
6	hexa	14	tetradeca
7	hepta	16	hexadeca
8	octa	20	icosi
9	ennea	24	icositetra

Note that you may occasionally see *nona* used for nine, which is derived from Latin, while the rest of the prefixes (like *ennea*) come from Greek. For simplicity – and when the Greek prefix may be hard to remember – numbers can be used for prefixes, especially for prefixes above 12, such as 14-flexagon for a flexagon with 14 pats.

As an example of putting all the pieces together to describe a flexagon's outward appearance, the *triangular bronze triangle hexaflexagon* in figure 16.4 has an overall shape of a triangle ("triangular"), is made up of leaves that are bronze triangles ("bronze triangle"), and has six pats in the main position ("hexa"). We can choose to simplify it to *bronze hexaflexagon* when unambiguous or when referring to the general class of bronze hexaflexagons that may differ in other features.

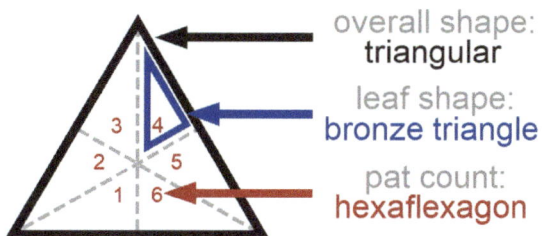

triangular **bronze triangle** hexaflexagon

FIG. 16.4 An example of naming a flexagon using its overall shape, leaf shape, and pat count.

To describe the internals of a flexagon, start with the base name, such as bronze hexaflexagon, and add the appropriate attributes. If it has three faces, you could use any of the following: bronze tri-hexaflexagon, bronze hexaflexagon (leaves: 9), bronze hexaflexagon (generator: P), bronze hexaflexagon (directions: /\/\/\/\/), or bronze hexaflexagon (pats: [[1,2],3,[4,5],6,[7,8],9]), depending on the feature you wish to emphasize.

As another example, consider different aspects of a flexagon that can have a count of five, some of which are shown in figure 16.5:

- A *pentagonal flexagon* has an overall shape of a pentagon.
- A *pentagon flexagon* has leaves that are pentagons.
- A *pentaflexagon* has 5 pats per face.
- A *penta-hexaflexagon* is a hexaflexagon with five faces you can pinch-flex between.
- A *flexagon (leaves: 5)* has five leaves in its template.

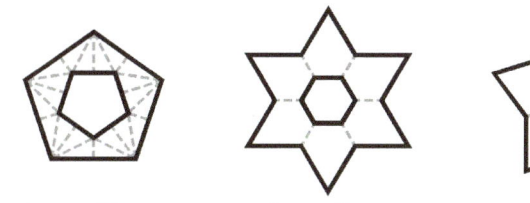

pentagonal flexagon   pentagon flexagon   pentaflexagon

FIG. 16.5 Some examples of the number five appearing in flexagons: five edges in the overall shape (pentagonal flexagon), five edges per leaf (pentagon flexagon), and five pats per face (pentaflexagon).

You may wonder if a hexaflexagon is named because the overall shape is a hexagon and a tetraflexagon is named because it's square shaped. Given just those two examples, this is reasonable. However, we now have a much larger variety of flexagons to consider. We have found that the number of pats per face is a better indicator of a flexagon's behavior than the overall shape is. Thus, our naming convention interprets the original hexaflexagon and tetraflexagon names as referring to the number of pats per face rather than the shape, which is consistent with the original names while generalizing nicely.

The portion of the name outside the parentheses describes the outward appearance, while the portion inside the parentheses describes the internal structure. The one exception is the optional prefix that describes the number of **pinch faces**, i.e., a set of leaf-faces that travel together during pinch flexing (or book flexing). Note that the outward appearance by itself doesn't uniquely identify a flexagon. Even including the number of pinch faces or leaves may not be sufficient. For example, as we saw in chapter 3, there are three different variations of a *hexagonal regular hexa-hexaflexagon* with 18 leaves, each of which have a different internal pat structure and therefore different behavior.

Summary

This chapter formalizes the flexagon naming convention we use throughout the book. The main part of the name describes the outward appearance of the flexagon, with optional details for describing the internal structure. Further examples of flexagons and their names can be found in appendix B, "Flexagon Charts".

Chapter 17

State Diagrams

Keeping track of where you are as you flex even a simple flexagon can be surprisingly challenging. A **state diagram** is a visual technique for tracking different states of a flexagon and how to travel between them.

Pinch state diagrams

We have seen the Tuckerman Traverse as a technique for pinch-flexing between all the faces in a flexagon, but it can still be tricky to know which face you'll see next or to figure out how to get to a particular face without simply continuing to flex until you eventually find it. One way to keep track of where you are is to draw a diagram that captures both the states you can visit and where you can flex between those states. We call a diagram that shows how to pinch-flex between states a **pinch state diagram**.

Figure 17.1 shows the pinch state diagram for the tri-hexaflexagon, which is the simplest pinch state diagram. The three states of the flexagon are represented by circles in the diagrams. The pair of numbers in the circles refers to the numbers you can see in that state. The first number is what appears on the top visible face, while the second number is what's on the back face, which you see when you turn the flexagon over. Arrows between two states represent flexing from one state to the other. Different types of flexes, such as the pinch flex or tuck flex, can be depicted using words along each arrow or by using different styles of lines for different flexes. Since this diagram is specific to the pinch flex, the unlabeled arrows are assumed to indicate the direction of a pinch flex.

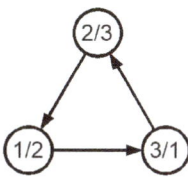

FIG. 17.1 The pinch state diagram for the tri-hexaflexagon.

Let's follow the arrows in the pinch state diagram from figure 17.1 to understand what it means. The diagram tells us that if we pinch-flex the flexagon with face 1 up and face 2 down (1/2), we will get the state where face 3 is up and face 1 is down (3/1). Pinch-flexing from this state will bring us to the face 2 up, face 3 down position (2/3). A pinch flex from this state brings us back to the original 1 up, 2 down state (1/2). Thus the diagram describes the cycle from 1/2 to 3/1 to 2/3 and back to 1/2.

DOI: 10.1201/9781003433538-20

The pinch state diagram also captures pinch flexes from the opposite faces. From state 1/2, turn over the flexagon so face 2 is up and face 1 is down (2/1). Pinch-flex from this state to get to 3/2. In figure 17.1, we see an arrow pointing from 2/3 to 1/2. We just flexed in the opposite direction of the arrow, which is equivalent to inverting everything – changing 1/2 to 2/1 and 2/3 to 3/2, as well as reversing the direction of the arrow. On the tri-hexaflexagon, we can pinch-flex from 2/1 to 3/2 to 1/3 and back to 2/1. If we imagine the inverted version of figure 17.1 – reversing the order of the face numbers and arrow directions – we can see that the diagram captures this cycle as well.

Next, take a look at the tetra-hexaflexagon from figure 3.1 and its pinch state diagram. In figure 17.2, you can see the same cycle from 1/2 to 3/1 to 2/3 that we saw with the tri-hexaflexagon, but there's also a cycle from 1/2 to 4/1 to 2/4. In particular, you can see that, from state 1/2, you can flex to either 3/1 or 4/1, depending on which hinge you pinch-flex from. There are also two arrows pointing at state 1/2. To properly follow this diagram, we need to dive more deeply into what it's communicating.

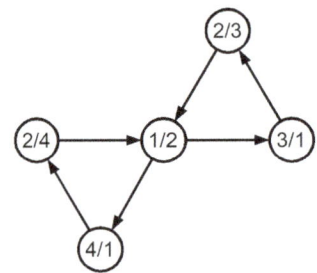

FIG. 17.2 The pinch state diagram for the tetra-hexaflexagon.

One thing this pinch state diagram communicates is that there are four different pinch flexes you can do from state 1/2 – two from face 1 (the arrows pointing away from the center circle) and two from face 2 (the arrows pointing toward the center circle). Start on face 1 with the set of hinges that pinch-flex to face 3. This is the hinge we would be at if we started in state 2/4 and pinch-flexed to state 1/2. Each of the options in the following list – A, B, C, and D – start from this same reference hinge, so you may want to mark it to help you keep track as you try these out. Here are the four different options for your second pinch flex after doing the initial pinch flex to 1/2:

A. *Pinch flex at the same hinge to reach 3/1.* Using the flex notation we first saw in chapter 7, we can say that the flex sequence PP (where P means "do a pinch flex") goes from 2/4 to 1/2 to 3/1.
B. Or *shift one hinge to the right and pinchflex to reach state 4/1.* Using flex notation, we say that P>P (where > means "shift one hinge to the right") goes from state 2/4 to 1/2 to 4/1.
C. Or *turn the flexagon over so you're in state 2/1, then pinch-flex to reach state 4/2.* Using flex notation, we say that P^P (where ^ means "turn over the flexagon but keep the same reference hinge") goes from 2/4 to 1/2 and back to 2/4.
D. Or *turn the flexagon over, shift one hinge to the right, and pinch-flex to go to state 2/3.* Using flex notation, we say P^>P goes from 2/4 to 1/2 to 2/1 to 3/2.

Figure 17.3 illustrates these four possible routes though the pinch state diagram, showing the corresponding flex sequence.

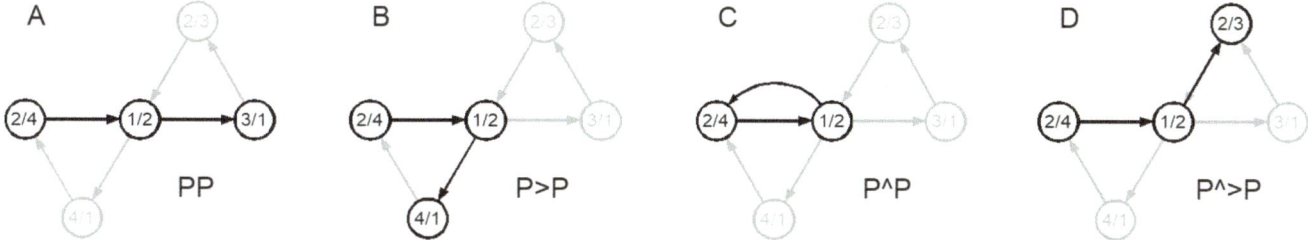

FIG. 17.3 How to travel through a pinch state diagram.

We can generalize these patterns to help us navigate more complex pinch state diagrams. If you have just followed an arrow in the diagram by pinch-flexing, here's how you can pick where you want to step to next:

- If you want to follow the arrow that goes in the same direction as the last arrow, pinch-flex at the same hinge.
- If you want to follow the arrow that takes a sharp turn (120°), shift one hinge to the right and pinch-flex.
- If you want to return to your previous state, turn over the flexagon and pinch-flex at the same hinge.
- If you want to take a slight turn in the diagram (60°), turn over the flexagon, shift one hinge to the right, and pinch-flex.

Note that turning over the flexagon effectively reverses the directions of the arrows in the pinch state diagram.

You should now be able to apply the same techniques to follow the arrows across the pinch state diagram in figure 17.4 for the penta-hexaflexagon from figure 3.3.

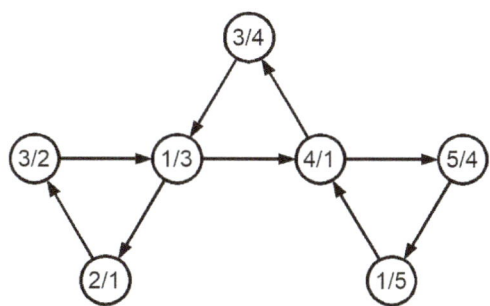

FIG. 17.4 Pinch state diagram for the penta-hexaflexagon.

As an example, consider starting in state 2/1. A pinch flex, P, takes you to 3/2. The sharp turn to 1/3 requires >P. Continue to state 4/1 with a pinch flex, P, at the same hinge. Take a slight turn to 1/5 using ^>P, turning it over, shifting one hinge to the right, and doing a pinch flex. Thus, by looking at the diagram, we know we can travel from 2/1 to 3/2 to 1/3 to 4/1 to 1/5 using the flex sequence P>PP^>P.

You can traverse an entire pinch state diagram using the Tuckerman Traverse: keep pinch-flexing at the same hinge until you can't pinch flex anymore, then shift to an adjacent hinge and repeat the process. This is a simple way to visit every state in the fewest flexes possible (though it's not the only way).

Recall from chapter 3 that there are three different hexaflexagons with six faces. Each one has slightly different behavior, which becomes visible once you see that they each have a different pinch state diagram. Figure 17.5 shows the diagram for variation 6a, the straight template from figure 3.4. Figure 17.6 presents the diagram for 6b, the hexagonal template from figure 3.5. And figure 17.7 is for 6c, the triangular template from figure 3.6.

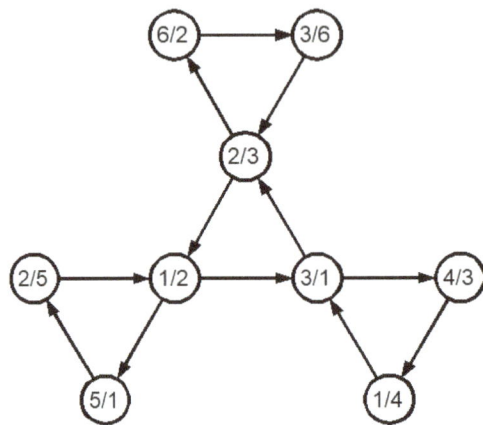

FIG. 17.5 Pinch state diagram for the hexa-hexaflexagon, variation 6a, the straight template.

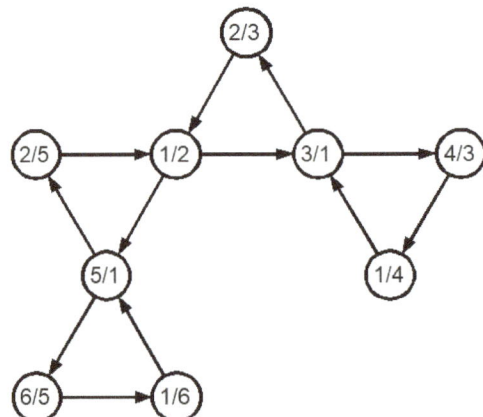

FIG. 17.6 Pinch state diagram for the hexa-hexaflexagon, variation 6b, the hexagonal template.

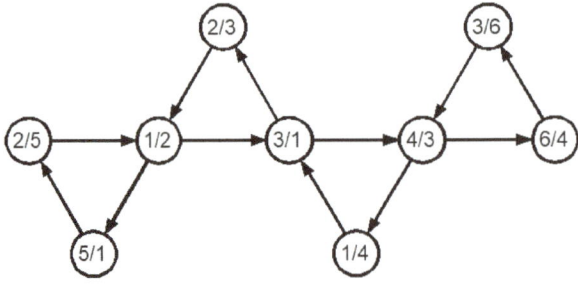

FIG. 17.7 Pinch state diagram for the hexa-hexaflexagon, variation 6c, the triangular template.

Pinch state diagrams and leaf angles

Pinch state diagrams can also tell you which of the three leaf angles is in the center of the flexagon, which informs you whether a given state is flat. To see how this works, let's revisit the bronze hexaflexagons from chapter 5. Recall that a bronze triangle has 30°, 60°, and 90° angles and a bronze hexaflexagon is a cup when the 30° angle is in the center, flat when the 60° angle is in the center, and a mountain-valley shape when the 90° angle is in the center.

First, let's take a look at the bronze tri-hexaflexagon from figure 5.2. You can travel between the same three states as with the regular tri-hexaflexagon, so, in theory, the pinch state diagram looks exactly the same. However, the cup state 2/3 has some distinctive behavior we need to consider. Recall that we can flex from flat 1/2 to mountain-valley 3/1 to cup 2/3, with 3's on the outside. If we could turn the cup inside out without ripping it, we could flex back to 1/2. Instead, we reversed course, eventually flexing from flat 2/1 to cup 2/3, but this time with 2's on the outside. Figure 17.8 is a modified version of figure 17.1, annotated with the angle that's at the center of each state and calling out the two cup arrangements. These cups represent the same state but turned inside out depending on how you get to the cup.

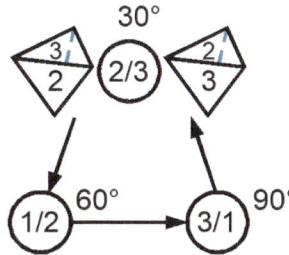

FIG. 17.8 The annotated pinch state diagram for the bronze tri-hexaflexagon showing two different arrangements of the cup position for state 2/3. The cup on the left has 2's on the outside and 3's on the inside, while the one on the right has 3's on the outside and 2's on the inside.

Now let's try the same thing with the bronze hexa-hexaflexagon from figure 5.10, marking up the pinch state diagram with the angle that's in the center and the cup states. Figure 17.9 shows the result.

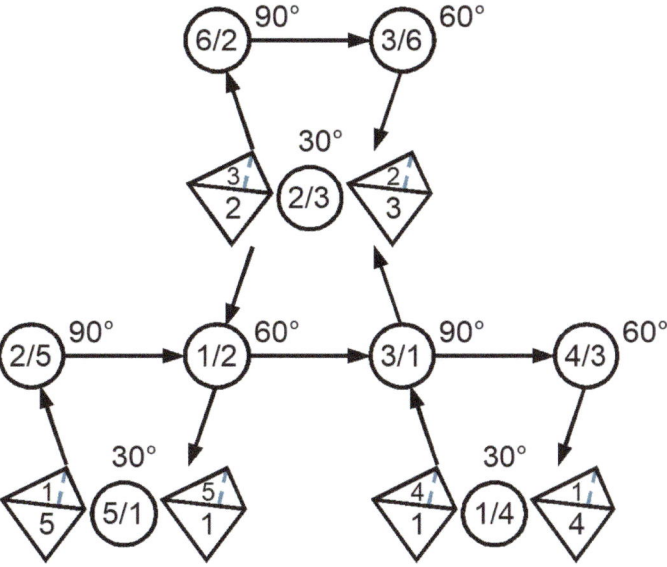

FIG. 17.9 The annotated pinch state diagram for a bronze hexa-hexaflexagon.

Try using the pinch flex to travel around this pinch state diagram to see which routes you can easily follow and which are more difficult because of the cup position. Can you figure out how to flex from flat 1/2 to mountain-valley 3/1 to cup 2/3 and then to flat 3/6? You should be able to fairly easily get to cup 2/3 with 2's on the inside and 3's on the outside, but how do you get to 3/6?

We saw earlier that traveling against the arrow is the same as inverting everything, so to accomplish that final step, you need to turn over the 2/3 cup so face 3 is on the front, then pinch-flex at the correct hinge to reach state 6/3. From there, you can pinch-flex to 2/6 (still going against the arrow) followed by a pinch flex to cup 2/3 with face 2 on the outside. From that cup, you can pinch-flex back to flat 1/2.

You can see that every cycle of three pinch flexes in figure 17.9, represented by three states in a triangle (such as 2/5, 1/2, and 5/1), has each of the three angles in the center (in this case, 90°, 60°, and 30°), so it cycles between flat, cup, and mountain-valley. If you follow any straight line in the diagram, you can see that it alternates between two angles, e.g., the line 2/5, 1/2, 3/1, and 4/3 alternates between 90° and 60°. From these two patterns, you can predict the angles for any state in a pinch state diagram, allowing you to figure out which states will be a cup, flat, or mountain-valley.

You can apply this same technique with any triangular leaf. For example, we also considered the silver octaflexagon, with leaf angles of 45°, 45°, and 90°. Each cycle of three pinch flexes will have two states where a 45° angle is in the center and one where the 90° angle is in the center. This means that roughly two thirds of the states will be flat.

Square tetraflexagons

Next, we'll build up a diagram of the states of a square tetraflexagon with arrows that show how to travel between them using the book flex and box flex. As with the pinch flex on the triangle hexaflexagon, these flexes keep the faces intact, so we just need to track the pairs of visible faces we can flex to. We'll use the square deca-tetraflexagon we saw in figure 8.6.

After folding the template, start in state 1/2, with 1's on the front face and 2's on the back face. Both pairs of hinges on each face support the book flex, which leads to four new states. From face 1, you can either open it up to face 4 or face 6, with 1's on the back face. From face 2, you can book-flex to either face 3 or 5, with 2's on the back. These are parts of two connected cycles that have state 1/2 in common. The state diagram in figure 17.10 captures these states and cycles.

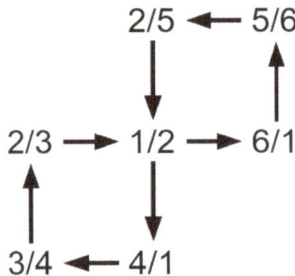

FIG. 17.10 State diagram showing the states accessible by just using the book flex on the square deca-tetraflexagon.

This way of arranging the diagram contains some additional information. Any time the diagram takes a right turn, you must rotate the flexagon 90° to the next hinge before flexing. If the diagram continues straight, you do the next book flex at the same hinge. For example, cycling between 1/2, 4/1, 3/4, and 2/3 requires rotating the flexagon 90° between each flex. But to go from 2/3 to 1/2 to 6/1, keep flexing at the same hinge.

What happens if we also include the box flex? First, let's just focus on the cycle involving faces 1 through 4. Midway between book-flexing between states 1/2 and 4/1, there's a flat position with face 1 on the outside. From here, you can open it up into a box with 1's on the outside and 3's on the inside. And from that state, you can open it up from the top to find state 7/1 or from the bottom for state 1/8. You see something similar for face 3. But if you try the same for face 2 or 4, the box position won't open to any new states.

In figure 17.11, a dashed blue line represents a box flex. The rectangle containing 1/3 indicates a box state with 1's on the outside and 3's on the inside. The 1/3 box opens up to either 7/1 or 1/8 from one side and 1/2 or 4/1 on the other. The single numbers in the middle of the central square represent the state partway through a box flex when there's only a single number visible.

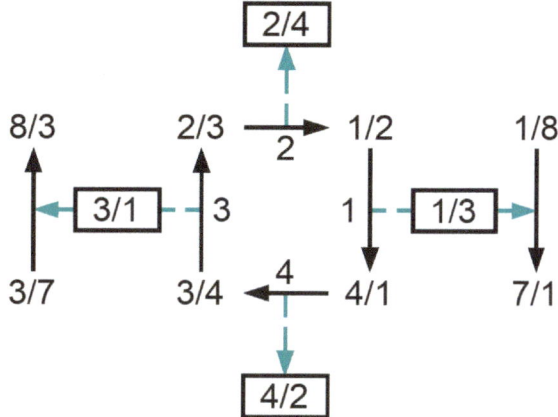

FIG. 17.11 State diagram for the square deca-tetraflexagon describing book flexes (solid black lines) and box flexes (dashed blue lines).

If you try book flexes from each of the four new states we just uncovered, you see that they're all tied together in a single cycle of book flexes: 1/8, 7/1, 3/7, and 8/3. Similarly, box flexes from the 1/2, 2/5, 5/6, and 6/1 cycle lead to a fourth cycle. Figure 17.12 puts that all together into a single state diagram. When doing the box flex, make sure the numbers inside match the diagram.

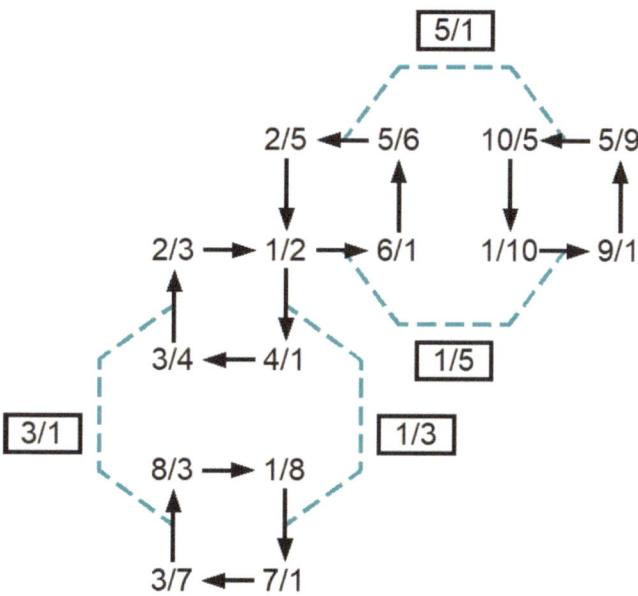

FIG. 17.12 Complete state diagram for the square deca-tetraflexagon for all book flexes and box flexes.

Pyramid shuffle diagram

Next, let's try explorations with a flex that mixes up the faces. To get a feel for some of the possibilities, make the flexagon in figure 17.13, which we'll explore using the pyramid shuffle. The small numbers in the corners of the leaves (ranging from 1 to 18) are **leaf IDs**, a unique number per leaf, which we use to track the states. Copy the leaf IDs to the back of each leaf. You may want to circle the leaf IDs to distinguish them from the larger leaf labels. The larger numbers in the middle of the leaves tell us how to fold the flexagon, as with other templates, while the IDs help distinguish the different states.

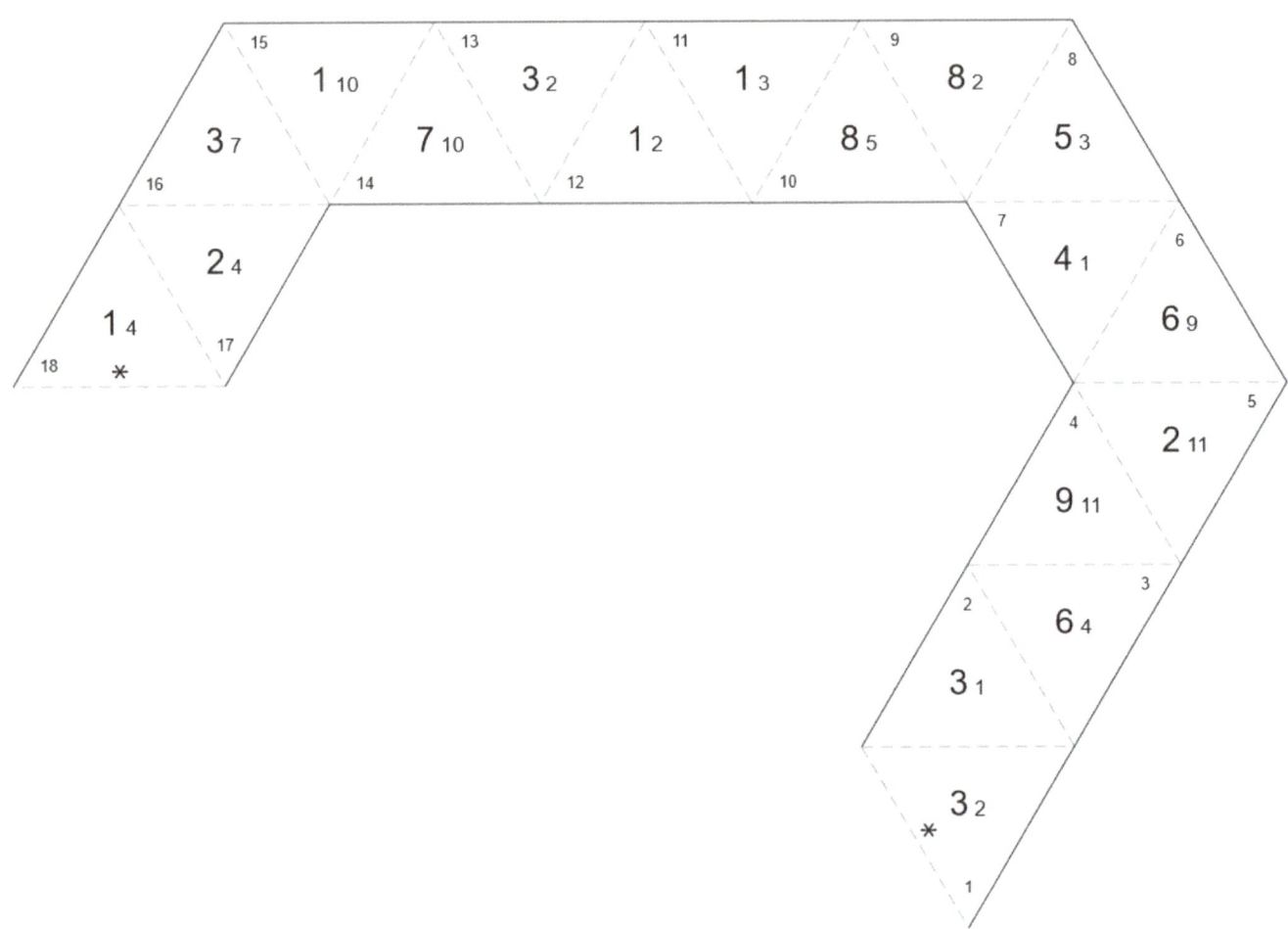

FIG. 17.13 Template for a hexaflexagon (leaves: 18) supporting various pyramid shuffles. Underlining the 6's and 9's can help you keep track of the numbers as the leaves rotate.

The first thing to realize is that you need to pay close attention to which of the six hinges you're starting from. With the pyramid shuffle, as with many flexes, every hinge on either face could possibly support a different pyramid shuffle. 6 hinges × 2 faces = 12 possibilities from a single state. Figure 17.14 shows the initial two faces including the internal pat structure and which hinges support the pyramid shuffle. Note that, for now, we'll figure out which hinges support a flex through trial and error, though we'll see how to predict this later in chapter 19.

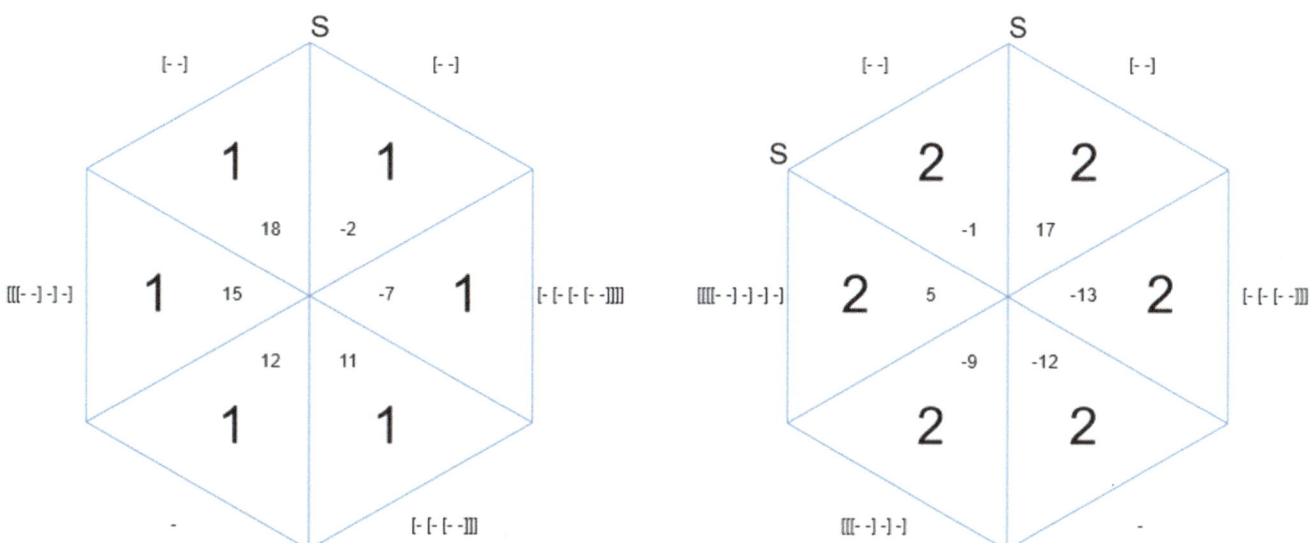

FIG. 17.14 Initial faces, pat structure, leaf IDs, and which hinges support the pyramid shuffle.

The second thing to realize is that most flexes other than the pinch flex mix up the faces, which means we can't just note a single number for an entire face. Assigning each leaf a unique ID allows us to describe each state in more detail. In order to track all the hinges and the ways the faces change, we start by writing down the visible leaf IDs, the small numbers shown on the template. The face labeled with all 1's (the top face of the flexagon after the initial folding) has leaf IDs of 2, 7, 11, 12, 15, and 18 going clockwise (ignoring the minus sign). The face labeled with all 2's (the bottom face of the flexagon after the initial folding) has the leaf with ID 1 on the opposite face from the leaf with ID 2, and then 5 opposite 7. Continuing counterclockwise, the IDs are 9, 12, 13, and 17. We'll call this state A, represented by the diagram in figure 17.15.

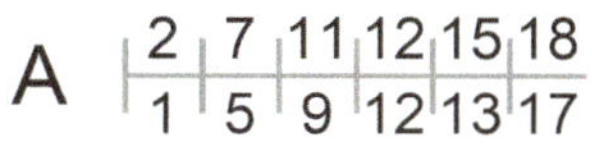

$$\text{A} \quad \frac{2 \mid 7 \mid 11 \mid 12 \mid 15 \mid 18}{1 \mid 5 \mid 9 \mid 12 \mid 13 \mid 17}$$

FIG. 17.15 Representation of the leaf IDs on opposite faces of state 1/2.

The lines between each of the leaf IDs represent hinges where it might be possible to perform a flex. Note that there's also a hinge between the first and last leaf IDs (such as 2 and 18). We'll put the line for this hinge at the start of the diagram, before the 2.

On this flexagon, you can perform a pyramid shuffle by using the hinge between 2 and 18 as the reference hinge. We'll call the new state you find state B. After you finish the flex, you can turn over the flexagon and do a pyramid shuffle at the hinge between leaves 13 and 15 to take you back to state A. You can verify this by checking that the leaf IDs in this state match state A. The diagrams in figure 17.16 capture where to do pyramid shuffles to travel between the two states. The red dashed line represents a pyramid shuffle, and the small letter at the end of the line tells you the state you get to after doing the flex.

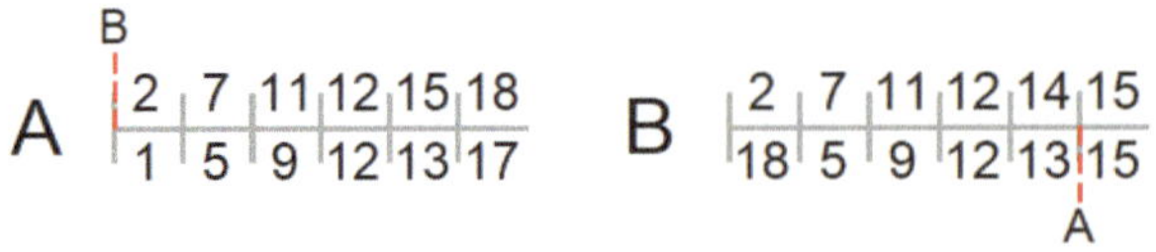

B

A $\quad\dfrac{|2\,|\,7\,|11|12|15|18}{|1\,|\,5\,|\,9\,|12|13|17}$ B $\quad\dfrac{|2\,|\,7\,|11|12|14|15}{|18|\,5\,|\,9\,|12|13|15}$

A

FIG. 17.16 Diagram that shows that you can do a pyramid shuffle from state A between leaf IDs 2 and 18 to get to state B, noting the corresponding pyramid shuffle that brings you back to state A.

Note that, in general, simply looking at the visible leaves isn't always sufficient to identify a specific state since it's possible to have the same visible leaves but different internal pat structure. We will develop the tools necessary to describe such a state in chapter 19. However, for the states we look at in this chapter, the visible leaves are unique for each state.

If we try to do a pyramid shuffle at each of the 12 possible hinges in state A, we find that three of the hinges allow the flex, leading to three new states. We can then write down the IDs for each of these states and try the pyramid shuffle at every hinge in these new states. After fully exploring every state we can flex to, we'll eventually find eight states. These states and flexes can be captured with the **state-transition diagram** in figure 17.17.

B

A $\quad\dfrac{|2\,|\,7\,|11|12|15|18}{|1\,|\,5\,|\,9\,|12|13|17}$ B $\quad\dfrac{|2\,|\,7\,|11|12|14|15}{|18|\,5\,|\,9\,|12|13|15}$

E D

C A

C $\quad\dfrac{|5\,|\,9\,|11|12|14|15}{|18|\,9\,|10|12|13|15}$ D $\quad\dfrac{|5\,|\,9\,|11|12|15|18}{|1\,|\,9\,|10|12|13|17}$

D

H A

G E

E $\quad\dfrac{|1\,|\,5\,|\,7\,|11|12|15}{|17|\,4\,|\,6\,|\,9\,|12|13}$ F $\quad\dfrac{|1\,|\,6\,|\,9\,|11|12|15}{|17|\,4\,|\,9\,|10|12|13}$

F

H

G $\quad\dfrac{|6\,|\,9\,|11|12|14|15}{|1\,|\,9\,|10|12|13|15}$ H $\quad\dfrac{|5\,|\,7\,|11|12|14|15}{|1\,|\,6\,|\,9\,|12|13|15}$

F

G E

FIG. 17.17 State-transition diagram showing how to do pyramid shuffles between eight different states.

To give us a better feel for how all these states tie together, we draw a circle for each of the eight states with lines between them representing the pyramid shuffles that take you between the states. This leads to the state diagram shown in figure 17.18.

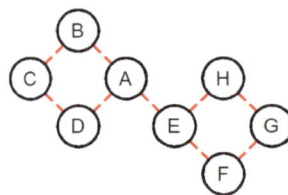

FIG. 17.18 State diagram summarizing the results from the state-transition diagram in figure 17.17.

This arrangement of circles was chosen to provide some additional information about how to flex between the states. Though not essential for using the state diagram, it may be useful to know that when you continue in the same direction in the diagram, e.g., from B to A to E to F, you carry out the pyramid shuffle on the same face, without turning it over. But whenever the lines switch from going up to going down (or vice versa), you need to turn over the flexagon before doing a pyramid shuffle. For example, if you want to go from state A to B to C, the line goes up from A to B then back down to C. This indicates that after you flex from A to B, you need to turn over the flexagon before flexing to state C. You can use the state-transition diagram in figure 17.17 to figure out which hinge to use.

General state diagram

We'll now extend these ideas so we can map out all the states of a flexagon reachable using multiple types of flexes. We will use the pinch flex (P), pyramid shuffle (S), tuck flex (T), and v-flex (V) on the hexaflexagon (leaves: 11) in figure 17.19.

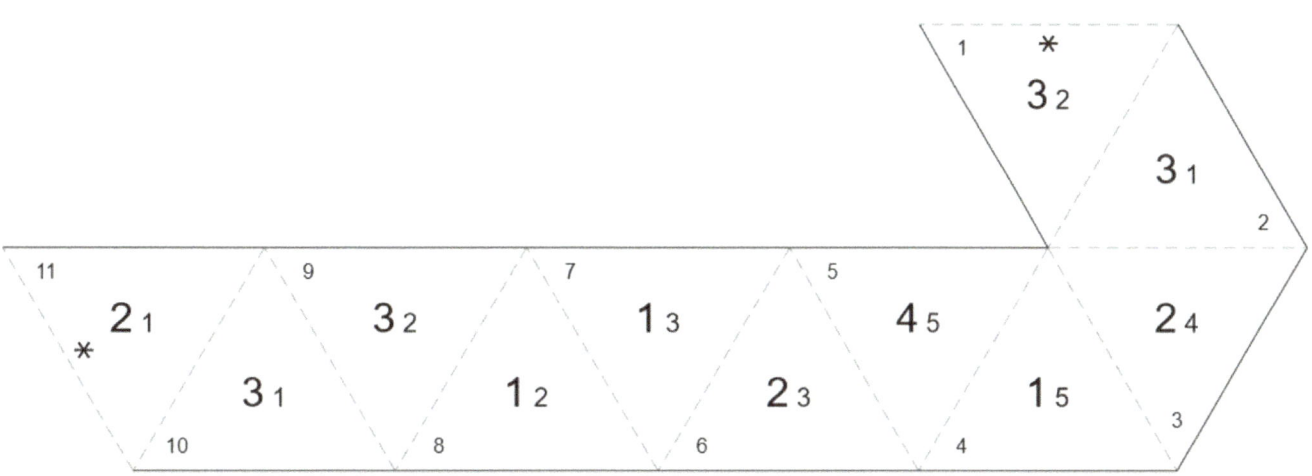

FIG. 17.19 Template for a hexaflexagon (leaves: 11).

As we did with the previous example, we write down the leaf IDs of the two faces from the initial 1/2 state and try P, S, T, and V at each hinge. From state A, there is one possible T at the hinge between IDs 2 and 4, one V between 10 and 11, and another V between 8 and 9 on the opposite face. To simplify the diagram, we only show one of the three possible hinges where you can do a pinch flex on a face, since they all lead to the same state. For example, doing P between 4 and 7, 8 and 10, or 11 and 2 are all equivalent, so we'll just choose the pair (4, 7). This results in the state-transition diagram in figure 17.20.

		P: pinch flex
		S: pyramid shuffle
		T: tuck flex
		V: v-flex

FIG. 17.20 State-transition diagram for the initial state of hexaflexagon (leaves: 11).

Recall that P, S, and V are their own inverses, which means that after you perform one of those flexes, you can turn the flexagon over and do the flex a second time to restore the flexagon to its original state. But T is one of the flexes where this isn't true in general. You may instead need to do an inverse tuck, T′, in order to return to the original state. Because of this, it's important to have the line for a tuck flex indicate whether it's a tuck or inverse tuck. In figure 17.20, the arrow pointing away from the center represents a tuck flex. An arrow pointing toward the center would represent an inverse tuck.

If we continue trying each of P, S, T, T′, and V at every hinge for every new state, we'll eventually find ten states. Figure 17.21 shows the state-transition diagrams for all ten states.

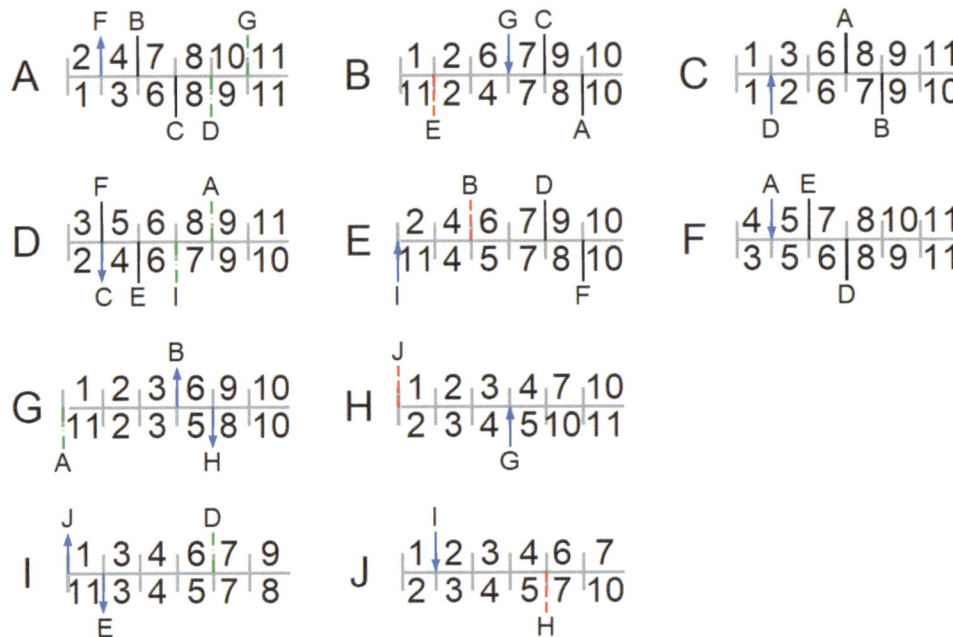

FIG. 17.21 State-transition diagram for hexaflexagon (leaves: 11).

We'll now draw a circle for each state and connect them with lines that indicate the type of flex that can take you between the states, as shown in figure 17.22.

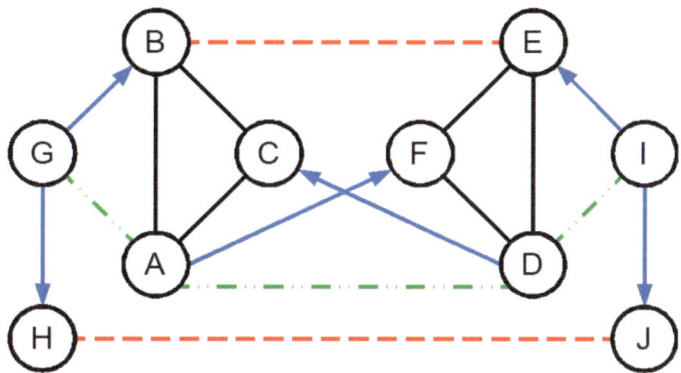

FIG. 17.22 State diagram for the 11-leaf hexaflexagon.

This state diagram makes it easier to answer various questions. For example, which states are reachable if you limit yourself to a single flex?

- **P:** Either ABC or DEF.
- **S:** Either BE or HJ.
- **T, T′:** Either AF, BGH, CD, or EIJ.
- **V:** ADGI.

We refer to a set of states that can be reached by one or more flexes as a **class**. For example, states A, B, and C make up a **pinch class** or **P class**. States D, E, and F make up a separate P class. If you use both the pinch flex and v-flex, you can travel between states A, B, C, D, E, F, G, and I, which we refer to as a **{P, V} class**. Note that when we refer to the class for a flex that isn't its own inverse, such as the tuck flex, we typically assume the inverse as well, enabling you to travel freely back and forth between all states in the class.

By examining the state diagram, you can figure out the flex classes for all pairs of flexes.

- {P, S} is ABCDEF or HJ.
- {P, T} is ABCDEFGHIJ.
- {P, V} is ABCDEFGI.
- {S, T} is AF, CD, or BEGHIJ.
- {S, V} is ADGI, BE, or HJ.
- {T, V} is ABCDEFGHIJ.

See figure 17.23 for state diagrams of several of these classes.

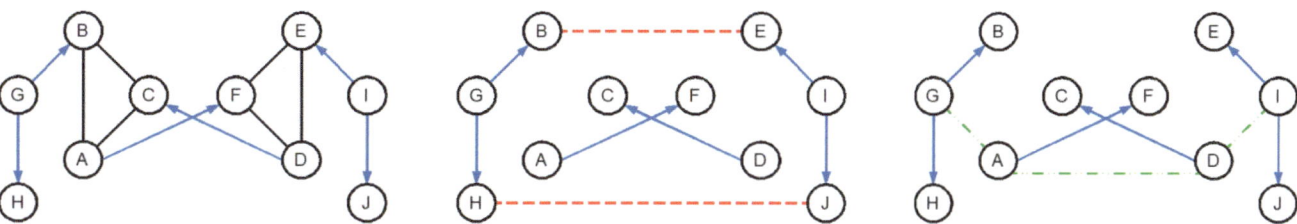

FIG. 17.23 State diagrams for the 11-leaf hexaflexagon showing just classes {P, T}, {S, T}, and {T, V}, from left to right.

Here are some observations:

- There are two P classes, each with three states. You can travel between the two pinch classes using S, T, or V.
- You can't visit every state using only a single type of flex.
- If you want to visit every state, you need to include T, since it's the only way to access states H and J from any of the other states.
- Either P or V can be used to travel between the four T classes, thus {P, T} or {T, V} allow you to visit every state.

It's interesting to note that if you include the slot-tuck-top-back (Ltb) and its inverse (Ltb'), you can find several additional states. Extending the diagrams with these flexes is left as an exercise for the reader.

Summary

Using a state diagram to keep track of how you can flex between the different states of a flexagon is a really helpful technique as you start exploring more complex flexagons. We saw the pinch state diagram for keeping track of faces as you pinch-flex, and generalized this concept when using the book flex and box flex on a square flexagon. We finished by tracking multiple types of flexes using a detailed state-transition diagram to supplement the more global state diagram.

Chapter 18

Flex Sequences

We have been using flex notation to describe simple sequences of flexes, and in this chapter we'll expand the notation and our uses of it. Some of the interesting ways we can use flex notation include the following:

- Describe a sequence of flexes.
- Define cycles and traversals.
- State that two sequences are equal or equal under certain conditions.
- Manipulate sequences to derive new equalities.
- Use generating sequences to describe and create new flexagons.

Sequences

We have discussed a wide variety of flexes, from the pinch flex and its variations to the v-flex to the pyramid shuffle to the morph-kite flexes. We have also given each of these a shorthand name, such as P, P3333, V, S, or Mkf. This shorthand is useful when you want to write out a sequence of flexes in a succinct way. Valid shorthand for a flex consists of a capital letter optionally followed by a series of lowercase letters and/or numbers.

We also define special flexes to indicate changing the reference hinge or turning over the flexagon. The **shift flexes**, < and >, are used for changing the reference hinge, shifting one hinge counterclockwise or clockwise respectively. And ^ is used to indicate turning the flexagon over while keeping the same reference hinge, called the **turn-over flex**.

We've used a tick mark, ', for indicating a flex's inverse, i.e., performing the flex in reverse. For example, T represents the tuck flex while T' represents its inverse.

It is also useful to be able to indicate that you would like to repeat a series of flexes a certain number of times. You do this by wrapping the flex sequence in parentheses followed by a number indicating the number of times to repeat the sequence. For example, (AB)3 says to repeat the sequence AB three times.

^	turn the flexagon over left to right
<	shift the reference hinge one hinge counterclockwise (to the left if the reference hinge is oriented at the top)
>	shift the reference hinge one hinge clockwise (to the right if the reference hinge is oriented at the top)
A'	the inverse of flex A, which undoes the effects of A
(A)n	repeat flex sequence A n times

DOI: 10.1201/9781003433538-21

Cycles

A **cycle** is a flex sequence that brings you back to your initial state.

A simple cycle is (P>)3, a series of three pinch flexes that takes you right back to where you began as shown in figure 18.1. Any triangle flexagon that supports a single pinch flex also supports this cycle, from the simplest tri-hexaflexagon to many of the more complex flexagons.

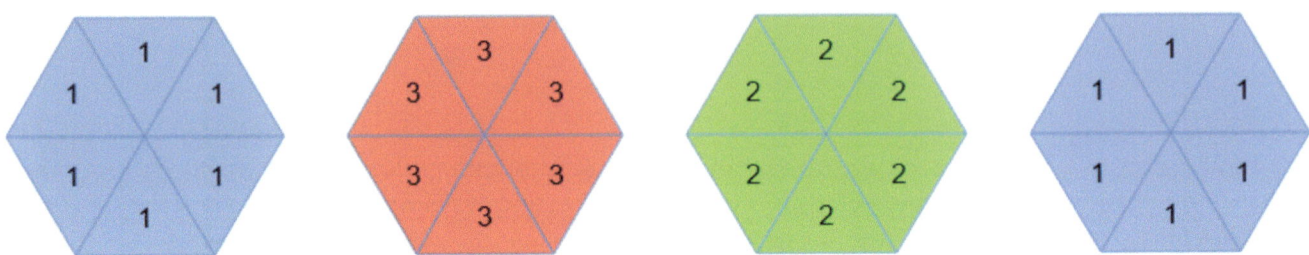

FIG. 18.1 Applying the sequence (P>)3 to a hexaflexagon.

The flip flex can also be a part of elegant cycles. We saw an example of this in chapter 12 where we did (F<)5 on a heptaflexagon, but other flexagons support a similar cycle. For example, the template in figure 18.2 folds up into an octaflexagon that supports a cycle of (F<)6. Figure 18.3 shows the states after each flip flex.

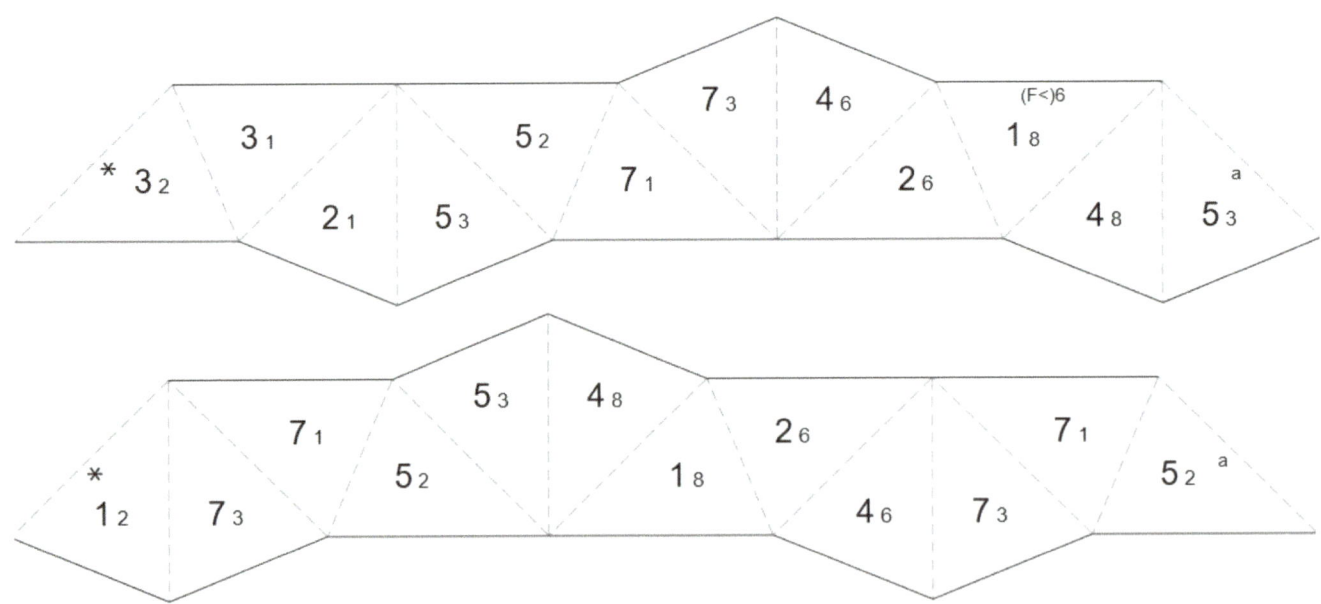

FIG. 18.2 Template for the octaflexagon (generator: (F<)6). Tape together the edges labeled with a before folding.

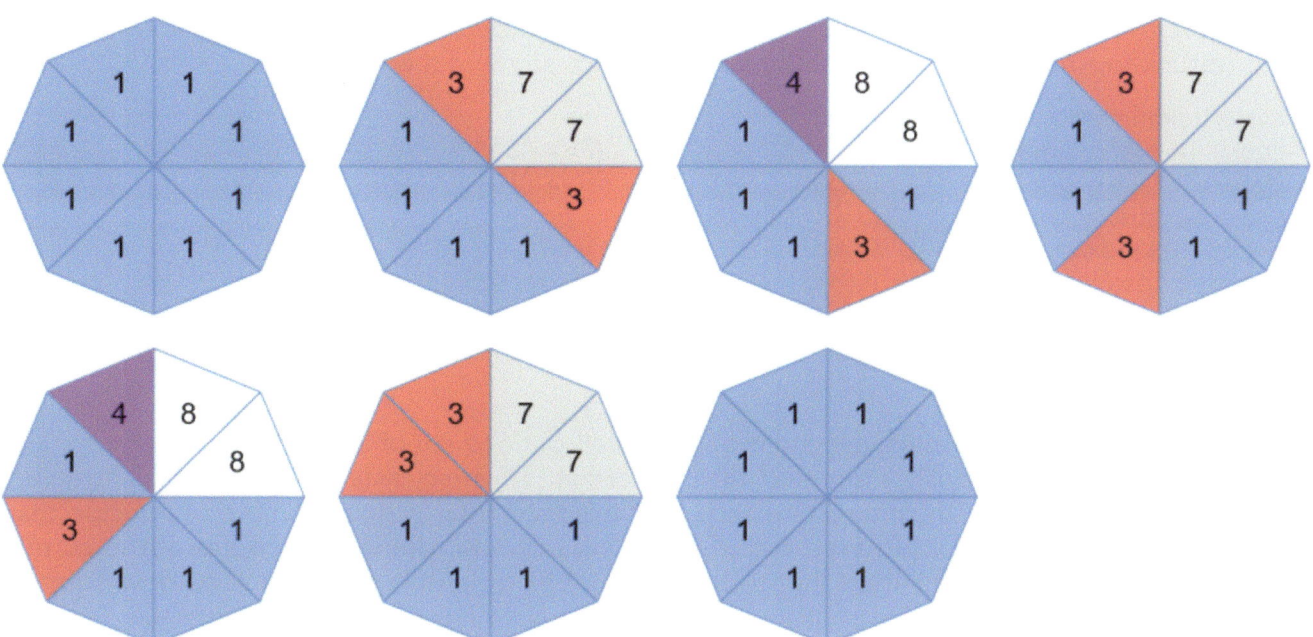

FIG. **18.3** Applying the sequence (F<)6 to an octaflexagon. Each image shows where to start the next flex, with the reference hinge at the top.

Whenever you can do a pyramid shuffle on a hexaflexagon, you can do the S-cycle, a sequence of pyramid shuffles and tuck flexes that cycle back to the starting state. The S-cycle is defined as (S>T'>^T<<^)2. You can do this even on the minimal flexagon for the pyramid shuffle. See figure 13.1 for a template. Since it can be tricky to keep track of the current hinge, figure 18.4 shows what the states look like before doing each pyramid shuffle, inverse tuck, and tuck in the sequence.

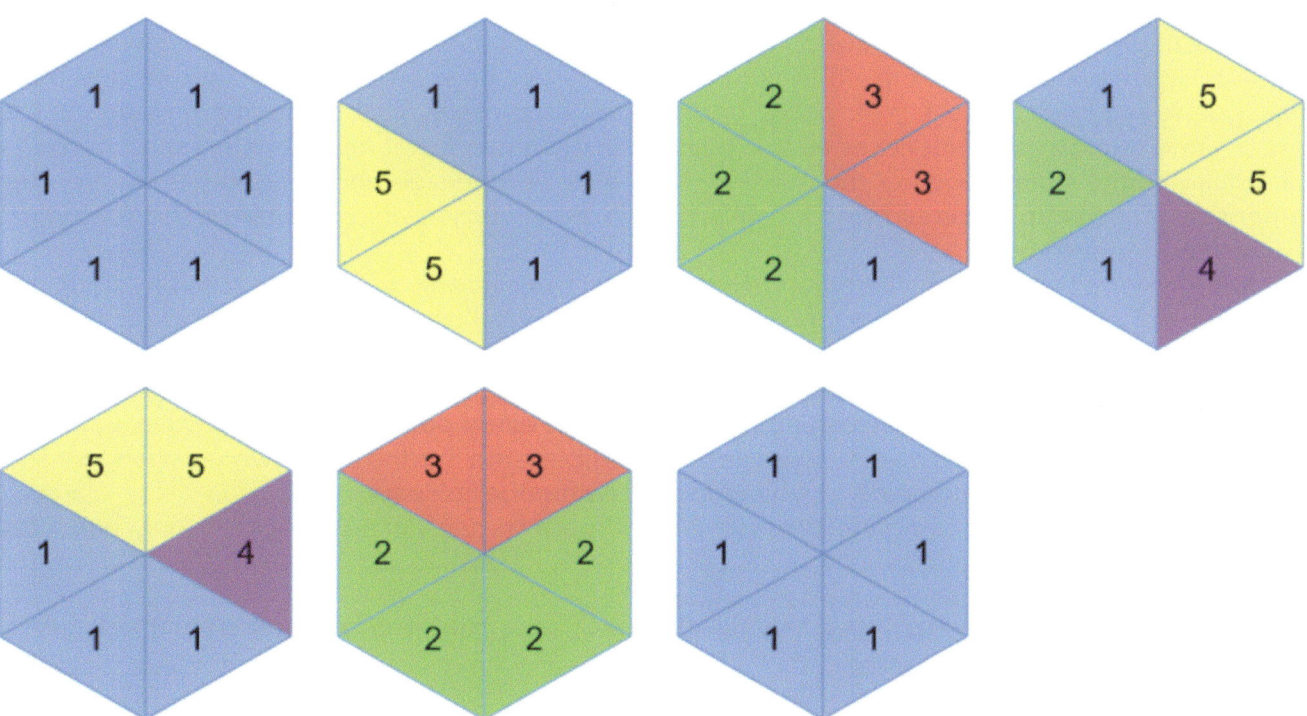

FIG. **18.4** Applying the sequence (S>T'>^T<<^)2 to a hexaflexagon. Each image shows where to start each S, T', and T from the sequence, with the reference hinge at the top.

Here are a few of the possible cycles, including ones described in this section and a couple for the Ltb (slot-tuck-top-back) flex. Multiple cycles are listed for some of the flexagons.

P-cycle hexaflexagon	(P>)3	
F-cycles heptaflexagon	(F<)5	(F>>>)3
F-cycles octaflexagon	(F<)6	(F>>>)10
S-cycle hexaflexagon	(S>T'>^T<<^)2	
V-cycles hexaflexagon	(V>)6	(V<)15
Ltb-cycles hexaflexagon	(Ltb<)3	(Ltb>>>)15

Traversals

A **traversal** is a cycle that visits every state out of a well-defined set of states.

For example, we previously discussed the Tuckerman Traverse, a technique for visiting every face using the pinch flex. One interesting aspect of this technique is that it's more of a recipe than it is a well-defined sequence of flexes. Recall that it says to keep doing a pinch flex at the current hinge until you can't anymore, then shift to the adjacent hinge and repeat. We need new notation to describe this:

$$\{A \mid B\} \quad \text{do sequence A if supported, otherwise do sequence B}$$

In other words, {A|B} means that if the internals of the flexagon allow us to do sequence A, then we do it. But if the structure doesn't allow it, we do sequence B instead. When spoken, you would pronounce {A|B} as "A or B".

This notation allows us to express the Tuckerman Traverse succinctly as ({P | >P})n, where n is the number of times to repeat. On a tetra-hexaflexagon, for example, the Tuckerman Traverse is ({P | >P})6, which means do a pinch flex at the current hinge if you can, otherwise shift one hinge to the right and do a pinch flex there, repeating six times.

Interestingly, the Tuckerman Traverse isn't the only way to traverse every face in the fewest steps. An alternate traversal is ({^>P | >P})n. You can try this out on a hexa-hexaflexagon or any other flexagon found in chapters 1, 3, 4, or 5. This traversal visits every pinch state in a flexagon in the exact same number of pinch flexes as the Tuckerman Traverse. But it has the interesting property of visiting all the states together that have a given label on one side and every other label on the other side. For example, it could visit 1/2, 1/3, 1/4, 1/5, etc. in order, which is different from what the Tuckerman Traverse does.

Note, however, that the Tuckerman Traverse doesn't necessarily visit every possible state of a flexagon. For example, if the flexagon supports the tuck or v-flex, these additional flexes will take you to states that aren't reachable by just doing pinch flexes. Recall that all states reachable by just using flex A is called an A class. Therefore, we say that the Tuckerman Traverse allows us to traverse every state in a pinch class, or P class, not that it traverses every possible state of the flexagon.

What would it look like to travel between every possible state in a flexagon? One challenge is that the number of states grows rapidly as you increase the number of leaves.

A simple example of a complete traversal is (P>)3 on a tri-hexaflexagon, since it only has three states. Another simple example uses the S-cycle from the previous section, (S>T'>^T<<^)2, on the minimal S hexaflexagon, since it only has six states.

A more challenging example uses the hexaflexagon template in figure 18.5. This flexagon has a total of 36 states and supports P, V, T, S, and Ltb. One flex sequence that hits every state is (Ltb>P>P>)12.

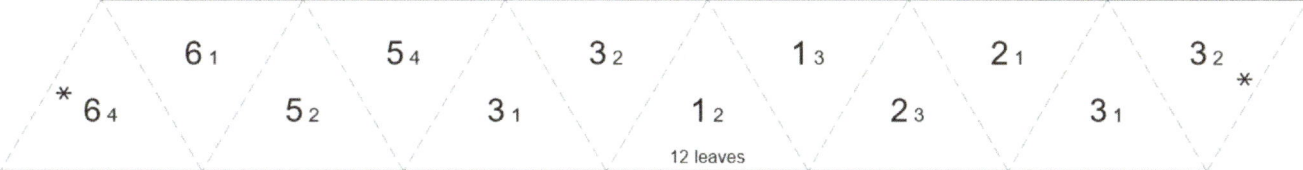

FIG. 18.5 Template for a hexaflexagon (leaves: 12) with 36 states.

Here are the traversals we've discussed:

Tuckerman Traverse	({P | >P})n
alternate traverse	({^>P | >P})n
S-cycle hexaflexagon	(S>T'>^T<<^)2
12-leaf hexaflexagon	(Ltb>P>P>)12

Manipulating sequences

Sometimes it's useful to be able to state that two flex sequences are equal, that is, they change a flexagon in the same way. Other times you may want to manipulate a flex sequence, substituting in equivalent sequences to arrive at new statements about sequences.

First, we introduce a new concept: the ***identity flex***, I, which doesn't change the flexagon.

In mathematics, it's often useful to have a way to capture the idea that nothing has changed. In flex notation, the identity flex serves that purpose. This gives us a way to succinctly say that a particular flex sequence is a cycle that returns to the original state. For example, (P>)3 is the cycle of pinch flexes you can do on a tri-hexaflexagon, so performing that cycle is like doing the identity flex because the flexagon looks the same at the end of the sequence. Note that the identity flex doesn't even change the reference hinge or turn over the flexagon.

Now we introduce notation for comparing flex sequences:

A = B	flex sequences A and B do exactly the same thing
A = B	{when A is supported}
A = B	{when B is supported}
A = B	{when A and B are supported}

If two sequences are exactly the same, you can always substitute one for the other. But if the two sequences are sometimes equal but require different internal structure, you can add the requirements after the equation inside curly brackets. For example, we say I = (P>)3 {when P is supported} to indicate that the two sequences have the same effect only if the internal structure supports a pinch flex. You can always do the identity flex at any hinge on any flexagon because it doesn't require any internal structure to the flexagon. But you can only do (P>)3 if the internal structure allows it. Depending on the flexes, you may need to specify when A is supported, B is supported, or both.

Next, we define some useful equalities involving the shift and turn-over flexes:

<' = >	>' = <	^' = ^
<> = I	>< = I	^^ = I

Finally, we provide some additional equalities for any set of flexes A, B, and C (see chapter 22 for why these are true):

IA = AI = A
(AB)' = B'A'
if A = B, then CA = CB and AC = BC

It's also important to note equalities that are not always true. For example, while A' = ^A^ for many flexes, such as the pinch flex and flip flex, there are flexes where this isn't true, such as the pyramid shuffle and slot-half flex. Also, it's generally the case that you can't swap the order of two flexes and expect the same result. In mathematics, we say that flexes are not *commutative*.

With those definitions and rules, we can now derive new flex sequences and prove how they behave. Let's consider the Mkf and Mkb morph-kite front and back flexes we discussed in chapter 14. By definition, we know that Mkf = ^Mkb^. We would like to start with the sequence Mkf Mkb' and come up with another sequence that's identical to it by applying known rules and identities to manipulate the sequence. First, we derive a definition for Mkb in terms of Mkf (where the arrow, =>, means *therefore*):

Mkf = ^Mkb^ => ^Mkf^ = ^^Mkb^^ => ^Mkf^ = Mkb

Now we can derive an alternate sequence for Mkf Mkb' through substitution:

Mkf Mkb' = ^Mkb^ (^Mkf^)'
          = ^Mkb^ ^' Mkf' ^'
          = ^Mkb ^ ^ Mkf' ^
          = ^ Mkb Mkf' ^

Thus, we have just shown that doing the sequence Mkf Mkb' is the same as doing the sequence Mkb Mkf' from the opposite face.

Generating sequences

A *generating sequence* is a flex sequence used to create the minimal flexagon that supports that sequence.

There are a couple additional pieces of notation useful for declaring generating sequences:

A+   create the minimal pat structure needed for A without doing A
A*   A+ A (create the minimal pat structure and perform A)

We have already used these concepts when working with the minimal flexagons for various flexes. For example, the generating sequence for the tri-hexaflexagon is P+, while the generating sequence for the minimal pyramid shuffle flexagon is S+.

With this new notation, we now have the ability to define more complex generating sequences, such as P*>St*^>T'*. (Note that this could also be written as P+P > St+St ^> T'+T', but using * makes the sequence more obvious.) This gives us a very general and powerful way to define new flexagons and to custom design them with the behavior we want. The majority of the templates in this book were created from generating sequences that illustrate the particular flex, set of flexes, or flex sequence being discussed.

Consider the three hexaflexagons with six faces that we discussed in chapter 2. Using the standard naming convention, these are all called hexa-hexaflexagons, with no easy way to differentiate between them. One way to be more precise is to use generating sequences to refer to the different variations. Here are generating sequences for them:

```
P* P* P+ > P P+      straight template (variation 6a)
(P*^>)4              hexagonal template (variation 6b)
(P*)4               triangular template (variation 6c)
```

Note that a generating sequence for a given flexagon isn't necessarily unique. For example, an alternate sequence for the straight template is (P*)3 (>P)2 P P*. You can try out both of those sequences on the pinch state diagram in figure 17.5.

This does present a question, however: given a generating sequence, how do we figure out the necessary template for it? We answer this question in chapters 19 and 20.

Other interesting sequences

We can now make precise statements about how various sequences of flexes compare to each other. For example, here are some equivalent sequences on a hexaflexagon:

PP = ^T' << T' <<^ T >> = (Tf<<)3
P^>P^>P< = (S<<)3

Or, if a flexagon has n pats, we can write more general formulas for them (where $n/2$ means half of n):

PP = (Tf<<)(n/2)
P^>P^>P< = (S<<)(n/2)

There are lots of interesting ways that various flexes are related to each other.

```
P        =    >>>V>>>T      {when >>>V>>>T is supported}
T        =    <P^SP^>       {when <P^SP^> is supported}
F        =    <^T<<T^S<     {when <^T<<T^S< is supported}
S        =    FSt'          {when FSt' is supported}
St       =    >Tf'<<Tf'>
S3<<<F   =    Fm<<Fm<       {when both are supported}
```

The flexes we saw in chapter 6 can be expressed in terms of a series of forced tucks when you do them twice at the same hinge using a simple approach: for every hinge you skip in the pinch variant, add a right shift. For example, (P345)2 can be expressed as Tf>>>Tf>>>>Tf>>>>>. Or, to make it more explicit, you can write it as Tf (>)3 Tf (>4) Tf (>)5.

```
(P222)2    =    Tf>> Tf>> Tf>>
(P333)2    =    Tf>>> Tf>>> Tf>>>
(P3333)2   =    Tf>>> Tf>>> Tf>>> Tf>>>
(P444)2    =    Tf>>>> Tf>>>> Tf>>>>
(P66)2     =    Tf>>>>>> Tf>>>>>>
(P345)2    =    Tf>>> Tf>>>> Tf>>>>>
```

We can validate these formulas manually by trying them out on actual flexagons. In the next chapter on pat notation, we'll see how to prove them mathematically.

Summary

The following table summarizes the notation discussed in this chapter:

<	shift reference hinge left (counterclockwise)
>	shift reference hinge right (clockwise)
^	turn flexagon over along axis perpendicular to reference hinge
A'	the inverse of flex A
(A)n	repeat flex sequence A n times
{A|B}	do sequence A if supported, otherwise do sequence B
I	the identity flex, the flex that doesn't change the flexagon
A = B	flex sequences A and B do exactly the same thing
A+	create minimal pat structure needed for A without doing A
A*	A+ A (create minimal pat structure and perform A)

Here are some identities and other rules:

$<' = >$ $>' = <$ $^' = ^$
$<> = I$ $>< = I$ $^^ = I$
$IA = AI = A$
$(AB)' = B'A'$
if $A = B$, then $CA = CB$ and $AC = BC$

Chapter 19

Pat Notation

Pat notation describes the internal structure of a flexagon. Some of the interesting ways to use it include the following:

- Define how a flex changes a flexagon's state.
- Determine if a given flexagon supports a given flex at a particular hinge.
- Predict how a flex sequence changes a flexagon's state.
- Use a generating sequence to create the pat structure necessary to perform a flex sequence.
- Prove if two flex sequences are the same or different.

In this chapter, we will specifically discuss notation that can describe the internals of triangle flexagons. We will focus on **isoflexagons**, flexagons where all the pats meet in the center. In chapter 20, we'll generalize this to any arrangement of triangular pats.

Structure

We will use the template from figure 19.1 to help illustrate how pat notation works. The pair of numbers in the middle of each leaf-face are the leaf labels we've already seen in previous templates, where the large number is for the front of the leaf-face and the smaller number is for the back of the leaf-face. In this template, the leaf labels range from 1 to 7. We also include a leaf ID – a different number for each leaf – in its corner, ranging from 1 to 14. Leaf labels are useful when folding templates or for tracking faces when we don't care about the identity of individual leaves. In contrast, leaf IDs are useful when we want to track every leaf. If we also want to track each face of each leaf, one leaf-face is marked with the positive ID number and the other leaf-face with the negative ID number.

DOI: 10.1201/9781003433538-22

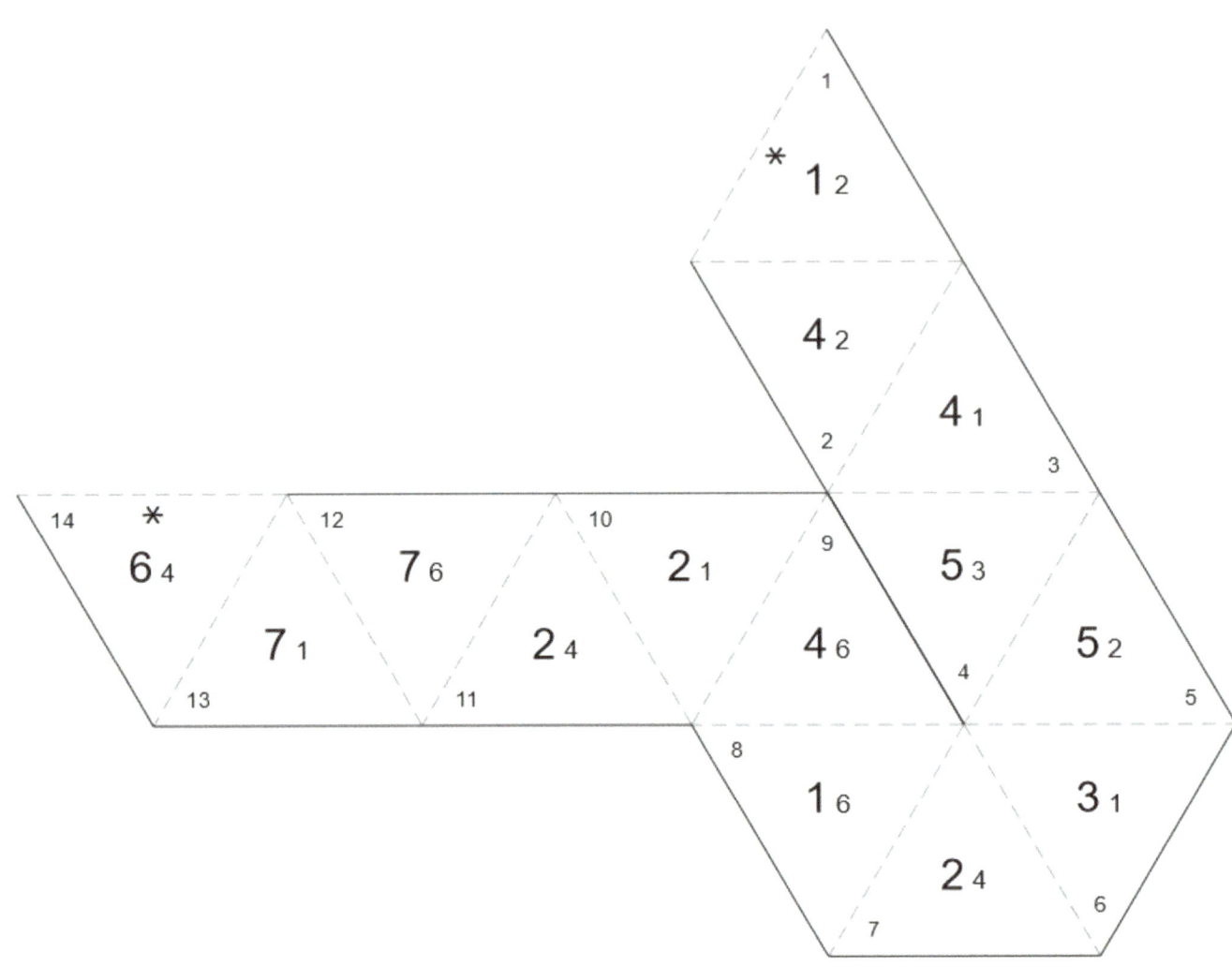

FIG. 19.1 Template for a hexaflexagon (leaves: 14) with leaf IDs.

Since we want to keep track of every leaf-face, after cutting out the template, write the equivalent negative leaf ID on the back of each leaf. For example, put –1 on the back of the first leaf (the one with the small 1 in the corner), –2 on the back of the second leaf (with the small 2 in the corner), and so on. You may want to underline the 6's and 9's to make them easier to read as you apply flexes.

Fold the template into a hexaflexagon by folding together adjacent pairs of leaf labels, starting with 7 on 7, then 6 on 6, and so on, as we do with other templates. But don't tape the ends together just yet because we want to be able to easily unfold the pats and examine them more closely. Start by looking at the face where all the labels (the large numbers) are 1's and the visible leaf IDs (the small numbers in the corner) are 1, –3, –6, 8, –10, and –13 going clockwise as shown in figure 19.2. We're going to focus on the leaf IDs.

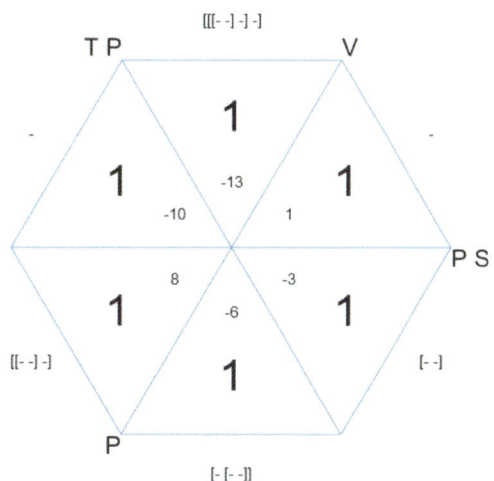

FIG. 19.2 Folded hexaflexagon (leaves: 14), with large leaf labels and small leaf IDs in the center.

Examine the pat with leaf ID 1. You'll notice that there's only a single leaf in this pat. In pat notation, we describe this pat simply as 1, the leaf ID that's visible on the top leaf-face. If we turn over the flexagon, we describe the resulting pat as –1, because the leaf ID on the opposite leaf-face is visible.

Now look at the pat that's connected to it with leaf ID –3 on the top. If you unfold it, you'll find a leaf with leaf label 4 and leaf ID 2 immediately below it, which is also facing up. The pat notation for this pat is [–3,2], which describes the order of the leaves in the pat from top to bottom.

Now examine the third pat, where you should find leaf ID –6 on top. If you fold back the –6, you will see two **subpats**: one with a single leaf (ID 6) and one with a pair of leaves. The bottom pair has a –4 on top with a 5 below it. Pat notation groups each leaf into the subpats we find as we unfold. Thus, this pat is written as [–6,[–4,5]].

Now look at the fourth pat, with 8 on the top. This one also has three leaves, but it has a pair of leaves folded on top of a single leaf, the opposite of the previous pat. This one is written as [[8,–9],–7].

The next pat is a singleton: –10.

The final pat has four leaves. If you unfold them and note both the pairs and the leaf IDs on the leaves, you should find that this pat is [[[–13,12],14],–11].

To describe the entire flexagon, we list the structure of each pat one at a time going clockwise. Thus, for our sample flexagon, we end up with the following, which we'll refer to as state f_1:

$$f_1 = 1, [–3,2], [–6,[–4,5]], [[8,–9],–7], –10, [[[–13,12],14],–11]$$

Caveat: Making a simple list of the pats like this is only sufficient if all the pats meet in the middle. We will revisit this assumption in the next chapter.

If we're only interested in the pat structure, but not the identities of the individual leaves, we can simply use a dash (-) to represent each leaf. Then f_1 looks like this:

$$f_1 = -, [- -], [- [- -]], [[- -] -], -, [[[- -] -] -]$$

^, >, and <

What happens if we turn over a single pat? If we turn over the first pat in state f_1, which consists of a single leaf, it switches from 1 to –1. If we turn over the second pat, which we described as [–3,2], it changes to [–2,3] because –2 is now on the top leaf and 3 is on the bottom leaf. When each individual pat is turned over, it swaps the IDs between positive and negative, and it reverses the order of the leaves and their nesting. For example, if we turn over the final pat, [[[–13,12],14],–11], we get [11,[–14,[–12,13]]].

What happens if we turn over the entire flexagon? When you examine the result, you'll see that every pat is turned over and the pat order is reversed. If we start from f_1 and turn it over, we get the following:

$$f_1 \wedge = [11,[–14,[–12,13]]],\ 10,\ [7,[9,–8]],\ [[–5,4],6],\ [–2,3],\ –1$$

By convention in pat notation, we enumerate pats starting with the reference hinge at the top and traveling clockwise around the flexagon. To shift where the reference hinge is, we can cycle the pats in the list. In f_1, the reference hinge is between IDs –13 and 1. To shift one hinge clockwise so the reference hinge is between IDs 1 and –3, we cycle the pats like this:

$$f_1 > = [–3,2],\ [–6,[–4,5]],\ [[8,–9],–7],\ –10,\ [[[–13,12],14],–11],\ 1$$

Similarly, when shifting the reference hinge counterclockwise, i.e., doing a < flex, we get the following:

$$f_1 < = [[[–13,12],14],–11],\ 1,\ [–3,2],\ [–6,[–4,5]],\ [[8,–9],–7],\ –10$$

Defining flexes

What happens if we apply a more interesting flex to our 14-leaf flexagon? Let's start with face 1 and try a pinch flex. You may want to loosely tape the ends together before the flex, then take the tape back off to examine the results.

Note that you can't start your pinch flex between leaf IDs –13 and 1, where the tape is. You have to shift to an adjacent hinge, for example, between 1 and –3, to do the pinch flex. After the pinch flex, you should be looking at the face labeled with 4's, with visible leaf IDs of 2, 3, –7, 9, –11, and –14. If you examine each pat, writing down the top leaf IDs and structure in each pat, you should end up with the following, which is state f_1 with the pinch flex applied:

$$f_1 P = [2,–1],\ 3,\ [–7,[[–5,4],6]],\ [9,–8],\ [–11,10],\ [–14,[–12,13]]$$

Now let's try using the same technique on the minimal hexaflexagon for the pinch flex, the tri-hexaflexagon. Figure 19.3 provides the template with leaf IDs. Figure 19.4 shows face 1 from state 1/2 followed by face 3 from state 3/1 after doing a pinch flex.

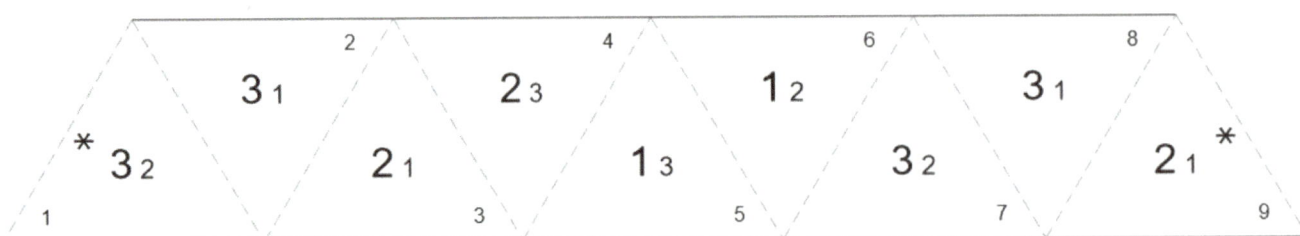

FIG. 19.3 Tri-hexaflexagon template with leaf IDs.

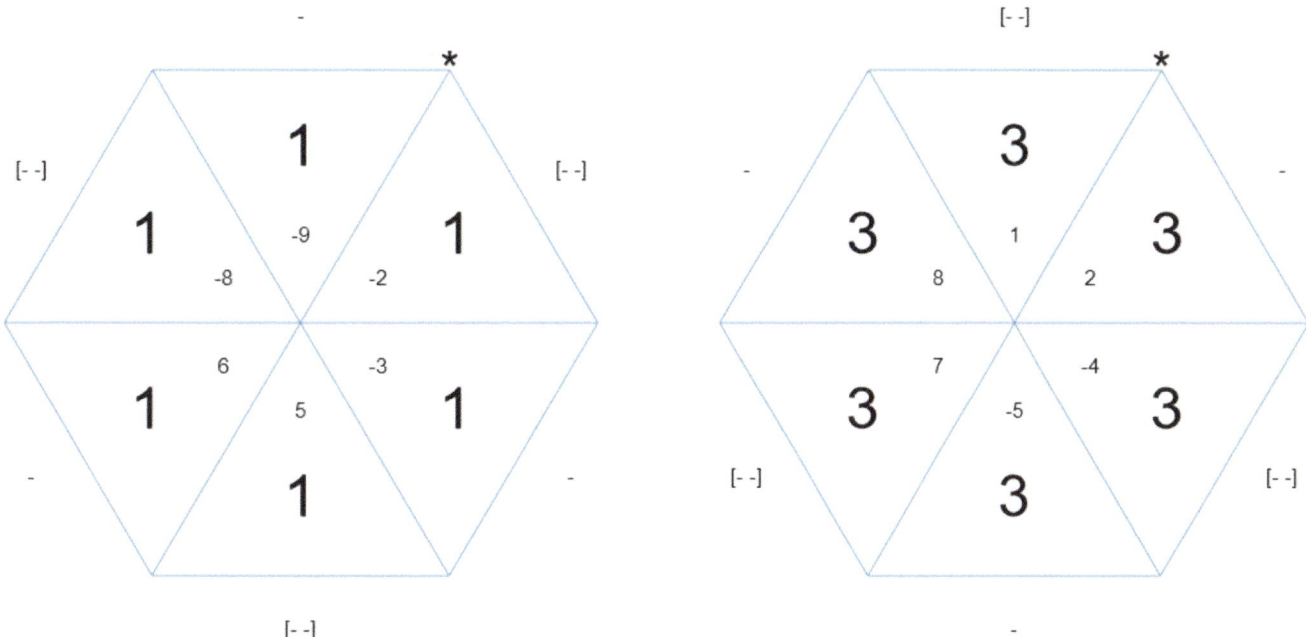

FIG. 19.4 Visible leaf IDs when looking at the initial face 1 (left) and the resulting leaf IDs after doing a pinch flex to reveal face 3 (right).

When you first fold the flexagon and examine it from face 1, you should find that the pat structure looks like this, which we'll call p_1:

$$p_1 = [-2,1], -3, [5,-4], 6, [-8,7], -9$$

After you do a pinch flex at the position marked by the asterisk (between pat −9 and pat [−2,1]), you will reveal face 3. The new arrangement of leaf IDs should look like this, which we'll call p_2:

$$p_2 = 2, [-4,3], -5, [7,-6], 8, [1,9]$$

These before and after states give us a precise definition of the pinch flex and can be used to predict exactly how the pinch flex modifies any other hexaflexagon, as we'll soon see.

Mathematically, p_2 is a permutation of p_1 (that is, a different arrangement of the leaves). This permutation describes a pattern that applies to any pinch flex on a hexaflexagon.

To see how this works, we'll apply a pinch flex to the hexaflexagon (leaves: 14) from figure 19.1 as an example, referring back to figure 19.2 for the folded flexagon. First, we compare the pat structure of that flexagon, f_1, to the initial pat structure of the minimal pinch flex tri-hexaflexagon, p_1. We previously described f_1:

$$f_1 = 1, [-3,2], [-6,[-4,5]], [[8,-9],-7], -10, [[[-13,12],14],-11]$$

Notice that f_1 starts with a single leaf, while p_1 starts with a pair of leaves. This tells us that we can't do a pinch flex between leaf IDs −13 and 1 (the current hinge) because f_1 doesn't have the required internal structure. That means that we have to shift to an adjacent hinge before we can do a pinch flex. Let's first apply the > flex to f_1 in order to change the current hinge and call the result f_2:

$$f_2 = [-3,2], [-6,[-4,5]], [[8,-9],-7], -10, [[[-13,12],14],-11], 1$$

Next, line up each pat from p_1 with the corresponding pat in f_2 in order to compare the structure between them:

p_1	[–2,1],	–3,		[5,	–4],	6,	[–8,	7],	–9
f_2	[–3,2],	[–6,[–4,5]]		[[8,–9,	–7],	–10,	[[[–13,12],14],	–11],	1

Ignoring the leaf IDs for the moment, you can see that for every leaf in p_1, f_2 has at least one leaf in the same location of the pat or subpat. This tells us that f_2 supports the pinch flex at its current hinge.

The next step is to figure out how the leaf IDs in p_1 correspond to the ones in f_2. If we look at leaf ID 1 from p_1, you can see that it corresponds to leaf ID 2 in f_2. Leaf ID –2 in p_1 corresponds to leaf ID –3 in f_2, which is the same as saying 2 corresponds to 3 (turning them both over). A more interesting case is leaf ID –3 in p_1, which corresponds to [–6,[–4,5]] in f_2. If we turn both of those pats over, we can see that leaf ID 3 in p_1 corresponds to [[–5,4],6] in f_2.

If we enumerate all the leaf IDs in p_1, lining them up with the corresponding structure in f_2, we get the following:

p_1	1	2	3	4	5	6	7	8	9
f_2	2	3	[[–5,4],6]	7	[9,–8]	–10	11	[–14,[–12,13]]	–1

Now substitute all those values into the description of p_2 to find state f_3, which is f_2 with a pinch flex applied to it.

p_2	2,	[–4,3],	–5,	[7,–6],	8,	[1,9]
f_3	3,	[7,[[–5,4],6]],	[8,–9],	[11,10],	[–14,[–12,13]],	[2,–1]

Finally, shift everything back to undo that shift we did earlier.

$$f_4 = [2,–1], 3, [–7,[[–5,4],6]], [9,–8], [–11,10], [–14,[–12,13]]$$

This is exactly the same as the state that we got by flexing our actual flexagon. Using the pat notation for the pinch flex on the tri-hexaflexagon and applying it to our sample flexagon allowed us to predict exactly how the pinch flex would change its internal structure.

This is a really powerful result and can be applied more broadly than just to the pinch flex and hexaflexagons. Here's the recipe for creating and applying a flex definition using pat notation:

1. Start with the minimal flexagon for a given flex.
2. Mark every leaf-face with a unique ID, with negative IDs on the back faces of the leaves.
3. Write out the initial state of the flexagon in pat notation, then apply the flex and write the end state in pat notation.
4. If an arbitrary flexagon has at least as much structure as the initial state from step 3, it supports the given flex at its current hinge.
5. If the flex is supported, use leaf ID substitution to predict how the flex will change the flexagon.

Here are definitions for selected flexes on a hexaflexagon, listing the input (in) and output (out) states:

>	in	1, 2, 3, 4, 5, 6
	out	2, 3, 4, 5, 6, 1
<	in	1, 2, 3, 4, 5, 6
	out	6, 1, 2, 3, 4, 5
^	in	1, 2, 3, 4, 5, 6
	out	–6, –5, –4, –3, –2, –1

P	in	[–2,1], –3, [5,–4], 6, [–8,7], –9
	out	2, [–4,3], –5, [7,–6], 8, [1,9]
S	in	[–2,1], –3, –4, –5, [[[–8,7],9],–6], 10
	out	[–2,[9,[1,–10]]], –3, –4, –5, [7,–6], 8
T	in	[[–2,3],1], 4, 5, [–7,6], –8, –9
	out	3, 4, 5, [–7,6], –8, [2,[–9,–1]]
V	in	1, [–3,2], [5,–4], 6, 7, [–9,8]
	out	[2,–1], 3, 4, [–6,5], [8,–7], 9

Each of these flexes can be applied to at least one hinge of f_1. You can use these definitions to predict if a hinge supports a given flex and, if so, exactly how the flexagon will be modified. You can compare the before and after states on the real flexagon to validate that the rules work. To define the inverse of a flex, swap the inputs and outputs.

Those selected definitions were all designed so that the leaf IDs flow sequentially in the corresponding template, but this isn't a requirement. All that's required is that the IDs are unique and are used consistently in the pat notation for the input and output. As we'll see, sometimes it's more convenient to have the leaf IDs arranged in other ways.

Angles

If you're flexing a regular hexaflexagon where all the leaf edges and angles are the same, simply tracking how the pats change is sufficient to understand how the flexagon changes. But if the leaves have different-sized edges and angles, you may also want to track how they change as you flex. To do this, you can label the angles and check how they change position as you flex.

We'll continue to assume all pats meet in the middle, as they do in most of the flexagons we've looked at so far, revisiting this assumption in the next chapter. Label the angles in the first leaf to the right of the reference hinge α (alpha), β (beta), and γ (gamma), where α is in the center of the flexagon and β is the angle immediately counterclockwise. Then perform a flex and recheck the angles of the first leaf to the right of the reference hinge to see how they have changed. Here's how α β γ changes after performing the flexes we just defined:

>	αγβ		P	βαγ
<	αγβ		S	no change
^	no change		T	no change
			V	γβα

In general, local flexes don't change the angles while global flexes may.

Generating sequences

In chapter 18, we introduced flex notation that can describe generating sequences. The basic idea is that you can declare the sequence of flexes you would like to be able to carry out on a flexagon, and somehow the necessary magic happens in order for the flexagon to have exactly the minimum amount of internal structure needed to support that sequence. With pat notation, we now have the tools for implementing that magic.

Let's start with a simple hexagon with a single leaf per pat, which we'll call h_1. We would like to apply the generating sequence P+>P+ to it, which will allow us to do a pinch flex at any of the six hinges.

$$h_1 = 1, 2, 3, 4, 5, 6$$

Obviously, our hexagon doesn't support that sequence, so what do we do? We need to figure out how to change our hexagon so it supports the first pinch flex in our sequence. Fortunately, the pinch flex definition tells us exactly what internal structure we need in order to do a pinch flex, so we can just change our flexagon to match, then apply > to the result. This supports the P+> portion of our generating sequence.

$$h_2 = [-2,1], -3, [5,-4], 6, [-8,7], -9$$
$$h_3 = h_2 > = -3, [5,-4], 6, [-8,7], -9, [-2,1]$$

But now that we need to be able to apply a second pinch flex, we're stuck. The pinch flex definition says that the first, third, and fifth pats each need to have at least two leaves. But the corresponding pats in h_3 only have a single leaf each. In order to support the generating sequence, we need to add the missing leaves as described in the flex definition to the pat notation for our current flexagon. As we add new leaf IDs for the new leaves, we'll start with leaf ID 10, the first number we haven't already used. For example, the first pat needs to contain two leaves instead of just one, so we change the pat from −3 to [−3,10]. The second pat, [5,−4], is a set that contains the single leaf we need, so we can copy it across as is. Continuing this recipe for every pat gives us the following result:

$$h_4 = [-3,10], [5,-4], [6,11], [-8,7], [-9,12], [-2,1]$$

That's it. We've just applied the generating sequence P+>P+ to our inflexible hexagon and described a new flexagon with exactly the minimum structure necessary to apply the sequence. If you look at the structure of the tetra-hexaflexagon, you'll see that this exactly describes its pat structure.

We could follow the same process for any arbitrary sequence, such as S+>T'+^>>V+. Recall that A* = A+ A, so if the sequence were instead S*>T'*^>>V*, we would apply each flex after first creating the internal structure for that flex.

This raises another question, however. Now that you can go from a generating sequence to pat structure, how can you go from pat structure to an unfolded template? That will be answered in the next chapter on atomic flex theory.

Proving equivalence

Next, we will leverage flexes defined using pat notation to prove that the equation Lh = Ltb T' is true. We start with the following flex definitions:

Lh	in	[[−2,[−4,3]],1], −5, −6, −7, [[9,−10],−8], −11
	out	[[−11,−1],10], −2, −3, [5,−4], [[−7,8],6], 9
Ltb	in	[[−2,[−4,3]],1], −5, −6, −7, [9,−8], 10
	out	−1, −2, −3, [5,−4], [[−7,8],6], [−10,9]
T	in	[[−2,3],1], 4, 5, [−7,6], −8, −9
	out	3, 4, 5, [−7,6], −8, [2,[−9,−1]]

Our strategy is to start from the input state for Lh, apply Ltb and T', and compare that result to the output state for Lh. If they match, then the two sequences are equal.

First, we start from the initial state for Lh and compare it against the input to Ltb:

Ltb	in	[[–2,[–4,3]],1]	–5	–6	–7	[9	–8],	10
Lh	in	[[–2,[–4,3]],1]	–5	–6	–7	[[9,–10],	–8],	–11

This gives us the following values for Ltb's leaf IDs:

Ltb	in	1	2	3	4	5	6	7	8	9	10
Lh	in	1	2	3	4	5	6	7	8	[9,–10]	–11

Plugging each of those correspondences into the output of Ltb gives us the result of applying Ltb to Lh's inputs.

$$A = –1, –2, –3, [5,–4], [[–7,8],6], [11,[9,–10]]$$

Next, we want to apply T', the inverse of T. Recall that in order to apply a flex's inverse, we need to swap the inputs and outputs of the flex definition. Then we line up our last result against the inputs for T' to get the following:

T'	in	3	4	5	[–7,6]	–8	[2,[–9,–1]]
A		–1	–2	–3	[5,–4]	[[–7,8],6]	[11,[9,–10]]

This gives us the following values for the leaf IDs:

T'	in	1	2	3	4	5	6	7	8	9
A		10	11	–1	–2	–3	–4	–5	[–6,[–8,7]]	–9

Therefore, applying T' to A gives us this result:

$$B = [[–11,–1],10], –2, –3, [5,–4], [[–7,8],6], 9$$

This is exactly the output of Lh, so we've proven that Lh has the exact same effect as the flex sequence Ltb T'.

General flex definitions

While the flex definitions we've explored thus far were specific to the hexaflexagon, the flexes themselves can be performed on flexagons with a pat count other than six. Let's look at how to generalize these definitions.

The pinch flex works on any flexagon with an even number of pats, as long as the number of pats is at least four. In the following definitions, n is the number of leaves (which, for the pinch flex, is the number of pats × 1.5, since half the pats need a single leaf and the other half need two leaves) and i refers to the next available leaf index, a pattern that keeps repeating until you reach the last set of pats.

If the number of pats is divisible by four, the general formula looks like this:

P	in	[–2,1], –3,	...	[i+1,–i], i+2,	...	[n–1,–(n–2)], n	
	out	2, [–4,3],	...	–(i+1), [i+3,–(i+2)],	...	–(n–1), [1,–n]	

If the number of pats is not divisible by four, the general formula is the same as the last definition with a minor change to the final pats:

P	in	[−2,1], −3,	...	[i+1,−i], i+2,	...	[−(n−1),n−2], −n
	out	2, [−4,3],	...	−(i+1), [i+3,−(i+2)],	...	n−1, [1,n]

We can use this to derive the definition for the pinch flex on the octaflexagon (with eight pats):

P	in	[−2,1]	−3	[5,−4]	6	[−8,7]	−9	[11,−10]	12
	out	2	[−4,3]	−5	[7,−6]	8	[−10,9]	−11	[1,−12]

Next, consider the pyramid shuffle, which requires at least five pats. The total number of required leaves is the number of pats + 4. Here is the general definition of the pyramid shuffle:

S	in	[−2,1],	...	−i,	...	[[[−(n−2),n−3],n−1],−(n−4)],	n
	out	[−2,[n−1,[1,−n]]],	...	−i,	...	[n−3,−(n−4)],	n−2

On a heptaflexagon, with seven pats, we can derive the following definition for the pyramid shuffle:

S	in	[−2,1], −3, −4, −5, −6, [[[−9,8],10],−7], 11
	out	[−2,[10,[1,−11]]], −3, −4, −5, −6, [8,−7], 9

Same visible leaves, different internal structure

In chapter 17, we noted that it was possible to have the same visible leaves but different internal pat structure. With pat notation, we have a way to precisely specify an example. The following pat notation describes two different ways to fold the template in figure 19.5:

[[[−2,[[−4,[7,[5,−6]]],3]],1],	8,	[−10,9],	[12,−11],	[−14,13],	−15
[[[−2,[7,[[[−4,5],3],−6]]],1],	8,	[−10,9],	[12,−11],	[−14,13],	−15

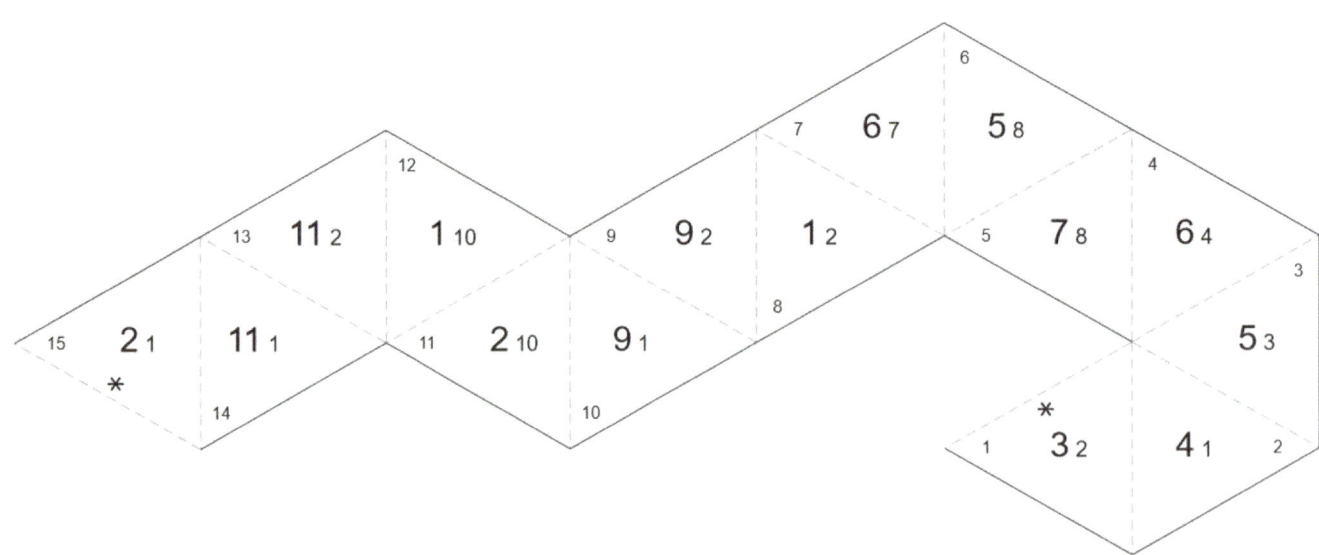

FIG. 19.5 Template for a hexaflexagon that allows you to tuck-flex between two states with the same visible leaf IDs but different pat structure. Copy the stars from the first and last edges onto the back of the template before folding.

If you use the standard folding technique (8 on 8, 7 on 7, and so on), you'll get the first folding. In both states, the thick pat has leaf ID –2 on the top face and leaf ID –1 on the bottom face. You can flex between those two states using a long series of tuck flexes. From face 1, start at the hinge marked with * (between leaf IDs –15 and –2) and perform the following flex sequence:

$$T\ T\ ^{\wedge}{<}T\ {>}T\ ^{\wedge}{>}T\ ^{\wedge}{>}T\ ^{\wedge}$$

This is a tricky flex sequence, so you'll need to pay close attention to the current reference hinge and where you are in the flex sequence. If you succeed in doing all the flexes, you should end up with a state that has the exact same visible leaves, with faces 1 and 2 on the outside, but the internal pat structure of the thickest pat has been rearranged. The resulting pat structure should look like this:

$$[\ [[-2,[7,[[[-4,5],3],-6]]],1],\ 8,\ [-10,9],\ [12,-11],\ [-14,13],\ -15$$

To get back to the original state, perform the following flex sequence by starting at the final reference hinge from the previous tuck flex sequence:

$$T\ {>}^{\wedge}T\ {>}^{\wedge}T\ {>}T\ {<}^{\wedge}T\ T\ ^{\wedge}$$

If, instead of doing the series of flexes, you wanted to fold the template directly to the second pat structure – the thick pat in particular: $[[-2,[7,[[[-4,5],3],-6]]],1]$ – you would start from the innermost structure, $[-4,5]$, and expand out. Thus, you fold leaf ID 4 on top of leaf ID 5 so you have $[-4,5]$. Then fold 3 under them so you have $[[-4,5],3]$. Continue on by folding 6 under them, 7 on top, 2 on top, and finally 1 under them all.

Relatively prime flexes

All the flexes in a set of flexes are **relatively prime** if none of them can be exactly expressed in terms of the others. As we saw earlier, Lh, Ltb, and T' are not relatively prime because Lh = Ltb T'. In contrast P, V, T, T', and S are all relatively prime. One implication of this is that each of those flexes can reach states that the others cannot, depending on the pat structure of the flexagon.

This concept is useful when exploring all the states of a flexagon. If Lh can be replaced by Ltb and T', then you know you don't need to try Lh in order to find every state.

Out of the flexes we've discussed so far, you can see the sets of relatively prime flexes on isoflexagons with a given number of pats in table 19.1.

TABLE 19.1   Relatively prime flexes on isoflexagons with a given number of pats.

# of pats	flexes	# of pats	flexes
4	P	8	F, Ltf, P, S, T, Tw, V
5	L3, Ltf, S	9	F, Ltf, S, T
6	F, Ltf, Ltb, Ltb', P, S, T, V	10	F, Ltf, P, S, T, Tw, V
7	F, Ltf, S, T	12	F, Ltf, P, S, T, Tw, V

Note that T is a stand-in for various tuck flex alternatives. Typically, it's used to refer to both the tuck and inverse tuck, since each of those flexes require a different pat structure. On the heptaflexagon and higher, it includes the variants such as T1 and T2 with their inverses. To explore more states, you could replace {T, T'} with the forced tuck, Tf, though you probably don't want to use this flex in practice. The forced tuck theoretically applies to the tetraflexagon and pentaflexagon as well.

You also may opt to exclude flexes that don't work well on the particular flexagon you're exploring. For example, the flip flex works well at three of the hinges on a bronze hexaflexagon, but requires severe bending on the other three hinges. And it requires severe bending on every hinge of the regular hexaflexagon. You may wish to prohibit the flip flex in those cases.

Pat notation provides a way to automate our explorations, since it describes exactly how a flex changes the pat structure without having to flex an actual flexagon. The one additional detail to take into account is that the same pat structure can be represented in multiple ways since a flexagon loops around while pat notation is linear. For example, the four pats ABCD could be shifted – BCDA, CDAB, or DABC – or turned over – D^C^B^A^ – or both – C^B^A^D^, B^A^D^C^, or A^D^C^B^. Thus, when checking if the state you found after a flex is unique, you need to check if it's a shifted or turned-over version of a state you've already found.

If we look at some of the hexaflexagons from chapters 1 and 3, try each of the relatively prime flexes (including Tf) at every hinge, and count all the distinct states, we come up with the results in table 19.2.

TABLE 19.2 The number of distinct states for several hexaflexagons.

	states		*states*
tri-hexaflexagon 3	3	hexa-hexaflexagon 6a	3420
tetra-hexaflexagon 4	17	hexa-hexaflexagon 6b	2358
penta-hexaflexagon 5	181	hexa-hexaflexagon 6c	627*

* Only 513 states if Tf isn't included.

You can see how many states support the various relatively prime flexes in some of the simple hexaflexagons in table 19.3.

TABLE 19.3 The number of states that support various flexes on several hexaflexagons.

	3	4	5	6a	6b	6c
F	0	0	37	528	1204	242
Ltf	0	0	50	2910	800	148
Ltb	0	0	38	2910	1376	91
Ltb'	0	0	26	1788	1014	60
P	3	5	25	3420	438	39
S	0	0	113	2910	1652	352
T	0	12	86	1788	1950	231
T'	0	16	142	3024	2346	399
V	0	13	50	1788	2310	117

We can see some interesting results in that table. While every state of the 6a hexa-hexaflexagon supports the pinch flex, only 39 of the 627 states of the 6c variant do. In fact, the pinch flex is the least commonly supported flex on variants 5, 6b, and 6c. The most commonly supported flex on variants 4, 5, 6b, and 6c is the inverse tuck.

Note that these hexaflexagons are specifically designed for the pinch flex. What do you find if you instead examine flexagons that are designed for other flexes such as the v-flex, tuck flex, flip flex, or pyramid shuffle?

Summary

Pat notation gives us a powerful tool that allows us to predict exactly how flexes change the internal structure of a flexagon. But there are still a few missing details, which will be filled in by atomic flex theory in the next chapter.

Chapter 20

Atomic Flex Theory

Atomic flex theory provides a fine-grained description of how flexagons operate. It expands on flex notation and pat notation with the following applications:

- Extend pat notation to flexagons where the pats don't all meet in the middle (**non-isoflexagons**).
- Provide a technique that can take a detailed description of a flexagon's structure and figure out the template that generates it, or the opposite, going from an unfolded template to the folded flexagon.
- Describe the individual steps to carry out in order to perform a flex.
- Show how flexes add and remove pats and cause a flexagon to shape-shift, which tells us which overall flexagon shapes we can flex between.

In the previous two chapters, we developed two languages we can use to describe the dynamic behavior of flexagons: *flex notation* and *pat notation*. Flex notation gives us a way to describe a series of flexes so that we can repeat the sequence and reason about it as a mathematical equation. Pat notation allows us to describe the internal structure of a flexagon so that we can show exactly how a flex or series of flexes changes the arrangement of leaves. In this chapter, we will expand both languages to the finest details of manipulating a flexagon so we can describe a wider variety of flexes, flexagons, and their dynamic behavior.

The first two parts of the book focused on the variety of flexes that start and end in the same flat position. But we also discussed morphing flexes that change the flexagon's shape and can be combined into more complex flexes, such as pocket flexes and morph-kite flexes. If you examine those flexes, you'll see that they are also made up of multiple, smaller flexes. What is the smallest unit we can break all flexes into?

Adding directions

In chapter 19, we made a simplifying assumption when we assembled adjacent pats into a simple list: we assumed we were describing an isoflexagon, where all the pats met in the middle. Figure 20.1 shows some examples of isoflexagons.

DOI: 10.1201/9781003433538-23

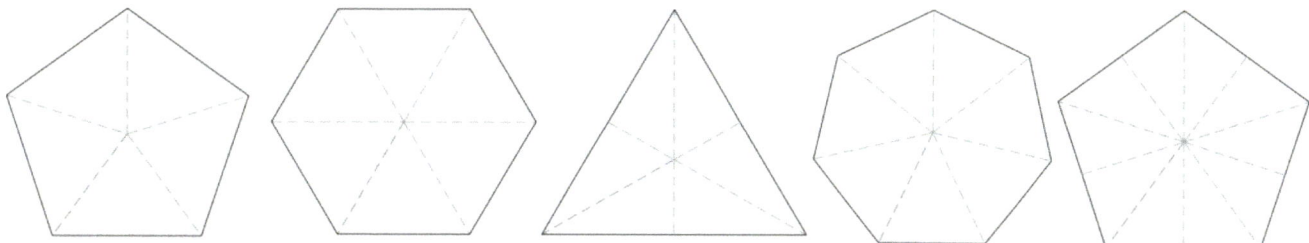

FIG. 20.1 Examples of isoflexagons.

One of the properties of an isoflexagon is that every hinge behaves the same as every other hinge. (Well, that's mostly true. For example, in a right triangle flexagon, such as the pentagonal decaflexagon in figure 20.1, the hinge adjacent to a right angle may make some flexes easier or harder than the same flex on a hinge not adjacent to a right angle. But this just means that you may need to bend the flexagon, not that the flex becomes completely impossible.)

However, we've also looked at flexagons where the pats *don't* all meet in the middle, such as the kite bronze octaflexagon and hexagonal silver dodecaflexagon. But there are many more non-isoflexagons, as exhibited by the examples in Figure 20.2.

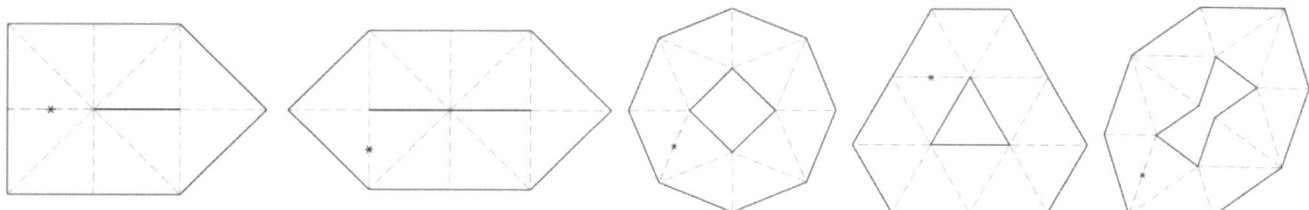

FIG. 20.2 Examples of non-isoflexagons: from left to right, the pentagonal silver decaflexagon, hexagonal silver dodecaflexagon, octagonal ring dodecaflexagon, hexagonal ring dodecaflexagon, and octagonal ring tetradecaflexagon.

With an isoflexagon, whether or not you can do a P, S, T, or V at a particular hinge depends solely on whether the internal pat structure supports it. But in general, whether or not you can flex at a particular hinge also depends on the ***directions*** between the pats.

Before we dig into how flexes depend on hinge directions, let's first define some notation we can use to describe those directions. Say that you are looking at the green pat in the leftmost image of figure 20.3. The previous pat is shown in gray, and you want to attach another pat to the green one. You can add it along either the upper or lower edge of the green pat, as shown in blue. We use \ to indicate that the next pat is connected along the upper edge, since the symbol looks like the upper edge. Similarly, we use / to indicate that the next pat is connected along the lower edge.

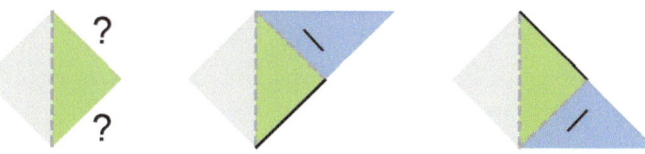

FIG. 20.3 The notation \ indicates the next pat is connected to the upper edge and / indicates the lower edge.

By convention, we travel clockwise around the flexagon, marking down the directions between pats. Thus, the directions of the six pats in a hexaflexagon are described as ////// (six forward slashes). In general, an isoflexagon with *n* pats is described using *n* /'s. In chapter 19, we assumed all the pats go in the same direction to keep the notation simple.

To see how we can use directions to describe a flexagon shape, take a closer look at the pentagonal silver decaflexagon, the leftmost image in figure 20.2. Start with the triangle immediately clockwise from the star and look at how the next triangle clockwise is connected to it. You can see that it's connected to the lower edge, which means its direction is /. If you keep stepping from triangle to triangle clockwise noting each direction, you'll find that the ten directions between triangles can be described as ///\//\///. The following list shows the directions for each of the flexagons in figure 20.2, starting from the star and going clockwise:

pentagonal silver decaflexagon	///\//\///
hexagonal silver dodecaflexagon	//\//\//\//\
octagonal ring dodecaflexagon	//\//\//\//\
hexagonal ring dodecaflexagon	///\///\///\
octagonal ring tetradecaflexagon	//\\//\//\\//

Morph-kite flexes

Let's take a closer look at the kite bronze octaflexagon that we used in chapter 14. In figure 20.4, each pat is annotated with / or \ depending on to which direction the next pat clockwise is connected. We find that we mostly have /'s, as with an isoflexagon, except for two pats that go the opposite direction. The end result looks like this: //\//\// (or shifted depending on which pat you start with).

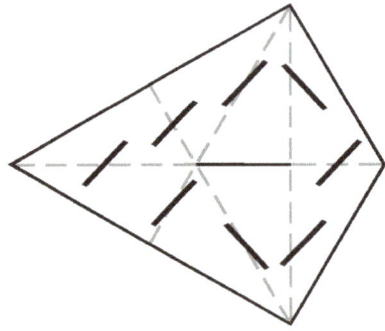

FIG. 20.4 The directions between pats in a kite bronze octaflexagon going clockwise around the flexagon.

Let's focus on the four pats on the right side that form a diamond, where the directions are \//\ going clockwise from the first \. These are the directions that the inverses of the morph-kite flexes operate on. A morph-kite flex transforms the directions between those four pats from \//\ to ////. This means that the entire flexagon goes from //\//\// to ////////. You can see this by doing Mkf' on the flexagon from figure 14.20, for example, which leads to all eight pats meeting in the middle of a non-flat star.

And if you look at the silver octaflexagon, where the directions of any four pats are ////, you'll see that the morph-kite flex transforms the directions from //// to \//\. This notation gives us a way to talk about how these flexes transform the overall shape of those flexagons. The entire flexagon transforms from //////// to //\//\//.

The important thing to note here is that the morph-kite flexes require not only specific pat structure but also specific directions between the pats. This means that the pat notation discussed in the last chapter isn't sufficient for fully describing the requirements and transformations of the morph-kite flexes. We need to include the pat directions as well.

Recall from chapter 19 that we define the inputs (in) and outputs (out) for a flex by adding leaf IDs to every leaf in its minimal flexagon. The minimal flexagon for Mkf was shown in figure 14.3. Start by numbering the leaf labeled 3/2 with ID 1, leaf 3/1 with ID 2, and so on across the template. This allows us to write down the full definition of the morph-kite front flex, Mkf, including the directions between the pats using / or \\:

Mkf	in	[−2,1] / −3 / −4 / −5 / −6 / −7 / [9,−8] / 10 /
	out	[1,−10] / [2,−3] \\ −4 / −5 / −6 / −7 / −8 \\ −9 /

This definition is explicit about both the pat structure and pat directions, and therefore tells us exactly when Mkf is allowed or not allowed.

We can generalize our Mkf definition to other numbers of pats much as we did in chapter 19:

Mkf	in	[−2,1] / −3 / ... *i* ... / [*n* − 1,−(*n* − 2)] / *n* /
	out	[1,−*n*] / [2,−3] \\ ... *i* ... / −(*n* − 2) \\ −(*n* − 1) /

Atomic pat notation

However, there's another way we could define Mkf that takes advantage of the fact that it's a local flex, only affecting a portion of a flexagon. In order to do that, we'll add some new concepts, which we call **atomic pat notation**:

- Use the symbol # to indicate where the reference hinge is.
- Use *a* to indicate the portion of the flexagon to the left that doesn't change and *b* to indicate the portion of the flexagon to the right that doesn't change.
- Use −*a* to indicate that *a* has been turned over without any other changes (such as when you use the ^ to turn over the flexagon).

Thus, instead of the reference hinge implicitly being between the first and last pat listed (as in chapter 19), you can place it between any two pats. And instead of explicitly listing every pat, even if it's not modified, you can simply list the pats that change. We can now write down simpler definitions of the morph-kite flexes like this:

Mkf	in	*a* [−2,1] / −3 / # [5,−4] / 6 / *b*
	out	*a* 1 \\ 2 / # [−4,3] / [−5,6] \\ *b*
Mkb	in	*a* 1 / [−3,2] / # −4 / [6,−5] / *b*
	out	*a* [1,−2] \\ [4,−3] / # 5 / 6 \\ *b*
Mkr	in	*a* [−2,1] / −3 / # −4 / [6,−5] / *b*
	out	*a* 1 \\ 2 / # [[−4,5],3] / 6 \\ *b*
Mkl	in	*a* [−2,1] / −3 / # −4 / [6,−5] / *b*
	out	*a* 1 \\ [4,[2,−3]] / # 5 / 6 \\ *b*

This has an additional benefit of being able to describe flexes that add half-twists to the surface of the flexagon. Consider the pocket flex, for example. If you trace your finger across the surface of the flexagon before a pocket flex, it travels across the front face but not the back. Now open up a single pocket and trace your finger along the surface of the flexagon. You will find that your finger travels across both faces of every pat. That means there's only a single surface, which is called a **Möbius strip**.

Using atomic pat notation, here is the definition of the pocket flex:

K   in   *a* [−2,1] / −3 / # [5,−4] / *b*
         out   *a* 1 \ 2 / # [−4,3] / −5 / −*b*

Note that after a pocket flex, the resulting flexagon contains *a* on one side and −*b* on the other. This indicates that the right side of the flexagon has been turned over, but the left side has not. In other words, this is telling us that the flex has added a half-twist to the surface of the flexagon.

You may have also noticed that the output of the flex has four pats where the input only had three pats. We created an extra pat! Adding a half-twist and adding an extra pat are intimately connected when flexing a flexagon, as we will soon see.

Atomic flexes

We now have all we need in order to define the lowest-level operations on any flexagon and assemble those operations into any of the flexes we have done on triangle flexagons.

We define the following four **atomic flexes**, listing the input state (after the equals sign) and output state (after the arrow), which we will soon explain in more detail:

> = *a* # 1 *b* → *a* 1 # *b*
^ = *a* # *b* → −*b* # −*a*
~ = *a* # *b* → −*a* # −*b*
Ur = *a* # [−2,1] / *b* → *a* # 1 \ 2 / −*b*

These four flexes together with their inverses form the **atomic flex axioms**. They can be composed together into any flex on a triangle flexagon. They describe how to unfold a flexagon into a template and how to fold a template into a flexagon. They also describe how the flexagon shape-shifts and adds/removes half-twists. This may seem surprising, given how basic these axioms are and how complex some of the flexes we've tried are.

The first flex, >, is the familiar clockwise shift flex. The second, ^, is the familiar turn-over flex. More explicitly, you turn the flexagon over left to right across an axis that goes through the current hinge. The third, ~, is called the **change direction flex** because it changes the directions of the pats, swapping / and \. It's similar to ^ except that you turn the flexagon over top to bottom across an axis perpendicular to the axis going through the current hinge. These three flexes (>, ^, and ~) simply rotate and shift the flexagon without otherwise changing it, as shown in figure 20.5.

 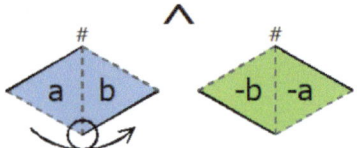 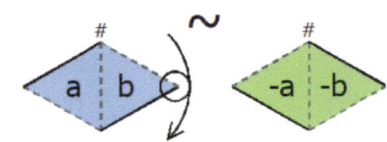

FIG. 20.5 The rotate and shift flexes.

The final flex, Ur, is the **unfold right flex**, illustrated in figure 20.6. It unfolds a single pat into two pats and reverses the direction of the pats to the right by adding a half-twist to the surface of the flexagon. Its inverse, Ur', folds two pats together into a single pat while removing a half-twist. In atomic pat notation, Ur starts with a # [–2,1] / b and ends with a # 1 \ 2 / –b. Following the steps in the reverse order, folding the pats together, is Ur'.

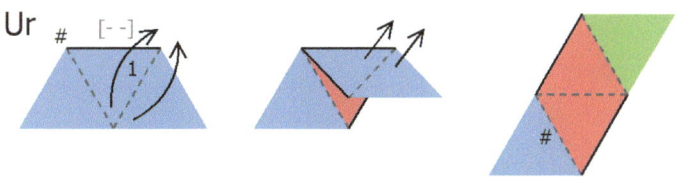

FIG. 20.6 Ur, the unfold right flex.

Note that we could have defined Ur in the opposite direction. Rather than [–2,1], we could have instead chosen [1,–2], giving us Ur2 = a # [1,–2] / b → a # 1 \ 2 / –b. This gives us flexagons that wind in the opposite direction. We call this an **enantiomorph**, a flexagon with opposite handedness. For consistency, the templates in this book all have the same handedness, but we could have chosen to wind them the opposite direction. With flexes such as the pinch, you wouldn't notice the difference. But other flexes, such as the pyramid shuffle, would need to be performed mirrored in order to work.

Half-twists

It's important to differentiate between *twists in the folded template* versus *twists in the surface of the flexagon.*

As we fold a *template*, the atomic flexes tell us that we add a half-twist with every fold. But once we've attached the first and last edges together, the number of times the template has been twisted is fixed and can't be changed without disconnecting a hinge.

However, we can independently consider the number of half-twists in the flexagon *surface* by treating each folded pat as a single leaf. We add three half-twists to the template as we fold a tri-hexaflexagon, but the resulting surface of the flexagon is a flat hexagon, with no twists. Physically, of course, the pats have thickness and there are thumbholes you can open in the surface. But ideally (mathematically speaking), the thickness of the surface is infinitesimally small. We do not count these as three-dimensional half-twists in the surface. Rather, they are discontinuities in the surface at the boundaries between pats. However, once we open a thumbhole, performing the Ur flex, we've added a half-twist (and dimensionality) to the surface, forming a Möbius strip. Every fold and unfold of the pats changes the number of half-twists in the flexagon surface, while keeping the number of half-twists in the template unchanged.

Thus, we can see that the atomic flexes tell us both how many half-twists we add to the *template* as we fold it and how many half-twists there are in the resulting flexagon *surface* as we flex it. Each flex adds and removes half-twists in the surface as we open up one pat and fold other pats back together again.

Angles and atomic flexes

Now let's see how the angles of the leaves change as we apply the atomic flexes. In particular, let's look at the leaf immediately to the right of the reference hinge. We use α (alpha) for the angle at the bottom, which would be in the center of an isoflexagon. The first angle counterclockwise is β (beta), and the final angle is γ (gamma).

Figure 20.7 illustrates what happens to the leaf angles as a result of the atomic flexes. You can see that ~ swaps α and γ, while ^ and Ur leave the angles unchanged. Interestingly, the shift flex, >, behaves differently depending on which hinge the next pat is attached to. If it's the lower one (/), then β and γ are swapped, but if it's the upper one (\), α and β are swapped. If we need to differentiate between the two, we use >/ to refer to the former and >\ to refer to the latter.

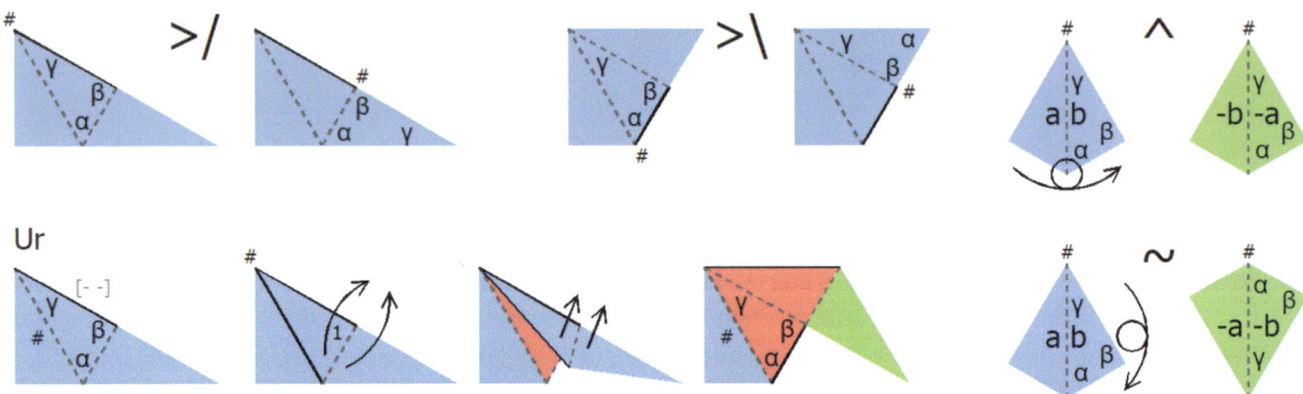

FIG. 20.7 Flex diagrams for the atomic flex axioms showing how the angles change.

This lets us refine our definitions of the atomic flexes to include how the angles αβγ of the leaf immediately to the right of the current hinge change, differentiating between the two different shifts. For simplicity, flex sequences typically use > for a clockwise shift, with the understanding that it resolves to >/ or >\ based on how it's connected to the next pat when you need to track the angles.

$$
\begin{array}{llll}
>/ & = & a \,\#\, 1\,/\, b & \rightarrow a\, 1\,/\,\#\, b & \alpha\gamma\beta \\
>\backslash & = & a \,\#\, 1\,\backslash\, b & \rightarrow a\, 1\,\backslash\,\#\, b & \beta\alpha\gamma \\
\wedge & = & a \,\#\, b & \rightarrow -b\,\#\,-a & \alpha\beta\gamma \\
\sim & = & a \,\#\, b & \rightarrow -a\,\#\,-b & \gamma\beta\alpha \\
\mathrm{Ur} & = & a \,\#\, [-2,1]\,/\, b & \rightarrow a\,\#\, 1\,\backslash\, 2\,/\,-b & \alpha\beta\gamma
\end{array}
$$

For reference, here are the definitions for a counterclockwise shift:

$$
\begin{array}{llll}
</ & = & a\, 1\,/\,\#\, b & \rightarrow a\,\#\, 1\,/\, b & \alpha\gamma\beta \\
<\backslash & = & a\, 1\,\backslash\,\#\, b & \rightarrow a\,\#\, 1\,\backslash\, b & \beta\alpha\gamma
\end{array}
$$

Building a flex vocabulary

We can compose our four atomic flexes into convenient building blocks for building up a description of more complex flexes. First, we derive <. Since it's the inverse of >, we simply swap the inputs and outputs to derive its definition.

$$
\begin{array}{ll}
< & = & >' \\
& = & a\, 1\,\#\, b \rightarrow a\,\#\, 1\, b
\end{array}
$$

Next, we define Ul, the **unfold left flex**, as being the same as Ur performed from the opposite face. We can derive its definition from the atomic definitions starting from the state a # [1,–2] \ b. See figure 20.8 for what this flex looks like.

Ul =   ~Ur~

input a # [1,–2] \ b
apply ~ $–a$ # [2,–1] / $–b$
apply Ur $–a$ # –1 \ –2 / b
apply ~ a # 1 / 2 \ $–b$

Ul = a # [1,–2] \ b → a # 1 / 2 \ $–b$

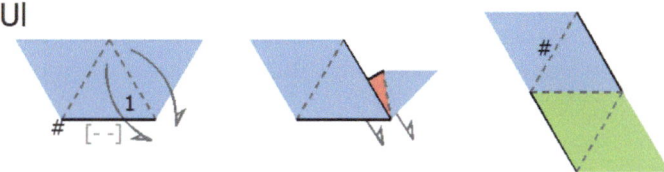

FIG. 20.8 Ul, the unfold left flex.

For Ul, we now have one definition using flex notation, ~Ur~, which gives us a recipe for how to perform this flex, and a second definition using atomic pat notation, which describes how the flexagon structure changes as a result. We derived the second definition using the first definition. We can continue to build each of the more complex flexes from these simpler flexes.

Next, we define the **exchange right flex**, Xr, which exchanges leaves between two adjacent pats. See figure 20.9 for the flex diagram for Xr. When you follow these directions, notice how it's just like one portion of a pinch flex:

1. Bring the backs of the middle two pats together.
2. Open from the front to reveal face 3.
3. Turn over top to bottom.

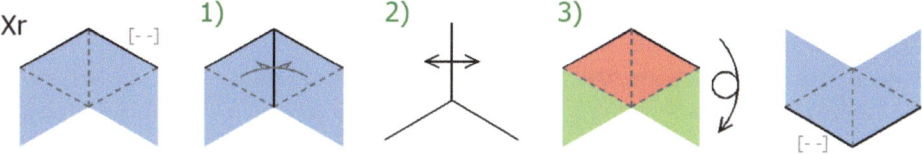

FIG. 20.9 Flex diagram for Xr, the exchange right flex.

Here's the definition of the exchange right flex using atomic flexes and atomic pat notation:

Xr = Ur < Ul' >
     = a 1 / # [–3,2] / b → a [1,–2] \ # –3 \ b

An alternate way to do Xr is to follow its flex decomposition step by step, as shown in figure 20.10.

1. Fold open the top layer from corner A.
2. Fold corner B behind the adjacent pat.
3. Rotate everything clockwise by 120°.

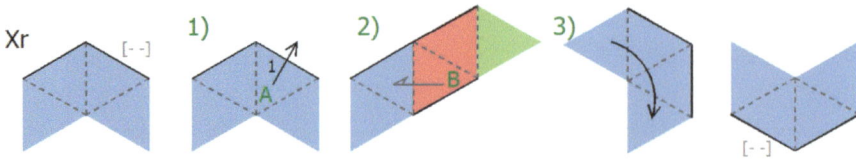

FIG. 20.10 Flex diagram for Xr by following its flex decomposition.

Note that Xr finishes on the original face, which requires turning over the flexagon in step 3 of figure 20.9. Defining it this way makes it easier to combine multiple exchange flexes together to build up a complete pinch flex since we can do all the exchange flexes from the original face before turning the entire flexagon over. To see how this works, let's look at the pinch flex on a hexaflexagon. Taking the pinch flex definition for the hexaflexagon from chapter 19 and adding in the directions between the pats, we get the following definition:

P	in	[−2,1] / −3 / # [5,−4] / 6 / [−8,7] / −9 /
	out	2 / [−4,3] / # −5 / [7,−6] / 8 / [1,9] /

Here's how this pinch flex is expressed using the exchange right flex:

$$P = (Xr \gg)3 \sim$$

We can use this definition to derive the atomic pat definition and verify that it matches the given inputs and outputs:

input	[−2,1] / −3 / # [5,−4] / 6 / [−8,7] / −9 /
apply Xr	[−2,1] / [−3,4] \ # 5 \ 6 / [−8,7] / −9 /
apply >>	[−2,1] / [−3,4] \ 5 \ 6 / # [−8,7] / −9 /
apply Xr	[−2,1] / [−3,4] \ 5 \ [6,−7] \ # −8 \ −9 /
apply >>	[−2,1] / [−3,4] \ 5 \ [6,−7] \ −8 \ −9 / #
apply Xr	−2 \ [−3,4] \ 5 \ [6,−7] \ −8 \ [−9,−1] \ #
apply >>	−2 \ [−3,4] \ # 5 \ [6,−7] \ −8 \ [−9,−1] \
apply ~	2 / [−4,3] / # −5 / [7,−6] / 8 / [1,9] /

While a pinch flex does an exchange flex on every other hinge, the different pinch variants instead skip multiple hinges. For example, here are the decompositions for some of the pinch variants on a dodecaflexagon:

P222222	=	(Xr >>)6 ~
P3333	=	(Xr >>>)4 ~
P444	=	(Xr >>>>)3 ~
P66	=	(Xr >>>>>>)2 ~

Next, we define the ***exchange left flex***:

Xl	=	Ul < Ur' >
	=	a 1 \ # [2,−3] \ b → a [−2,1] / # −3 / b

We can then define the v-flex on a hexaflexagon using the two exchange flexes:

$$V = <\; Xr>>Xr>>Xl' >>\sim\; >$$

Here's a definition of the pocket flex in terms of simpler flexes:

$$K = Xr\sim^\wedge > Ul^\wedge$$

We can use that definition to derive our earlier definition for the pocket flex:

input	$a\ [-2,1] / -3 / \# [5,-4] / b$
apply Xr	$a\ [-2,1] / [-3,4] \backslash \# 5 \backslash b$
apply ~	$-a\ [-1,2] \backslash [-4,3] / \# -5 / -b$
apply ^	$b\ 5 / \# [-3,4] / [-2,1] \backslash a$
apply >	$b\ 5 / [-3,4] / \# [-2,1] \backslash a$
apply Ul	$b\ 5 / [-3,4] / \# -2 / -1 \backslash -a$
apply ^	$a\ 1 \backslash 2 / \# [-4,3] / -5 / -b$

$$K = a\ [-2,1] / -3 / \# [5,-4] / b \rightarrow a\ 1 \backslash 2 / \# [-4,3] / -5 / -b$$

Then we can build up the morph-kite flexes from our simpler flexes:

Mkf	=	K > Ul' <
Mkb	=	^Mkf^
Mkr	=	<< Ur >>>> Xl << Ur'
Mkl	=	^Mkr^
Mkfs	=	> K < Ur << Ul' >> Ur'
Mkbs	=	^Mkfs^

Now we compose various flexes from all these pieces:

F	=	Mkf Mkb'	
St	=	Mkf Mkfs'	
S	=	< Mkfs Mkb' >	{when < Mkfs Mkb' > is supported}
T	=	< Mkr Mkf' >	{when < Mkr Mkf' > is supported}
Fm	=	< Mkr Mkb' >	
S3	=	< Mkr Mkl' >	

Note that some of the above definitions have stricter requirements than the actual flex. For example, the pyramid shuffle only requires three pats, while the morph-kite flexes require four.

If we look carefully at the various slot flexes from chapter 15, we can reinterpret them as consisting of three pieces: the morph-kite front flex (Mkf), a flex that slides pats through a slot, and a final flex that returns the flexagon to a flat state. We call the middle flex the kite-to-kite slot flex, abbreviated Lkk, since it starts and ends in a kite position. This gives us the following definitions for Ltf and Lk, the slot flexes we tried on the pentaflexagon, hexaflexagon, and octaflexagon:

Lkk	=	> Ul Ur <<<< Ul' < Ul' >>
Ltf	=	Mkf Lkk Mkf' <
Lk	=	Mkf Lkk Mkbs' <

If we create new definitions for flexes called the morph-kite-half flex (Mkh) and morph-kite-tuck flex (Mkt), we can provide breakdowns for the hexaflexagon-specific slot flexes. This list includes a bonus flex, the tuck-top-front flex (Ttf), which you may be able to reconstruct based on its atomic decomposition.

```
Mkh  =  Xr~ >>> Xl <<<
Mkt  =  < Ur ^<<< Ur' <<^ Xl <<<
Lh   =  Mkf Lkk Mkh'
Ltb  =  Mkf Lkk Mkt'
Lbb  =  Mkf Lkk >>> Mkt' <<
Lbf  =  Mkf Lkk >>> Mkf' <<
Ttf  =  Mkh Mkf'
```

Here are some more specific atomic definitions of a few of the flexes:

$$
\begin{array}{lll}
\text{Tf} & = & a\ 1\ /\ \#\ [[-3,4],2]\ /\ b \\
 & \rightarrow & a\ [3,[1,-2]]\ /\ \#\ 4\ /\ b \\
\text{S} & = & a\ [[[3,-2],-4],1]\ /\ -5\ /\ \#\ [7,-6]\ /\ b \\
 & \rightarrow & a\ [-2,1]\ /\ -3\ /\ \#\ [7,[-4,[-6,5]]]\ /\ b \\
\text{F} & = & a\ [[3,-4],[1,-2]]\ /\ -5\ /\ \#\ [7,-6]\ /\ 8\ /\ b \\
 & \rightarrow & a\ 1\ /\ [-3,2]\ /\ \#\ -4\ /\ [[-7,8],[-5,6]]\ /\ b \\
\text{St} & = & a\ [3,[1,-2]]\ /\ 4\ /\ \#\ [7,[5,-6]]\ /\ 8\ /\ b \\
 & \rightarrow & a\ 1\ /\ [[-3,4],2]\ /\ \#\ 5\ /\ [[-7,8],6]\ /\ b \\
\text{S3} & = & a\ [[[3,-2],-4],1]\ /\ -5\ /\ -6\ /\ \#\ [8,-7]\ /\ b \\
 & \rightarrow & a\ [-2,1]\ /\ -3\ /\ -4\ /\ \#\ [8,[-5,[-7,6]]]\ /\ b \\
\text{Fm} & = & a\ [[3,-4],[1,-2]]\ /\ -5\ /\ -6\ /\ \#\ [8,-7]\ /\ b \\
 & \rightarrow & a\ 1\ /\ [-3,2]\ /\ -4\ /\ \#\ [8,[-5,[-7,6]]]\ /\ b
\end{array}
$$

Angles and flexes

If we want to be able to figure out what a flexagon looks like after a flex, we also need to look at how the leaf angles change as they're rotated and flipped. If we can figure out the angles of the first leaf after the reference hinge, we can infer the rest because we know each leaf is mirrored across the hinge described by the directions. In chapter 19, we figured this out by trying it on an actual flexagon, but now let's look at how we can use the atomic flex axioms to derive this information.

Let's try this with the pinch flex. We previously saw that the pinch flex can be decomposed into $(Xr>>)n$ ~, where $Xr = Ur<Ul'>$. Take a look at a single exchange flex applied to a pair of adjacent pats, paying close attention to the directions of the hinges when shifting the current hinge.

	1 / # [−3,2] /	αβγ
Ur	1 / # 2 \ 3 /	αβγ
</	# 1 / 2 \ 3 /	αγβ
Ul'	# [1,−2] \ −3 \	αγβ
>\	[1,−2] \ # −3 \	γαβ
~	[2,−1] / # 3 /	βαγ

The end result is that the pinch flex turned αβγ into βαγ, effectively swapping the first two angles. If we label the angles of a hexaflexagon and do a pinch flex, we can validate that this is exactly what happens.

Unfolding a flexagon

We now have the tools for creating an unfolded template from a generating sequence. For reference, we'll be using both of the unfolding flexes, Ur and Ul:

$$Ur = a \# [-2,1] / b \quad \rightarrow a \# 1 \setminus 2 / -b$$
$$Ul = a \# [1,-2] \setminus b \quad \rightarrow a \# 1 / 2 \setminus -b$$

Let's try deriving the template from the generating sequence S+ on a hexaflexagon. We start with a description of the folded flexagon, then apply flexes that unfold it into a template. From the atomic pat definition for S, here's a description of the folded S+ hexaflexagon:

$$S1 = a \# [-2,1] / -3 / -4 / -5 / [[[-8,7],9],-6] / 10 / b$$

You can see that the input pattern for Ur, [-2,1] / b, occurs right at the beginning of our flexagon. Applying Ur unfolds the first pat, giving us the following:

$$S2 = a \# 1 \setminus 2 / 3 \setminus 4 \setminus 5 \setminus [6,[-9,[-7,8]]] \setminus -10 \setminus -b$$

Now we need to unfold the pat with four leaves. If we shift the current hinge so it's right before [[[-8,7],9],-6], we see that the input matches the input for Ul, [1,-2] \ b. Applying the sequence >>>>>Ul to S2 gives us the following:

$$S3 = a \ 1 \setminus 2 / 3 \setminus 4 \setminus 5 \setminus \# 6 / [[-8,7],9] \setminus 10 / b$$

Again, we need to shift the current hinge before unfolding with Ul, giving us the following:

$$S4 = a \ 1 \setminus 2 / 3 \setminus 4 \setminus 5 \setminus 6 / \# [-8,7] / -9 \setminus -10 \setminus -b$$

We can now apply Ur to give us a description of the unfolded template:

$$S5 = a \ 1 \setminus 2 / 3 \setminus 4 \setminus 5 \setminus 6 / \# 7 \setminus 8 / 9 / 10 / b$$

We can draw the unfolded template using the directions between the pats as seen in figure 20.11.

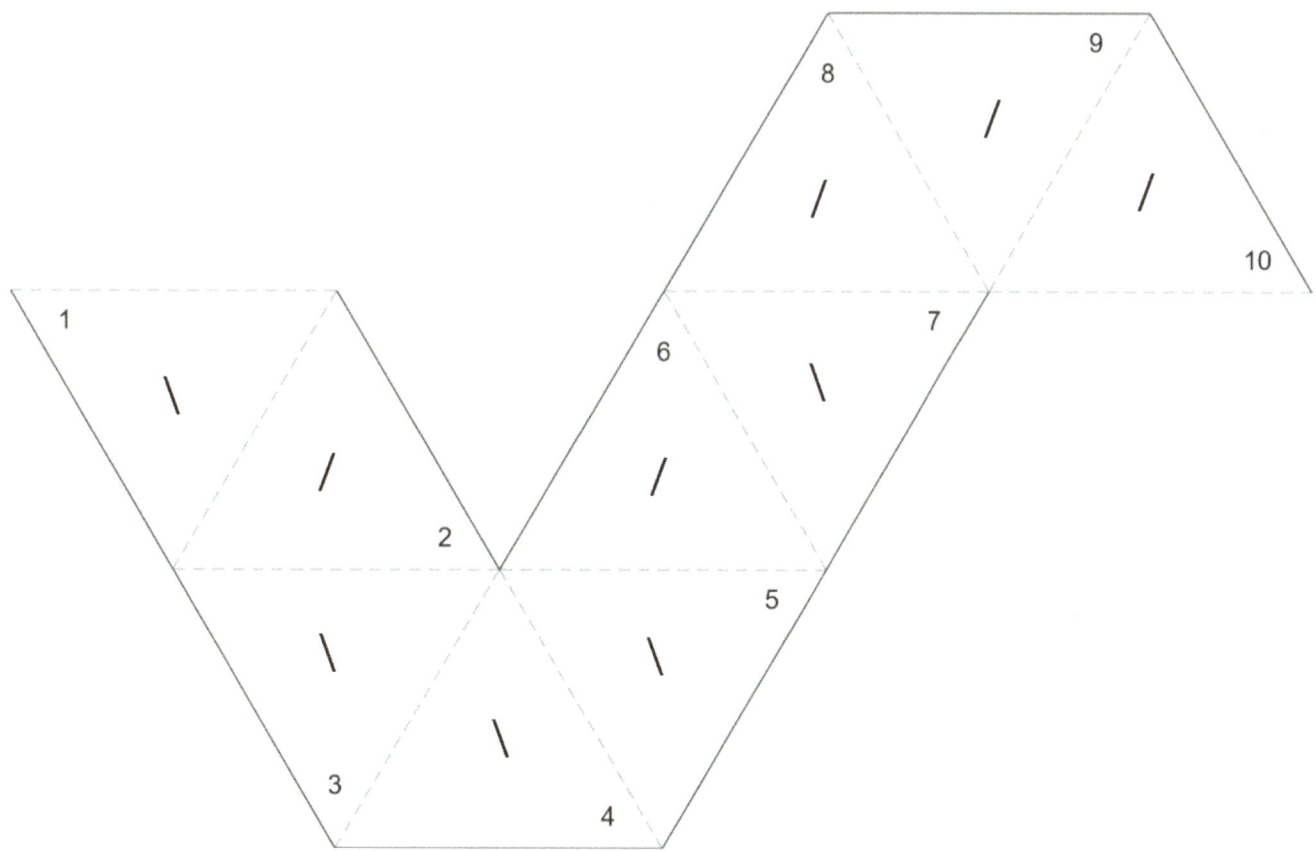

FIG. **20.11** Unfolded template for the hexaflexagon (generator S+).

You can follow each step of our atomic pat notation in reverse to fold it back up into a hexaflexagon:

1. S4 contains [–8,7], which means you should fold 8 on top of 7 so that you see –8 on top.
2. S3 contains [[–8,7],9] indicating that you should fold 9 underneath the [–8,7] pat.
3. S2 contains [6,[–9,[–7,8]]], so fold [[–8,7],9] under 6.
4. S1 contains [–2,1], so fold 2 over 1.
5. Finally, tape 1 and 10 together to make a hexagon.

To make it easier to refold, every time you unfold, mark the two leaf-faces that were folded together with the same label. This allows you to make your flexagon simply by folding the adjacent pairs of matching labels together.

This technique allows you to go from any arbitrary generating sequence to atomic pat notation to an unfolded template. Or the opposite: use Ur' and Ul' to fold a template into a flexagon.

Shape-shifting

We've looked at what the atomic flex axioms enable, such as predicting the effect of flexes and unfolding templates, but it's also useful to look at what they prohibit. The unique behavior of flexagons comes from this trade-off.

Consider a portion of a flexagon described by 1/2/, as shown in figure 20.12. There is no way to use the atomic flex axioms to fold 1 and 2 together, since the directions between pats don't match the requirements for Ur' or Ul', even after applying >, ^, or ~. Thus, if you have an isoflexagon such as a hexaflexagon, the first atomic operation in a flex can't be a fold. You instead have to start by unfolding at least one pat. Therefore, all definitions for flexes on isoflexagons, such as the pinch or flip flex, start by unfolding at the atomic level.

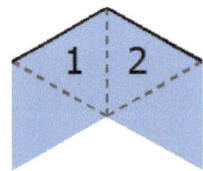

FIG. 20.12 A portion of a flexagon represented by 1/2/.

The atomic flex axioms lead us to two important conclusions:

Every pat is always connected across two different edges to pats that are mirror images. If you could fold 1 against 2 in figure 20.12, the resulting pat would be connected to adjacent pats across the same edge, but we just saw that this is prohibited. By starting with leaves that are connected to mirror-image pats across two different edges and recursively applying the atomic folding operation to fold up a pat, we know that all pats are connected across two *different* edges to mirror-image pats.

Every fold removes a half-twist and every unfold adds a half-twist. The definition of Ur (unfold right) takes *a* and *b* as input (the two ends of the portion of the flexagon being manipulated) and creates *a* and *–b* as output, where *–b* means that the back face of leaf *b* is now on top. Given that the flexagon is a closed loop, the only way for this to happen is to add a half-twist to the flexagon surface. Adding a half-twist to a flat band turns it into a Möbius strip. Ur' (fold right) adds a half-twist in the opposite direction from Ur, effectively canceling out the half-twist created by the unfold. To figure out how many twists there are in the flexagon surface, simply count the number of times pats have been folded together. Note that the surface of the flexagon typically has fewer half-twists than the template it's folded from.

Now we've seen that every pat is connected to adjacent mirror-image pats across two different edges and that we can use folding and unfolding to track how many half-twists are in the template and the surface of the flexagon. This helps us understand how a flexagon can morph into other shapes as we flex it.

A question we might ask is whether we can morph a flat hexaflexagon into a flat octaflexagon through a series of flexes. How can we answer that?

Consider that our hexaflexagon has six pats with directions //////, while the octaflexagon has eight pats with directions ////////. We want them both to be flat, so the flexagon surfaces have no half-twists. But the atomic flex axioms tell us that the only way to add a pat, Ur, also adds a half-twist, while the only way to remove a pat, Ur', reverses that half-twist.

Thus, in order to add two pats to our hexaflexagon, we also have to add two half-twists. That means that the only way to morph from a hexaflexagon to an octaflexagon is to add a full twist, so the result won't lie flat. That means we can't flex from a flat hexaflexagon to a flat octaflexagon.

This is a general result, called the **Morph and Twist Theorem**: If we morph to another shape with a different number of pats, it will have a different number of half-twists in the flexagon surface. If we morph to a shape with the same number of pats, it will have the same number of half-twists.

Summary

In this chapter, we defined atomic pat notation by extending pat notation with several new concepts:

- Use / and \ to indicate the direction between pats.
- Use # for marking the current hinge.
- Use a and b to refer to the pats at either end that don't change or get turned over.

With this notation, we defined the four atomic flexes that make up the atomic flex axioms: >, ^, ~, Ur, along with their inverses:

$$
\begin{aligned}
> \quad &= \quad a \,\#\, 1 \, b & &\rightarrow a \, 1 \,\#\, b \\
\wedge \quad &= \quad a \,\#\, b & &\rightarrow -b \,\#\, -a \\
\sim \quad &= \quad a \,\#\, b & &\rightarrow -a \,\#\, -b \\
\text{Ur} \quad &= \quad a \,\#\, [-2,1] \,/\, b & &\rightarrow a \,\#\, 1 \,\backslash\, 2 \,/\, -b
\end{aligned}
$$

We then used these flexes to build up more complex flexes, deriving their definitions in atomic pat notation. This gives us a vocabulary for breaking flexes into smaller pieces so we can accurately describe and reproduce them.

Next, we showed how to go from atomic pat notation to an unfolded template. And finally, we used that knowledge to understand how a flexagon morphs as we flex it, changing the number of half-twists as we add and remove pats.

Chapter 21

Defining Flexagon and Flex

We've now seen lots of examples of flexagons and flexes, but what exactly are flexagons and flexes? What makes them different from other ways of folding paper?

The magic of flexagons comes from how *unfolding a pat* and *folding together adjacent pats* are restricted and how they're complements of each other, especially as they occur across multiple pats, combining in novel and unexpected ways.

To make this more precise, we start with a definition of a basic flexagon and basic flex, which is what atomic flex theory describes. Next, we explore flexagon variants that twist those rules in interesting ways. Note that this chapter focuses on triangle flexagons.

In a **basic flexagon**, every leaf is connected to exactly two mirror-image copies of itself, and it's folded using the atomic flex axioms.

A **basic flex** is composed of a sequence of atomic flexes.

Almost every triangle flexagon in this book is a basic flexagon, and almost every flex is a basic flex. These simple rules give us a huge variety of interesting triangle flexagons. We describe ways of breaking these rules in the second half of this chapter.

Two mirror-image copies

Every leaf and every pat is connected to exactly two mirror images of itself in many different situations: a template (except that the first and last leaves aren't yet attached), intermediate states as you fold the template, the flexagon, and intermediate states as you apply various flexes.

Figure 21.1 shows examples of non-flexagon arrangements.

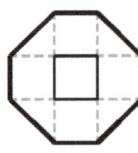 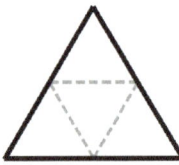

FIG. 21.1 Examples of arrangements that are not flexagons: the first is because leaves are connected to leaves that aren't mirror images, and the second is because the leaves aren't all connected to exactly two other leaves.

Atomic flex axioms

The atomic flex axioms restrict how you can fold a template or flex a flexagon. While you can fold any two pats of a flexagon together, flexagon structure is only created by folding together *adjacent*

DOI: 10.1201/9781003433538-24

pats. And it's not just any adjacent pats, since you can't fold together pats if their directions are //, as we saw in the "Shape-shifting" section in chapter 20. Furthermore, while you can fold together either the backs or fronts of adjacent pats, only one of those produces valid structure for a basic flexagon.

To illustrate this, let's look at what the axioms have to say about folding together the two pats shown in figure 21.2, where we see the pat structure $a \# 1 \backslash 2 / -b$. We can use Ur' to fold together 1 and 2 into $a \# [-2,1] / b$. This tells us to fold pat 2 under pat 1, not pat 2 over pat 1.

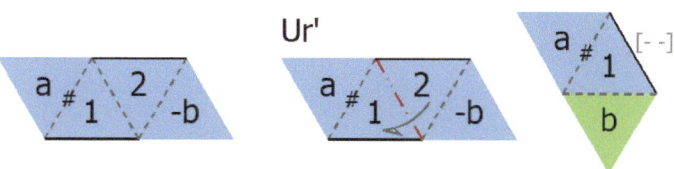

FIG. 21.2 Applying Ur' to fold together pats 1 and 2.

As noted when the axioms were introduced, we could have chosen the definition of Ur to contain either [−2,1] or [1,−2], but not both. It's not possible to combine the axioms in some way that allows for both foldings in the same flexagon. For example, if we apply ^, the directions are no longer correct for Ur' unless we also apply ~, which leads us back to Ur'.

$$
\begin{array}{ll}
 & a \# 1 \backslash 2 / -b \\
\wedge & b -2 / -1 \backslash \# -a \\
\sim & -b \, 2 \backslash 1 / \# a \\
<< & -b \# 2 \backslash 1 / a \\
\text{Ur'} & -b \# [-1,2] / -a \\
\wedge< & a \# [-2,1] / b
\end{array}
$$

Another interesting consequence of the atomic flex axioms is that a basic flexagon can't have zig-zagged stacks within a pat. This means that if you have three adjacent leaves 1/2\3/, you can create valid pat structure by folding them into [1,[3,−2]] or [[−2,1],3] but not [1,−2,3], which zigzags within the stack of leaves. This is why pat notation always has exactly one or two items in each nested pair, never more. Yes, you may temporarily create a zigzag while doing a flex, but that's only incidental, not an integral part of the pat structure.

Physical constraints

When determining if a flex works at a given hinge in a specific flexagon, there are three aspects of the flexagon you need to consider: the internal pat structure, the directions between pats, and the three-dimensional geometry of how the pats move. Atomic flex theory describes both pat structure and directions, but not the three-dimensional geometry.

As we noted when discussing the tuck flex and flip flex, sometimes a flex can be done with perfectly rigid leaves, sometimes you need to bend the leaves a bit, and sometimes you need to bend them so much that a paper model will rip. We feel that flexing and flexagon theory are more interesting if you allow some bending. But since the line between reasonable bending and too much bending is fuzzy, we don't attempt to model the three-dimensional aspects of flexing in our definition.

Breaking the rules

Throughout this book, we've used basic rules to create templates from generating sequences and pat structure, as well as to define flexes and predict how flexagons change as flexes are applied. But there are more general flexagons that don't strictly follow these definitions, or at least they seem like they don't. We'll look at the following examples of how to break the rules while still following the basic mechanics of flexagons:

1. Modify a portion of the flexagon not involved in a flex.
2. Add/remove parts of leaves as long as the flex is unaffected.
3. Fold together nonadjacent pats.
4. Attach a leaf to more than two other leaves.
5. Make nested flexagons, called matryoshkas, which flex.
6. Connect leaves across points rather than edges.
7. Create flexagon surfaces with twists and knots.
8. Use an infinite number of leaves.

Rule-break 1

First off, any leaves and hinges not involved in a flex don't have to follow the rules. If you're using a local flex, such as a flip flex, the rest of the flexagon could be anything as long as it permits enough freedom of movement to perform the flex.

Rule-break 2

Secondly, you can modify the leaves as long as it doesn't interfere with the flexing mechanism too much. For example, you can trim off parts of the leaves as long as the hinges still work, or add to the leaves as long as the extra bits don't get in the way of the flexes. In fact, there are times when trimming the leaves can make it easier to flex, such as when leaves need to slide through a slot during a slot flex.

Consider figure 21.3, which has two different leaf shapes: some are triangles (angles of 40°, 60°, and 80°) and others are kites (angles of 40°, 100°, 100°, and 120°). It folds into a hexagon with nine pats and supports the P333 just like the enneaflexagon in Figure 6.2. This is because the P333 flex only requires six of the nine pats and three of the nine hinges, leaving us free to modify other portions of the flexagon. We changed the three unaffected leaves into kites so that the overall shape is a flat hexagon. This means that flexing from state 1/2 to 3/1 goes from one flat hexagon (with nine 40° angles in the center) to another flat hexagon (with three 120° angles in the center). State 2/3 is not flat, though you can arrange it so that six 60° angles are in the center, with tented leaves rising off the face.

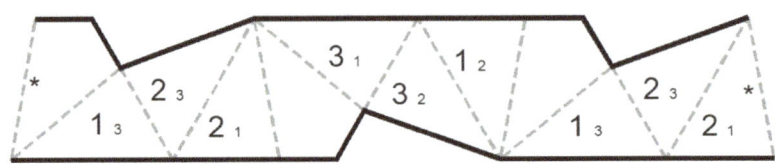

FIG. 21.3 Template where not all leaves are the same shape. This folds into a hexagon but supports the P333.

Rule-break 3

You may have noticed that some flexes, such as the silver tetra flex, involve folding together nonadjacent pats, which seems to violate the definition of a basic flex. But if you look closely at how the flex operates, you can see that it's only incidental that nonadjacent pats are folded together. The actual flex only requires folding together adjacent pats. Similarly, you may find that a stack of leaves zigzags when flexing, even though this is just incidental to the actual structure.

Rule-break 4

Figure 21.4 contains an example of a template where some leaves are attached to more than two other leaves. To assemble this flexagon, start by folding x's against x's and y's against y's, which requires folding pairs of adjacent pats together. After that, fold as normal. The resulting flexagon behaves like a tetra-dodecaflexagon, with the x's and y's not appearing during basic flexes. The advantage of this template is that you can fold it from a straight strip of paper, even though the template of the corresponding tetra-dodecaflexagon is hexagonal rather than straight.

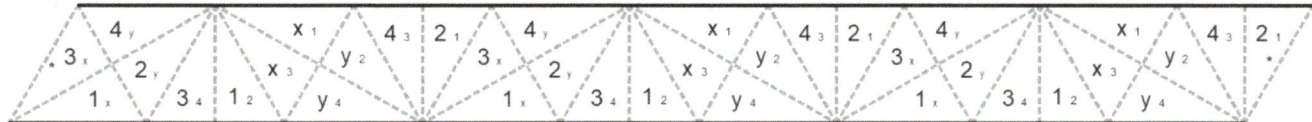

FIG. 21.4 Template where some leaves are attached to more than two other leaves. This behaves like a tetra-dode-caflexagon where the x's and y's remain hidden.

Because folding together the x's and y's gives you the template for the bronze dodecaflexagon (generator: P*P*), you can do all the same flexes on this flexagon that you can do on the corresponding basic dodecaflexagon. This includes the basic pinch flex, the reverse pass-through from chapter 5, the P444 from chapter 6, the tuck from chapter 11, as well as some of the flexes described in chapter 29 such as the rhombic-morph flex. But if you have any hope of finding the x's and y's, you would have to try something that's not a basic flex (Schwartz and Rutzky, 2009, pp. 265–266).

Rule-break 5

A **matryoshka** is created when you fold together pats to make a smaller flexagon you can flex, like nested dolls, but some of the pats don't seem to follow the atomic flex axioms. Typically, folded pats that don't follow the axioms restrict the hinges to the point that they can no longer flex, but sometimes there's still enough freedom to flex. A basic flex on a matryoshka only operates on the pat structure that follows the axioms, sometimes opening back up to the original flexagon structure. You can find an example in chapter 30.

Rule-break 6

Our next example follows the rules but breaks our assumptions of how to interpret the rules. Instead of connecting leaves across an *edge*, leaves connect across a *point*. These are called **point flexagons**, to contrast with **edge flexagons** that connect across an edge like most flexagons in this book. See figure 21.5 for the template for a point flexagon with six faces. For practical reasons, the physical model uses narrow hinges rather than points, with the first and last narrow hinges taped together. It's flexed across a pair of hinges, where the same number appears on the top and bottom of a single pat.

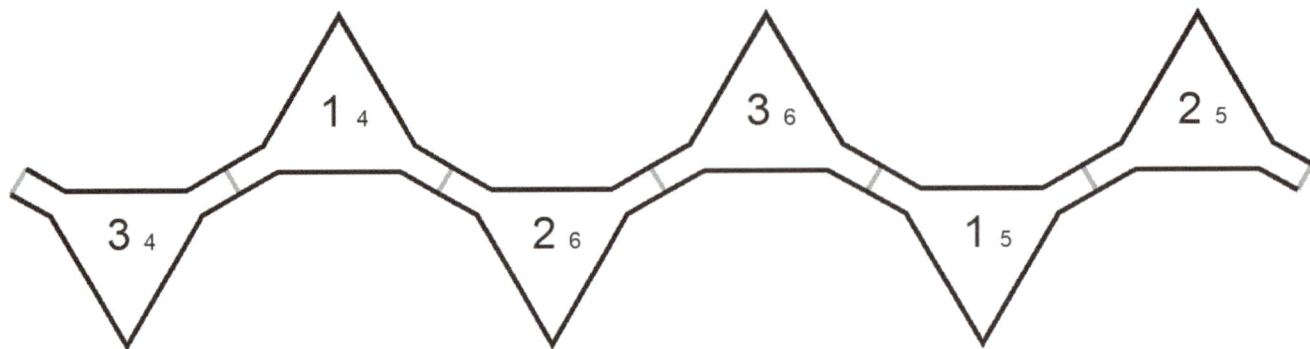

FIG. 21.5 Template for a hexa-hexa point flexagon, where leaves are connected across points rather than edges.

Point flexagons are the answer to our earlier question about how to create a flexagon with only a single copy of the underlying pattern. In order to flex a single copy, you need to trim off enough of the leaves and hinges so that the hinges are reduced from edges to points. This provides enough freedom to flex the resulting flexagon.

One intriguing detail you'll find if you draw the state diagram for this point flexagon is that it looks very similar to the state diagram we saw in figure 17.5 for the corresponding hexa-hexaflexagon, except that there are a few unexpected shortcuts you can take. From the state with 1's on the outside, you can open it up and flip around pats to see state 2, 3, 4, or 5. This is similar to moving between states 1/2, 1/3, 1/4, or 1/5 on a hexaflexagon. However, from face 4, you can flex to face 1, 3, 5, or 6. This is equivalent to states 4/1, 4/3, 4/5, and 4/6, even though faces 5 and 6 are inaccessible from face 4 in a hexaflexagon connected across edges.

Rule-break 7

Adding twists and knots to the flexagon surface also breaks our assumptions for how to follow the rules, while still fitting the definition: every leaf is connected to two other leaves and the pats are folded using the atomic flex axioms.

After folding a template, you can sometimes add twists or even knots to the flexagon surface before connecting the final edges. Chapter 31 on silver bracelets discusses examples.

Rule-break 8

Mathematically, a flexagon can have an infinite number of leaves either through infinitely nested pat structure or by having an infinite chain of pats. Both of these approaches can be useful in theory. We'll see infinitely nested pats in chapter 22 and infinite chains of pats in chapter 23. However, since infinite paper models are fairly impractical, we won't provide any infinite templates in this book.

Not a flexagon

The term *flexagon* was coined from *flexible polygon*. Specifically, this referred to the structures found in hexaflexagons and square tetraflexagons. The generalization of this structure is what this book is about.

However, there are lots of other possible ways to build something that can be considered a "flexible polygon". As a result, the term *flexagon* is sometimes applied to constructs that are different from either the original meaning or our definition of a *basic flexagon*.

At the time of this writing, if you do an image search for *flexagon*, one of the most common results is for a mechanism typically called a kaleidocycle. A **_kaleidocycle_** is a ring of tetrahedra connected across edges, where a tetrahedron is a three-dimensional shape with four triangular sides, like a pyramid with a triangular base. This is very different from a flexagon, with no hidden structure and no "flexible polygons," so calling them flexagons seems to be a misnomer.

There are many other types of foldings that are similar to flexagons including cootie catchers (aka fortune tellers), fleptagons (Conrad and Hartline, 1962, pp. 218–220), bregdoids (Conrad and Hartline, 1962, pp. 214–217), flexahedra (Pook, 2003, pp. 154–163), flexatubes (Pook, 2003, pp. 89–90), Hooke's joint flexagon (Pook, 2003, pp. 88–89), and flapagons (Pook, 2009, pp. 308–310). Some of these behave like a "flexible polygon", but given that these are significantly different from the mechanisms of a basic flexagon, we won't discuss them further.

Summary

Connecting every leaf to two mirror-image copies and applying the atomic flex axioms gives us all the rich, dynamic behavior of the triangle flexagons we've seen in this book. But as long as you preserve the basic nature of the underlying hinges, there are many ways to modify these basics to make a wide variety of other interesting, dynamic mechanisms.

Chapter 22

Groups and Flexagons

A group is a powerful and general mathematical abstraction that follows certain rules that allow us to reason about various operations in lots of interesting ways. Examples of groups include addition, polynomial equations, special relativity, molecular chemistry, and even moves on a Rubik's Cube. When a mathematician sees any new set of things that can be manipulated, one of their first questions is whether it forms a group. Notice that we've already used the word *set*, which itself is a mathematical term that just means a collection of objects, as opposed to a *group*, which we'll define in a moment.

Since a collection of flexagon states is a set, and flexes are a way to manipulate those states, it's natural to ask if flexagons and flexes form a mathematical group. In order to answer that question, let's first look at the definition of a group.

Informally, you can think of a **group** as a set and an operation on that set that obeys these four rules:

1. There's an operation that doesn't change anything.
2. Every operation has an inverse operation, which essentially lets you undo it.
3. If you want to carry out a series of three operations, it doesn't matter if you start with the first two or the last two.
4. You can always perform any sequence of operations.

More formally, we define a set G as a collection of objects known as elements, where the elements are referred to using lowercase letters a, b, c, etc. We define an operation on the elements of the set, represented as an asterisk *. If G and * have the following properties, they form what's known mathematically as a group:

1. *Identity:* There's a unique identity e in G such that, for every a in G, $a * e = e * a = a$.
2. *Inverse:* For every a in G, there exists b in G such that $a * b = b * a = e$, where e is the identity.
3. *Associative:* For all a, b, and c in G, $(a * b) * c = a * (b * c)$.
4. *Closed:* If a and b are in G, so is $a * b$.

To make these properties more concrete, let's consider one of the groups we're most familiar with: adding numbers. The set is all numbers and the operation is addition. If you start with the numbers 1 and 2 and apply the addition operator – in other words, add them together – you end up with 3. Numbers and addition are a group because they follow the definition of a group:

1. There is an identity, 0, because $a + 0 = 0 + a = a$.
2. Every number has an inverse, because $a + (-a) = 0$.

DOI: 10.1201/9781003433538-25

3. It's associative, because $(a + b) + c = a + (b + c)$.
4. It's closed, because whenever you add two numbers, the answer is another number.

On the other hand, the set of integers (whole numbers and their negatives) with the operation multiplication is *not* a group. It follows only three of the rules:

1. There is an identity, 1, because $a \times 1 = 1 \times a = a$.
2. It's associative, because $(a \times b) \times c = a \times (b \times c)$.
3. It's closed, because whenever you multiply two integers, the answer is another integer.

But the inverse condition is not satisfied. The inverse of a whole number is a fraction, not a whole number. For example, ½ is the inverse of 2. Also, 0 does not have an inverse.

Flex sequences as elements

How can we align the *group* way of thinking about things with the way flexagons work? Our first attempt might be to think of all the states of a flexagon as the set and applying different flexes as the operation, but this isn't how groups work. For one thing, there are lots of different flexes, but a group only has a single operation, not a collection of them. A second issue is that an operation is something done to an ordered pair of elements, not a single element. Flexing can be thought of as "take state A, apply flex B, and get state C", while a group wants "element A operated with element B creates element C", where A, B, and C are all the same kind of thing. Figure 22.1 illustrates these two different approaches.

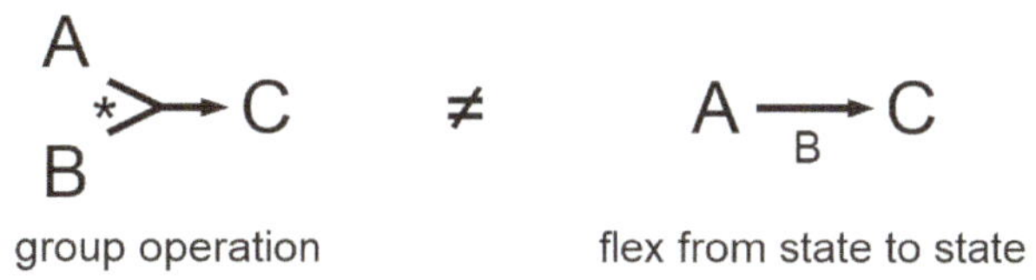

FIG. 22.1 Illustration of how to think about groups versus the way we typically think about flexes.

Instead, let's pivot how we think about flexagons and flexes to match the way a group works. Rather than using *flexagon states* as the elements of our set, let's use *flex sequences*. Thus, our elements are sequences such as P>F and T'^>S. And rather than *flexes* being the operation, *combining flex sequences* will be our operation. In other words, the *operation* consists of doing one sequence followed by the second sequence. For example, combining the two flex sequences P>F and T'^S gives us a third flex sequence P>FT'^S, as shown in figure 22.2.

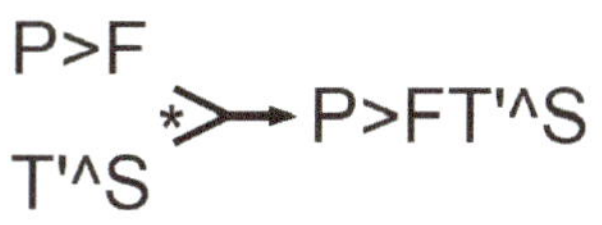

FIG. 22.2 Combining two flex sequences into a new flex sequence.

To be more precise, we're not actually interested in the *flex sequence* as much as the *transformation* the flex sequence represents. Thus, the atomic pat notation description of how the flex sequence changes the flexagon structure is the important part. A flex sequence is simply convenient shorthand for describing the transformation. Just note that sometimes two flex sequences actually represent the same transformation, and therefore the same group element.

Does this new framing give us a group? Let's look at how the four requirements of a group align with this approach.

1. *Identity:* The identity is I, the flex that leaves a flexagon as is. For any flex A, AI = IA = A.
2. *Inverse:* Every flex sequence A has an inverse A′, which represents doing sequence A in reverse. Thus, for any flex A, AA′ = A′A = I.
3. *Associative:* Is it associative? In other words, if you have flex sequences A, B, and C, is it always true that A * (B * C) = (A * B) * C? Consider how we've defined flexes in terms of the pat structure before the flex and the transformed pat structure after the flex. While we typically include square brackets to provide extra information about how the leaves are folded together (e.g., [1,[[2,3],4]]), we could simply list the leaf IDs without the brackets (e.g., [1,2,3,4]). This makes it clear that flexes are simply permutations, which are associative.
4. *Closed:* This brings us to the final requirement of a group: it must be closed, i.e., applying two flex sequences in a row must always give us another valid element in our set. Is this always true?

To answer that, let's examine a specific flexagon: the tetra-hexaflexagon, with four faces. A face on this flexagon is a hexagon with six hinges, one hinge between each pat. If you restrict yourself to just pinch flexes, it has five states. In one of these five states, you can do a pinch flex at any hinge. But in the other four states, you can only do a pinch flex at three of those hinges, not the other three, as shown in figure 22.3. After you do a pinch flex, you need to shift to one of the adjacent hinges (or the hinge on the opposite face) in order to do another pinch flex. This means that there are times where you can do the sequence P>P, but not PP. In other words, you can always combine P and >P, but you can't always combine P and P. Therefore, it's not closed.

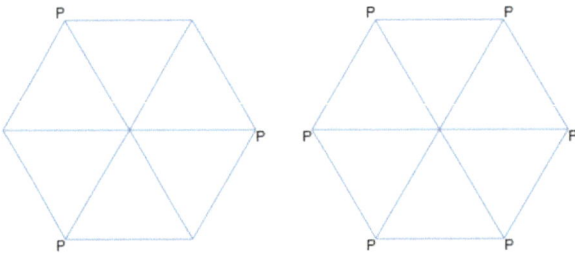

FIG. 22.3 How hinges on different faces of the tetra-hexaflexagon support the pinch flex.

This is a very general issue. For any given flexagon, you typically can't do all flexes at all of the hinges because the internal pat structure doesn't support them. That means you may only be able to perform flex sequence A followed by flex sequence B some of the time.

It should be noted that we've satisfied the definition of a **groupoid** because we have an identity, an inverse, and associativity. But is there a way to achieve closure as well, so we have a group?

To investigate, consider the three cases in figure 22.4: isoflexagons (where all pats and hinges are symmetrically equivalent) with non-morphing flexes (so all flexes preserve the directions between pats), non-isoflexagons with non-morphing flexes, and the case that includes morphing flexes.

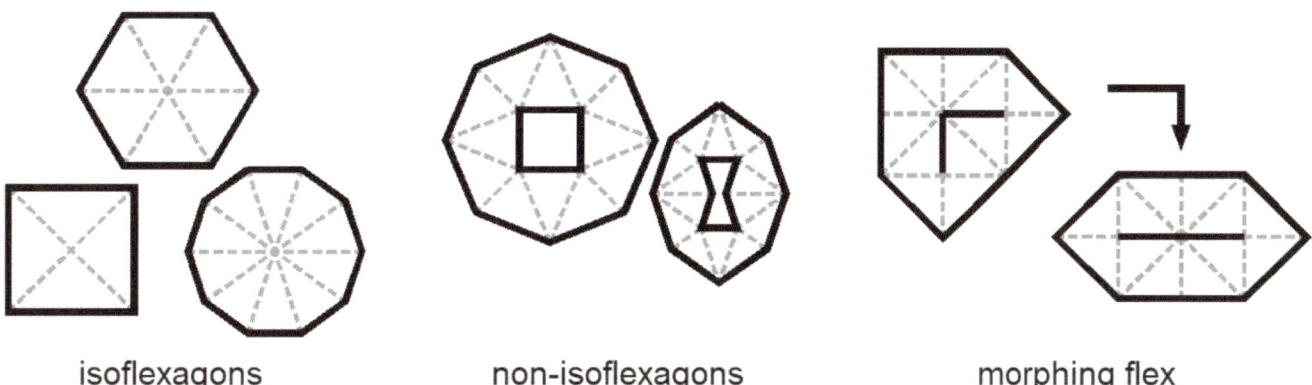

isoflexagons non-isoflexagons morphing flex

FIG. 22.4 Three cases to investigate: isoflexagons, non-isoflexagons, and morphing flexes.

Fortunately, there is a way to salvage this approach for isoflexagons. Rather than attempting to model a specific flexagon, such as the tetra-hexaflexagon, we instead model a theoretical flexagon with infinitely nested pats. We call this a **universal flexagon**. A universal flexagon always supports every flex at every hinge.

To understand universal flexagons, consider the pat notation for a single pat. We represent a pat with a single leaf with a dash, -. Just as we can create new leaves by applying a generating sequence, we can replace a leaf with two leaves folded together, represented by [-, -]. Each of those leaves can be replaced with a pair of folded leaves, making it [[-, -], [-, -]]. And each of those leaves can likewise be replaced with two leaves, giving [[[-, -], [-, -]], [[-, -], [-, -]]]. In theory, we can continue this process forever, making all our pats infinitely nested.

On a universal isoflexagon, any flex that can be done on the flexagon can always be done at every hinge because the required pat substructure is always present. This makes our flex composition operation closed, giving us a group. While this obviously doesn't apply to the finite flexagons we can actually make, it gives us a useful tool for reasoning.

Recall from chapter 20 on atomic flex theory that you can't flex from a flat hexaflexagon to a flat octaflexagon because the only way to add pats (going from six to eight in this case) also adds twists to the flexagon surface. This implies that there's a different universal flexagon for each different number of pats in an isoflexagon. Thus, there's a universal hexaflexagon, a universal octaflexagon, a universal tetraflexagon, a universal dodecaflexagon, and so on. No series of flexes can transform between these different universal isoflexagons.

We can adjust our strategy slightly for non-isoflexagons. Recall that flexes on non-isoflexagons require both a given pat structure and a specific set of hinge directions. The universal flexagon approach gets around the first requirement, but not the second. To deal with varying hinge directions, we'll construct all our flex sequences so that the current hinge never changes. This means that each flex sequence includes the shifts necessary to change the reference hinge, perform the flex, and shift back to the original hinge. For example, rather than considering the two flexes {>, A}, we might rephrase the collection of flexes as {>A<, >>>A<<<} so each flex sequence starts and ends at the same hinge. In this case, we may have excluded >>A<< because A couldn't be applied at the given hinge. Note that we may also be able to leverage the flexagon's symmetry to reduce the number of flex sequences. This strategy permits a group for universal non-isoflexagons and non-morphing flexes because any pair of flex sequences is now closed.

What about the final case, where we include morphing flexes? Recall that the various flexes have specific requirements for the pat directions, and morphing flexes change the directions between pats. Thus, once you've morphed the flexagon, the set of flexes you can apply is now different. This means that, in general, you can't get closure if morphing flexes are included in the operations, so you don't have a group.

In summary, we can now describe flexagons and flexing using group theory by adopting the following strategy:

- The elements of the set are *flex sequences* (or, more precisely, atomic pat transformations).
- The operation is *combining* flex sequences.
- There are four cases:
  - Group: universal isoflexagons, non-morphing flexes.
  - Group: universal non-isoflexagon, non-morphing flexes, defining all flex sequences so they start and end at the same hinge.
  - Groupoid: morphing flexes.
  - Groupoid: finite flexagons.

Why is this interesting? In short, it allows us to utilize powerful rules that have been proven for groups. For example, all groups follow this rule:

$$(AB)' = B'A'$$

For flexagons this means that if you want to undo a sequence of two flexes, do the inverse of the second flex followed by the inverse of the first flex. Groupoids also obey this rule whenever you can perform the flex sequence AB, so this applies to plain old finite flexagons as well as universal flexagons.

Finite subgroups

We can also talk about **subgroups**, subsets of the overall flexagon group that obey the group rules by themselves.

One of the simplest subgroups does nothing but shift hinges. This allows you to make any hinge the current hinge, but you can't turn it over or perform any flex that rearranges the leaves.

To see what this looks like using the language of groups, let's examine a triangle with three hinges. We use e to represent the identity flex I, which does nothing, and a to represent >, the shift right flex. Group notation uses a^n as shorthand to say that a is repeated n times, e.g., a^2 is the same as $a * a$. We label the leaves of our flexagon 1 to 3 clockwise from the current hinge so we can keep track of the leaves as we shift the current hinge. We refer to each of the three possible states using the series of flexes needed to reach it from the original state. Hence, e is [1,2,3], a is [2,3,1], and a^2 is [3,1,2], as illustrated in figure 22.5. Note that a^3 returns us to the original position, so it's the same as e.

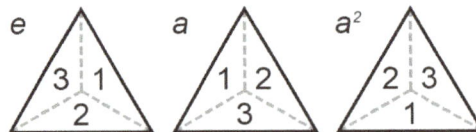

FIG. 22.5 The states of the group on a triangle that only supports >.

Next, we build table 22.1 to show the results of doing any pair of flexes, one after the other. The result of doing the flex listed for a row followed by the flex listed for a column is placed in the corresponding table cell. For example, if you look at where the row for a^2 intersects the column for a, you'll find e, which indicates that $a^2 * a = e$. You can verify this by applying >> followed by > to see that it returns to the original state. The resulting combination of operations is called group Z_3, where Z is used for a **cyclic group** (a group that cycles) and the 3 represents the number of elements.

TABLE 22.1 Composing the elements $\{e, a, a^2\}$. This is group Z_3.

	e	a	a^2
e	e	a	a^2
a	a	a^2	e
a^2	a^2	e	a

An interesting property of this table is that each of the symbols (e, a, and a^2) appears exactly once in each row and column. This is a general property of groups. Note that we created this group of three elements from a single flex, >, repeating it till it cycled back to the original state. We say that $\{>\}$ is the **generator** of the group.

We can also use multiple flex sequences as generators, such as > and ^. If e is the identity flex I, a is >, and b is ^, then figure 22.6 shows the six possible states we can reach by combining these flexes. We use $-n$ to indicate the back face of leaf n.

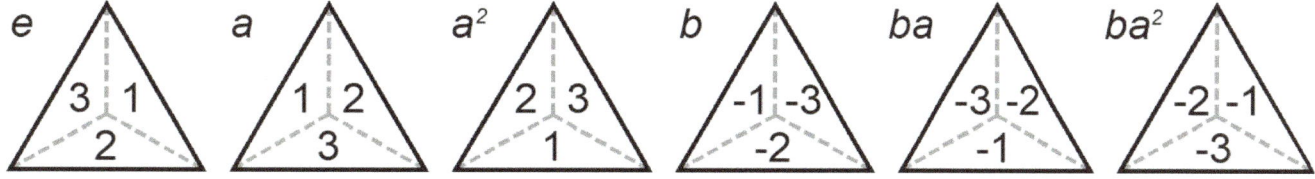

FIG. 22.6 The states of the group on a triangle with generators $\{>, \wedge\}$.

Table 22.2 shows how these flexes compose. Again, note that each flex sequence appears exactly once in each row and column. This is called group D_6, for reasons we'll see later.

TABLE 22.2 Composing the elements $\{e, a, a^2, b, ba, ba^2\}$, where $a = >$ and $b = \wedge$. This is group D_6.

	e	a	a^2	b	ba	ba^2
e	e	a	a^2	b	ba	ba^2
a	a	a^2	e	ba^2	b	ba
a^2	a^2	e	a	ba	ba^2	b
b	b	ba	ba^2	e	a	a^2
ba	ba	ba^2	b	a^2	e	a
ba^2	ba^2	b	ba	a	a^2	e

So far, our examples simply use the symmetry of a triangle, rotating and turning it over, which isn't specific to flexagons. But we can also create subgroups from nontrivial flexes, such as the flexes from any of the cycles listed in chapter 18. For example, the cycle (P>)3 leads to the subgroup $\{P>\}$. This means that the only flex sequence you can do is P>, i.e., a pinch flex followed by shifting one hinge to the right. Ignoring the current hinge for the moment, we see that $\{P>\}$ creates a cycle of three states – 1/2 to 3/1 to 2/3 to 1/2 – just like {>} on a triangle. In fact, if you set a = P>, you can reuse table 22.1 to understand how the flexes combine. This is because {>} and {P>} represent the same group and hence follow the same pattern, even though they use different flexes.

We can add additional generators to our cycle. When cycling through P> on a flexagon with three faces, at any point you could turn over the flexagon and shift one hinge to the right (^>) before continuing with another P>. This means we can create a group with the generators {P>, ^>}. If we choose a = P> and b = ^>, we can reuse table 22.2, which we created for the generators {>, ^} on a triangle.

These examples show one of the powers of groups: the same patterns can be found across groups even when they represent different types of transformations. In this case, > on a triangle and P> both cycle between three states. In general, a cycle of n states is represented as Z_n in group theory, so a three-state cycle is Z_3, as we saw in table 22.1. Thus, the two groups {^>} and {^}, which both just alternate between two states, are both Z_2. Table 22.3 shows Z_2, where a could be ^>, ^, or any other flex sequence that alternates between two states.

TABLE 22.3 Group Z_2 with elements $\{e, a\}$.

	e	a
e	e	a
a	a	e

In group theory, combining the rotations of an n-sided polygon with turning over the polygon creates the **dihedral group** D_{2n}. (Note that some sources use D_n. We'll use D_{2n} since the corresponding group has $2n$ elements just like Z_n has n elements.) Thus, the group with generators {>, ^} on a triangle, which we saw in figure 22.6 and table 22.2, is the group D_6. If we ignore the current hinge, the group with generators {P>, ^>} is also group D_6. Notice how both {>, ^} and {P>, ^>} have one flex sequence that rotates the flexagon and one that turns it over.

Of course, we have seen that it's important to track the current hinge since we usually can't perform a particular flex at every hinge. So, when we look more closely at the group {P>}, we see that we don't return to the original state and hinge until we've repeated P> *nine* times, not just three. This means that when we track the current hinge, {P>} is actually Z_9 and {P>, ^>} is D_{18}.

Next, let's try a pair of flex sequences that don't involve turning over the flexagon. Using the octaflexagon (generator: (F<)6) from figure 18.2, try a = (F<)2 and b = (F<)3. You'll find that $a^3 = e$ (group Z_3) and $b^2 = e$ (group Z_2). Mapping out all the combinations of the two sequences gives us table 22.4. In group theory, we refer to this group as the direct product of Z_2 and Z_3, written as $Z_3 \times Z_2$, which is equivalent to Z_6. One important difference between Z_6 and D_6 is that $ab = ba$ in Z_6 but not D_6. This means that the order in which you perform flexes doesn't matter for Z_6 but does for D_6. We say that operations in Z_6 are *commutative*. Cyclic groups are commutative while dihedral groups with $n > 4$ are not.

TABLE 22.4 Composing the elements $\{e, a, a^2, b, ba, ba^2\}$, where a = (F<)2 and b = (F<)3, which is the group $Z_3 \times Z_2 = Z_6$.

	e	a	a^2	b	ba	ba^2
e	e	a	a^2	b	ba	ba^2
a	a	a^2	e	ba	ba^2	b
a^2	a^2	e	a	ba^2	b	ba
b	b	ba	ba^2	e	a	a^2
ba	ba	ba^2	b	a	a^2	e
ba^2	ba^2	b	ba	a^2	e	a

In the next example, we create a group from the cycle (F<)5, supported by the heptaflexagon shown in figure 12.6. Therefore, the generator {F<} creates group Z_5, as shown in table 22.5 where $a = F<$.

TABLE 22.5 Group Z_5 with elements $\{e, a, a^2, a^3, a^4\}$.

	e	a	a^2	a^3	a^4
e	e	a	a^2	a^3	a^4
a	a	a^2	a^3	a^4	e
a^2	a^2	a^3	a^4	e	a
a^3	a^3	a^4	e	a	a^2
a^4	a^4	e	a	a^2	a^3

Similar to earlier examples, this means that the generator {F<, ^<} creates group D_{10}. Why did we choose ^< and not ^ or ^> or something else? Because only ^< preserves the pat structure so that the flex operations are *closed*. The cycle (F<)5 on a heptaflexagon requires the pat structure [--] / - / [[--][--]] / [[--][--]] / [[--][--]] / [[--][--]] / - /, and doing either F< or ^< keeps this pat structure unchanged even as the leaves are moved around. But other flex sequences, such as ^>, don't.

This observation that every generator preserves the minimal pat structure can be solidified into a theorem that's useful for finding and describing other finite subgroups.

Structure Preservation Theorem: *Given a finite flexagon group G with elements/flexes $\{e, a, a_2, ..., a_n\}$ and pat structure P created from the generating sequence e+ a+ a_2+ ... a_n+, every flex from every state preserves P.*

Proof: By definition, P is the minimal pat structure that supports every flex sequence in our list of elements from the original state e. We need to show this is true from every state.

Assume there's an element b in group G that transforms P into Q such that P ≠ Q. Since P and Q must have the same number of leaves, b must remove some structure from at least one pat (or subpat) and add structure to at least one other pat (or subpat).

Since we have a group, we know that we can perform any flex sequence on Q. But this means that the additional structure in at least one of the pats in P versus the corresponding pat in Q isn't necessary, therefore P isn't the minimal pat structure. This is a contradiction, so we know our assumption that P ≠ Q is false.

Therefore, every flex sequence in a finite group always preserves the group's minimal pat structure.

Recall that the challenge in finding group structure in flexagons is that flex sequences typically aren't closed. Aiming for cycles that preserve common pat structure helps for finding useful flex sequences to act as group generators.

Now when we describe the group created by a set of flex sequences, we can include the minimal pat structure for the associated flexagon, which is preserved after any sequence of the generator flexes, as shown in table 22.6.

TABLE 22.6 Examples of groups showing the pat structure that is preserved after every flex in the group.

generator	group	pat structure	pat count
{P>, ^>}	D_{18}	[--] / - / [--] / - / [--] / - /	6
{F<, ^<}	D_{10}	[--] / - / [[--][--]] / [[--][--]] / [[--][--]] / [[--][--]] / - /	7
{F>>>, ^>>>}	D_6	[--] / - / - / [--] / - / [[--][--]] / - /	7
{V>, ^>}	D_{12}	- / [--] / [[--][--]] / [--] / - / [--] /	6
{V<, ^<}	D_{30}	[--] / [[--][--]] / [--] / - / [--] / [[--][--]] /	6
{Ltb<}	Z_3	[[-[--]]-] / - / - / - / [--] / [[--]-] /	6

Note that there's no non-trivial combination of ^ and > that preserves the pat structure needed by Ltb< (why?), so we can't compose it with turn-over-and-shift flexes as we can with the other cycles.

As we saw with the generator {(F<)2, (F<)3}, you can use multiple flex sequences that involve more than ^, >, and <. You can also use different types of flexes within the same sequence. And you can also use more than two sequences. Table 22.7 shows several examples of more complex generators.

TABLE 22.7 Examples of groups created from more complex generators.

generator	group	pat structure	pat count
{Ltb>P>P>}	Z_{12}	[[[- -][- -]] -] / - / [- -] / - / [- -] / - /	6
{^>P>, PP^}	D_4	[[- -] -] / - / [[- -] -] / - / [[- -] -] / - /	6
{<<(S^<)2, (>>SS)2 ^<}	D_6	[[[- [[- -] -]] -][- [- -]]] / - / [[[- [[- -] -]] -] -] / - / [[[- -] -] -] / - /	6
{>^T<<^S>T', ^<}	D_{10}	[[- -][- -]] / [- -] / - / [- -] / - / [- -] / - / [- -] /	8
{(F<)2, (F<)3, ^<}	D_{12}	[- -] / - / [[- -][- -]] / [[- -][- -]] / [[- -][- -]] / [[- -][- -]] / [[- -][- -]] / - /	8

For our final example, take another look at the square tetra-tetraflexagon from figure 7.3. Start on face 1 in state 1/2 with the text (B>)4 in the lower left corner. If you perform the sequence (B>)4, you'll find you've returned to state 1/2 but the text has moved to the upper right corner. This means that the flexagon has rotated 180°, so you have to perform B> *eight* times to return to the original state and hinge. Thus {B>} is group Z_8.

In group theory, Z_n, the cyclic groups, represent the rotational symmetries of a polygon, while D_{2n}, the dihedral groups, are cyclic groups with an added reflection operation. It makes intuitive sense that these are the groups we found because of how flexagons and flexes are analogous to polygons and their symmetries. A flexagon is a ring of leaves where each leaf is connected to two neighbors similar to the way the edges of a polygon are connected to two adjacent edges. And flexes typically rotate hinges, while ^ behaves like reflection. Can you figure out how to use atomic flex theory to check if this intuition is correct?

Infinite subgroups

So far, all the examples we've explored are finite subgroups with a finite number of different states. But there are also infinite subgroups.

Consider the subgroup {P, ^, >}, where you're allowed to do a pinch flex at the current hinge, turn it over, or shift the flexagon one hinge to the right. This gives you enough freedom to perform any possible Tuckerman Traverse.

But what happens if you keep pinch-flexing at the same hinge without ever switching hinges? Let's look at the pat structure created by the generating sequences P*, (P*)2, and (P*)3 on a hexaflexagon to see what they teach us:

```
P*        - / [- -] / - / [- -] / - / [- -] /
(P*)2     - / [- [- -]] / - / [- [- -]] / - / [- [- -]] /
(P*)3     - / [- [[- -] -]] / - / [- [[- -] -]] / - / [- [[- -] -]] /
```

Every additional P* adds more leaves. This is because the pinch flex requires the odd-numbered pats to have at least two leaves, but after each P*, the odd-numbered pats only have a single leaf. This means that repeating the pinch flex at the same hinge always takes you to a new state, never looping back to a state you've visited before. Therefore, there are an infinite number of elements in this subgroup. Very few flex sequences cycle back to the original state, so most subgroups have an infinite number of elements.

Summary

Having a way of talking about flexes as groups (or groupoids) gives us a powerful way to describe and reason about flexagons by leveraging theory that applies to a wide variety of mathematics. Proving statements about flex sequences and reasoning about subgroups are just a couple examples.

Because flexagons are rarely closed with respect to a given set of flex sequences, it can be hard to know when you can perform a sequence and easy to get lost among the possible states. But if you stick to the generators of a finite group on a flexagon that supports them, you can always perform any of the flex sequence elements.

If you're interested in further exploring the mathematics of flexagons and groups, here are a few questions to ponder:

- The largest group we saw had 30 elements. Can you make bigger finite subgroups?
- We created groups with up to three flex sequences as generators. Can you figure out how to make groups with more than three generators? How about if you avoid combinations of the same flex sequence, e.g., something like {<<(S^<)2, (>>SS)2^<} rather than {(F<)2, (F<)3}?
- Are there any finite subgroups that aren't cyclic or dihedral?
- How would you classify the group structure of various universal flexagons?

Chapter 23

Flexagon Computers

Flexagons and flex sequences are powerful enough that we can use them to write "programs" that can compute anything a general-purpose computer can. In technical terms, we say they're **Turing complete**.

It may seem surprising that you can turn a flexagon into a computer. But think about the key components of a computer: they *store data*, they *examine and conditionally change* the stored data, and they have *input* and *output*. Similarly, a flexagon stores data in the form of its pat structure, a flex conditionally changes pat structure only if it matches a particular pattern, and you can provide input through a flex sequence with the resulting visible leaves and internal pat structure as the output.

Consider the flex sequence ({P | >P})12, which describes how to reveal every pinch face on a hexa-hexaflexagon using the Tuckerman Traverse. At each step, it tells you to do a pinch flex at the current hinge if you can, otherwise shift one hinge to the right and do a pinch flex there. This ability to make decisions based on pat structure is powerful, but is it powerful enough to build a computer?

Emulating Rule 110

One way to formally prove that a system is Turing complete is to show that it can run another system that is already known to be Turing complete. The system we'll choose to emulate is the Rule 110 one-dimensional cellular automata (CA) because it's fairly straightforward to implement, it illustrates how you can build a flexagon computer, and it has been proven to be Turing complete (Cook, 2004).

Rule 110 takes a list of cells that are on or off and applies rules for changing the state of each cell. For brevity, 0 = off and 1 = on, so a list of cells might be 011010001. When figuring out what the new state of a cell is, you look at the state of its neighbor to the left, the cell's state, and the state of the neighbor to the right. The rules are applied to every cell at the same time to create the next generation. Here are the rules for Rule 110:

pattern	111	110	101	100	011	010	001	000
new center	0	1	1	0	1	1	1	0

For example, if a cell's left neighbor is 0, its state is 0, and the right neighbor is 1, we look up 001 in the rules and find that the current cell changes to 1. If we assume cells to the left and right of the following table are 0, here's what four generations look like when starting with 0001110110:

DOI: 10.1201/9781003433538-26

generation

1		0	0	0	1	1	1	0	1	1	0
2		0	0	1	1	0	1	1	1	1	0
3		0	1	1	1	1	1	0	0	1	0
4		1	1	0	0	0	1	0	1	1	0

Figure 23.1 shows several dozen generations of Rule 110, with each row representing the generation that follows the row above it. The proof that it's Turing complete leverages "spaceships", emergent patterns that can travel across cells and interact with other patterns.

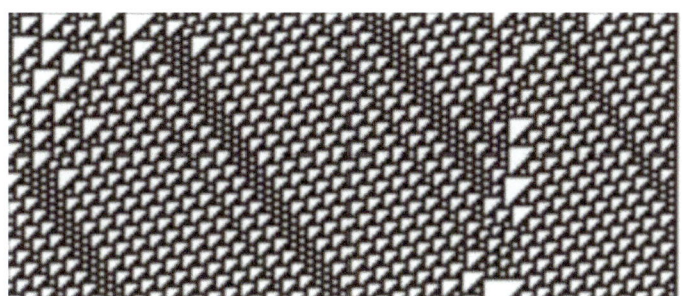

FIG. 23.1 Multiple generations of cells from running Rule 110, where black represents on and white is off.

Our basic strategy for running this rule on a flexagon is to track whether a cell is on or off by checking if you can perform a tuck flex on a given pair of adjacent pats. Applying the tuck flex changes the state from off to on, while applying the inverse tuck changes the state from on to off. This means that a cell is on if you can do a tuck flex at the cell's hinge, and it's off if you can instead do an inverse tuck. If you want to check the state without changing it, you follow the flex with its inverse. This gives several basic operations at the hinge representing a cell:

T	change cell from off to on
T′	change cell from on to off
TT′	check if cell is off without changing it
T′T	check if cell is on without changing it

We can use this technique to store the states of multiple cells across a series of pats. For example, to initialize a flexagon with the states 1011, we use the generating sequence T*>>T′*>>T*>>T*>>, since T turns a cell on and T′ turns it off. Because the tuck flex impacts two pats, we move two hinges between each flex to get to the next cell. See figure 23.2 for a flexagon that shows states 1011.

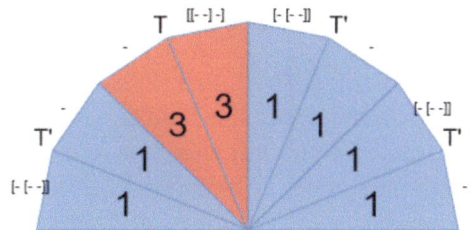

FIG. 23.2 A portion of a flexagon showing the states 1011 using the tuck and inverse tuck flexes to store on/off.

This also gives us the tools for describing a rule that changes a single cell. First, we'll show how to emulate the rule that says 111 → 0, which means if the neighbor to the left is on (1), the current cell is on (1), and the neighbor to the right is on (1), then turn the current cell off (0). We do this by defining the flex R111 with the following definition and breakdown:

$$R111 = <<T'T>> \; T'T \; >>T'T<< \; T'$$

<<T'T>>	step to the left neighbor, check if on, return to center
T'T	check if current cell is on
>>T'T<<	step to the right neighbor, check if on, return to center
T'	turn the current cell off

As with any flex, we can convert the R111 flex into atomic pat notation that specifies the pat structure needed in order to apply the flex and the result. This is an alternate way to demonstrate that R111 can only be applied if all three cells are currently on, and in that case, it will turn off the current cell. See figure 23.3 for what R111 looks like on a flexagon.

R111	in	[3,[1,−2]] / 4 / [7,[5,−6]] / # 8 / [11,[9,−10]] / 12 /
	out	[3,[1,−2]] / 4 / 5 / # [[−7,8],6] / [11,[9,−10]] / 12 /

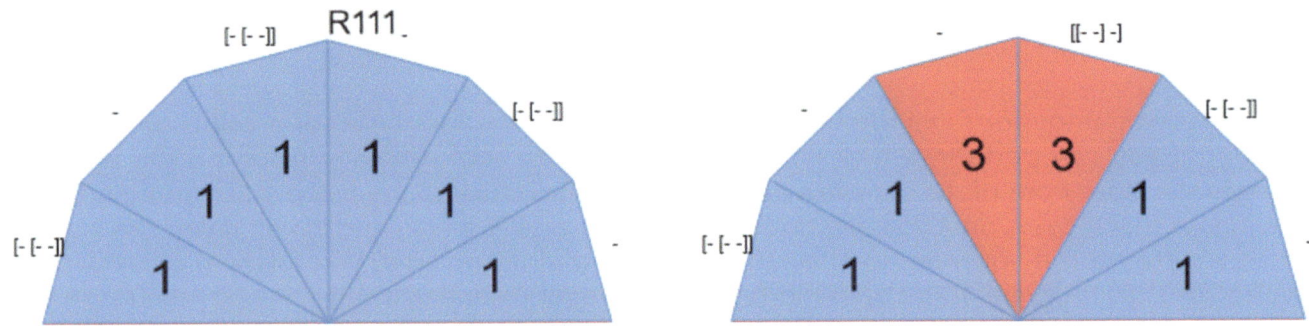

FIG. 23.3 The first flexagon shows states 111. The second flexagon shows the result of applying the flex R111, which is equivalent to running the 111 rule to turn the center cell off.

If you look at the rest of Rule 110, you'll see that most of the remaining rules don't change the current cell, so we don't need to define flexes for any of those cases. The only other ones that change the state are 101 → 1 and 001 → 1. Notice that we get the same result whether the left neighbor is 0 or 1, so we don't actually need to check it. We only need to check that the current cell is off and the right neighbor is on. If so, then we turn the current cell on. We define the Rx01 flex that accomplishes this like this:

$$Rx01 = TT' \; >>T'T<< \; T$$

It may seem that we now have everything we need. We can simply try R111 and Rx01 at every second hinge across the entire flexagon to apply the rule to every cell when the structure matches. However, recall that cellular automata assume that the rule is applied to every cell *simultaneously*, but we have to apply flexes one at a time. If we're not careful, this could mean that some transitions would be wrong because one of a cell's neighbors changed before we checked them.

To understand why, consider four cells with states 1011, focusing on the two cells in the middle (01). The second cell and its neighbors are 101, while the third cell and its neighbors are 011. Looking at the table for Rule 110, we see that the second cell changes to 1 and the third stays at 1, so the next generation is 1111. However, if we change the second cell *before* we check on the third cell, the four cells will now be 1111, which means the third cell and its neighbors are 111. In this case, Rule 110 tells us to change the cell to 0 instead of leaving it as 1, so we end up with 1101 instead of 1111.

We can fix this by saving a cell's original state before we change it so its neighbor can use its previous state allowing us to modify all cells independently. We'll use an extra pair of pats to store each cell's previous state to its right. Then for each generation of the rule, we run all the rules before copying the final results, which sets us up for the next generation. Figure 23.4 shows what this adjusted technique looks like for cells 1011.

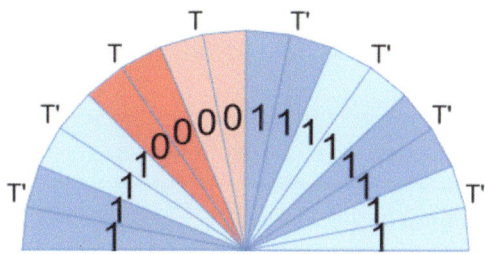

FIG. 23.4 A portion of a flexagon representing the states 1011 along with copies of the state. Dark blue represents a cell in state 1, with the adjacent light blue representing a copy. Dark red represents 0, with the adjacent light red representing a copy. A rule examines the copy to the left, the current cell, and the copy six hinges to the right.

To handle copying cell states, we define flex sequences to check the cell's state and copy it if needed. We define Coff to copy the off state and Con to copy the on state:

$$\text{Coff} \quad = \quad \text{TT}' >> \text{T}'<<$$
$$\text{Con} \quad = \quad \text{T}'\text{T} >> \text{T}<<$$

When we initialize a flexagon with a set of cell states, we need to initialize both the cell and its copy. The first step is to run a generating sequence that creates the needed pat structure for the whole flexagon, with four pats per cell. If n is the number of cells, we can use the following generating sequence:

$$\text{Init} \quad = \quad (\text{T*}>>\text{T*}>>)n$$

We define Ioff for initializing a cell to off and Ion for initializing it to on:

$$\text{Ioff} \quad = \quad \text{T}'>>\text{T}'>> \,|\, >>>>$$
$$\text{Ion} \quad = \quad \text{T}>>\text{T}>> \,|\, >>>>$$

Then we can provide input using a series of those flexes based on the original cells' states. For example, the following sequence initializes the flexagon with states 0110:

Ioff Ion Ion Ioff

Now that we have additional pats representing the copied state, we need to tweak the definitions of R111 and Rx01 so they check the proper hinge. When checking the left neighbor's state, they can simply step two hinges to the left to check the copy. But when checking the right neighbor's state, they skip over their own copied state and the neighbor's current state to get to the neighbor's copied state, which means they need to step six hinges to the right. This gives us the following:

$$R111 \quad = \quad <<T'T>> \ T'T \ (>)6 \ T'T \ (<)6 \ T'$$
$$Rx01 \quad = \quad TT' \ (>)6 \ T'T \ (<)6 \ T$$

Next, we define Rall to apply the rules to every cell and Csall to copy every cell. Note the use of the *or* symbol, |, which tells us to only apply a flex if the needed structure is present. Also note that we include I, the identity flex, in the list of flexes in order to cover the case where we don't need to apply a rule or copy state, allowing the sequence to always succeed.

$$Rall \quad = \quad ((R111 \ | \ Rx01 \ | \ I) >>>>)n$$
$$Csall \quad = \quad ((Coff \ | \ Con \ | \ I) >>>>)n$$

Running Rule 110 on a flexagon now consists of the following steps:

1. *Create computer:* Apply the Init flex to create the flexagon structure.
2. *Provide input:* Apply a sequence of Ioff and Ion flexes describing the original cells' states.
3. *Run program:* Repeat the following steps as many times as desired:
   a. Apply Rall to run rules on every cell.
   b. Apply Csall to make copies of the cells' states.

To see these steps in action, look at figure 23.5. The first image shows state 1011, followed by the result of running Rall, and the final result after copying the new state using Csall. Notice that these steps properly compute the end result of 1111.

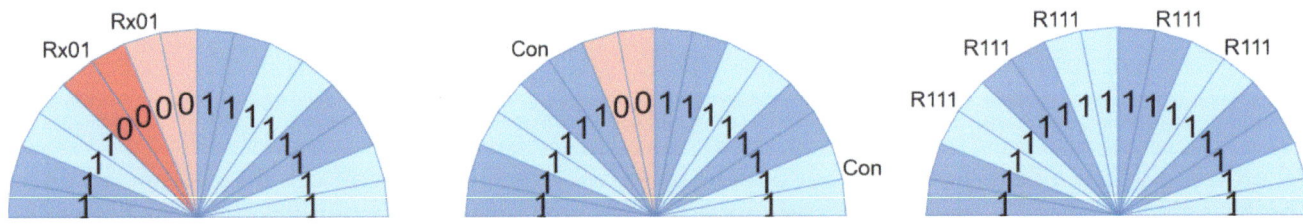

FIG. 23.5 The first flexagon represents state 1011, showing that you can apply flex Rx01 at the hinge representing state 0. While you can also apply Rx01 on the copy, the Rall flex sequence will skip that hinge. The second flexagon shows the result after applying Rall to run the rules. The third flexagon shows the result after applying Csall to copy the final states. It shows that you can run rule R111 on the result to compute the next generation.

One additional detail we glossed over is that a **Turing machine**, the original Turing complete system, requires an infinitely long tape for input and output. We'll define an **infinite tape flexagon** as a flexagon with an infinite number of pats with which to build our theoretical computer. Perhaps it spirals around to infinity without ever joining the ends, or the pats have infinitesimal width, or we take a different approach and create a flexagon where the pats zigzag back and forth forever (like the bracelet flexagon we'll see in chapter 31). In any case, we can place an infinitely long series of cells for Rule 110 on our infinitely long flexagon. However, given that infinite flexagons are somewhat impractical and all actual computers are finite, in practice, we're fine to use finite flexagons.

Theorem: *Flexagons and flex sequences are Turing complete.*

Proof: Using the strategy outlined in this chapter, we can build a flexagon with flex sequences that emulate Rule 110. Rule 110 is Turing complete, therefore this flexagon system is Turing complete.

It's interesting to see how examining flexagons through the lens of group theory versus computation are complementary. Flexes on a *universal flexagon* (infinite depth, finite pats) form an infinite group that contains many subgroups, where groups help you find order within complexity. In contrast, flexes on an *infinite tape flexagon* (finite depth, infinite pats) are Turing complete, allowing you to compute anything, harnessing the complexity behind seemingly simple flexes. And, in fact, we showed Turing completeness by making use of the fact that you can't always do a flex at a given hinge, which is why flexes on a finite flexagon don't form a group.

Summary

Our technique for emulating Rule 110 is fairly straightforward. We used the tuck flex and the pat structure it requires to represent a single bit – 0 or 1 – in a pair of adjacent pats. Then we defined flex sequences that effectively make decisions based on the states of various bits to change those values.

Here are some additional questions to ponder: What if you wanted to access a bit that's not an immediate neighbor? Can you think of other ways to utilize pat structure and flexes? Maybe using cycles and sequences from elsewhere in the book? What other problems might be interesting for a flexagon computer?

Chapter 24

Conrad and Hartline's Flexagon Theory

One of the earliest works on flexagons was a paper written by Anthony Conrad and Daniel Hartline in 1962 and published as a very long technical report by the Research Institute for Advanced Studies, a division of the Glenn Martin aerospace company. It covers much of the basic theory of flexagons – including how to make many types of flexagons, flexagon templates, and flexagon maps (diagrams) – and quite a lot about the combinatorics of flexagons. Their research is limited to isoflexagons and the pinch flex, however, the only type of flexagon and flex that were considered valid at that time.

Slitting and doubling

Conrad and Hartline began their exploration of flexagons with the tri-hexaflexagon. The basic idea with which they create new templates for flexagons is by adding leaves using a concept they call **slitting**.

Start with a regular hexagon divided into six regular triangles, or leaves. Now, imagine a leaf is thick enough that you can cut a slit through its thickness from one internal edge to the other internal edge, leaving the third, outer edge intact, as seen in figure 24.1. The pat now has two leaves, creating a thumbhole. Do this for every second triangle around the hexagon and you get a tri-hexaflexagon.

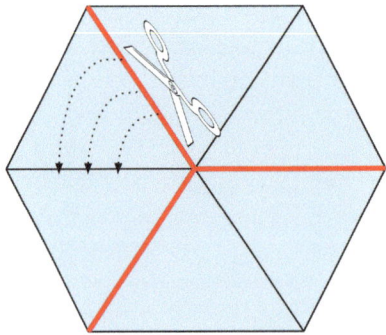

FIG. 24.1 Slitting a hexagon to turn it into a tri-hexaflexagon. The scissors show where to start one slit, with the dotted arrows indicating the direction of the slit. The bold red lines indicate where to start each of the three slits, continuing each slit counterclockwise from there.

If you instead slit all six of the leaves, you get a tetra-hexaflexagon. Of course, instead of slitting, which is physically impossible (or at least very impractical), you can just double the single leaves that you want to slit connected across the correct edges. Once you have the basic hexaflexagon

DOI: 10.1201/9781003433538-27

templates, you can create more by doubling the strip or by taking longer strips and shortening them by pasting unwanted faces together.

Slitting one leaf is equivalent to using a generating sequence of Ur+ to add a new leaf using the techniques from chapter 20. Doubling and pasting are the same techniques we used in chapter 3, where we first discussed making hexaflexagons with different numbers of faces.

Conrad and Hartline's template notation

Conrad and Hartline devised a flexagon notation that describes the structure of a template and how to label it. Here we briefly summarize their method when applied to flexagons whose leaf shapes are triangles.

Their starting point for describing the structure of a flexagon template is what's known as **Tukey triangles** (after John Tukey from the original Flexagon Committee). In figure 24.2, you start with a triangle whose edges are marked with arrows in a cycle, entering the triangle from below. If you leave it in the same direction as the arrow you just crossed (in this case, going to the right), you indicate this using a +. If you leave the triangle in the opposite direction, you indicate this with a −.

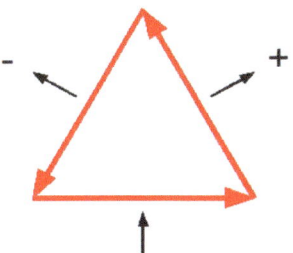

FIG. 24.2 A Tukey triangle.

Consider a straight-strip tri-hexaflexagon template as shown in figure 24.3, made from a series of Tukey triangles. Enter the first leaf through the left-hand edge, go across it and through the hinged edge into the next leaf. Keep repeating until you exit the template. Since each time we leave a triangle, we go through the edge that points to the base of the arrow of the edge through which we entered, the triangle is labeled with a minus sign. Using this notation, all the triangles in the tri-hexaflexagon are labeled with a minus sign giving us − − − − − − − − −. The entry and exit points of the whole template are the ends that are taped together to form the flexagon. Note that if we were to instead enter the tri-hexaflexagon template from the right-hand side, all the triangles would be pluses. Figure 24.4 shows the Tukey triangle template for the tetra-hexaflexagon, which gives us + + − − + + − − + + − −.

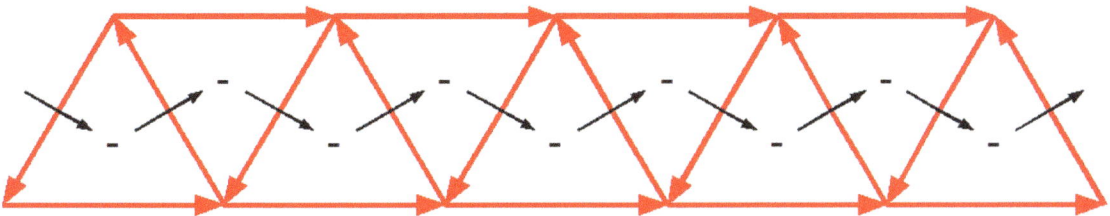

FIG. 24.3 A Tukey triangle tri-hexaflexagon template.

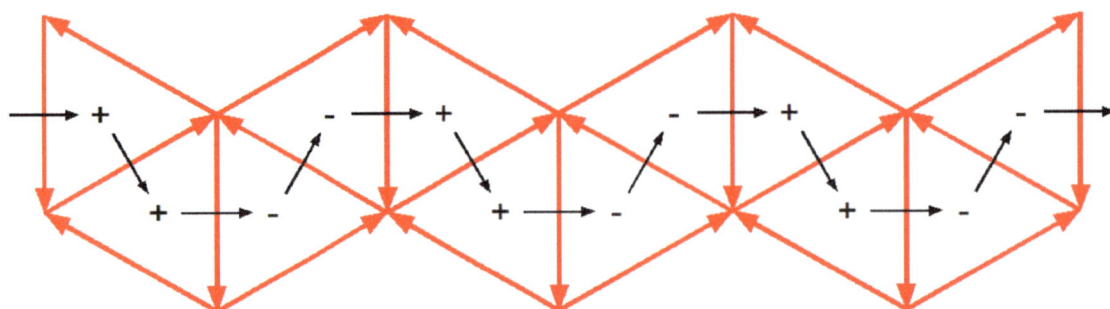

FIG. 24.4 A Tukey triangle tetra-hexaflexagon template.

There are some interesting features we can learn from observing these two templates:

- Whenever there is a deviation from a straight strip of triangles somewhere along the template, there is a change of sign.
- Entering the template from the opposite side swaps the pluses and minuses.
- Every sector in each template has the same pattern, so the notation for the whole template is just the notation of a single sector of a hexaflexagon repeated three times, since it has three sectors. For the tri-hexaflexagon the pattern on each sector is − − − and for the tetra-hexaflexagon it is + + − −.
- You can slit every other triangle on a tri-hexaflexagon template to turn it into a tetra-hexaflexagon template. We represent this in the notation by substituting every second minus triangle with two plus triangles.

It turns out that these observations apply more generally. This gives us notation for describing the structure of a template and how to derive new flexagons from it. So when we write − − − − − − − − − (or + + + + + + + + +), we immediately know it's the straight-strip tri-hexaflexagon template, and when we write + + − − + + − − + + − − (or − − + + − − + + − − + +), we know it's the wavy tetra-hexaflexagon template. In fact, as long as we know the number of sectors in a flexagon, we only need to write the notation for one of them. Thus we write − − − for the tri-hexaflexagon, + + − − for the tetra-hexaflexagon, and + − − + + for the penta-hexaflexagon. The 6a, 6b, and 6c hexa-hexaflexagons variations are + + + + + +, + + − − − −, and + − + − − −, respectively. Remember that because Conrad and Hartline only considered the pinch flex, all the sectors are the same.

An important thing to note is that we can cut any flexagon along any of its hinges to get its template. Another way of putting this is that a flexagon template is cyclic. We can remove any number of triangles on one end and join them to the other end. So there can be a few templates for the same flexagon that differ only in where we tape the ends together; but for all purposes, they are the same template. This is not to be confused with completely different templates that occur for higher-order flexagons (like the three different templates for the hexa-hexaflexagon). This is reflected in the Conrad and Hartline notation by the fact that all cyclic permutations of the notation refer to the same template. For example, + + − −, + − − +, and − − + + all refer to the one tetra-hexaflexagon template, depending on where it starts and ends.

Another important thing to note is that, while Conrad and Hartline's + and − notation captures the directions between leaves, it is not the same as the symbols \ and / that we used in the atomic flex theory. For example, the Conrad and Hartline notation for a hexagon is + − + − + −, whereas in atomic flex theory it is //////.

Templates from maps

The plus-minus labeling we just saw tells us what a flexagon template looks like, but it doesn't tell us how to label or color the faces. Instead of starting from templates and then coloring them, Conrad and Hartline suggest a scheme to color any of the basic hexaflexagons from a given flexagon pinch flex diagram, which they call a **_flexagon map_**. We'll briefly show a simplified version of their method applied to the hexa-hexaflexagon map shown in figure 24.5, which is the map for variation 6b from figure 3.5. The map is a different way of showing the information from a pinch state diagram, in this case the one in figure 17.6.

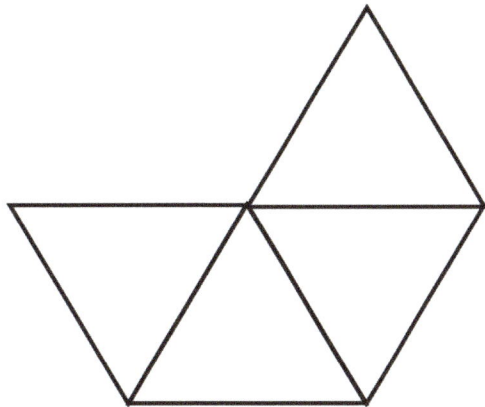

FIG. 24.5 One of the three hexa-hexaflexagon pinch flex state diagrams (maps).

First, draw a network of straight lines that connect the midsections of the edges of the triangles in the map by continuing a straight line as long as possible and breaking it wherever stuck until returning to the point where you started. This network of lines is known as the Tukey triangle network (the dotted lines in Fig. 24.6).

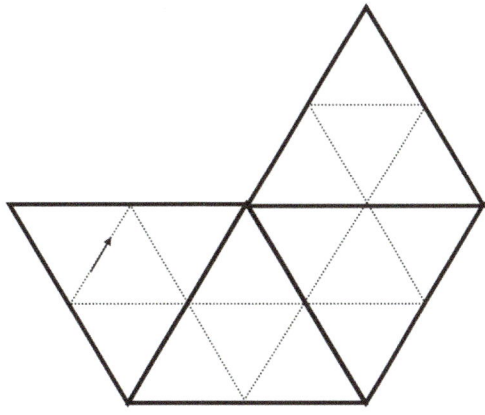

FIG. 24.6 The Tukey triangle network on the hexa-hexaflexagon map.

Number the vertices (corners and intersections) clockwise from 1 to *n*, where *n* is the number of vertices or states. For our hexa-hexaflexagon, this is 6. Then number the edges between two adjacent vertices with the lower vertex number, so the edge between 1 and 2 is labeled 1, the edge between 2 and 3 is labeled 2, and so on. When choosing between the highest and lowest number, in this case 6 and 1, pick the highest number, as shown in figure 24.7.

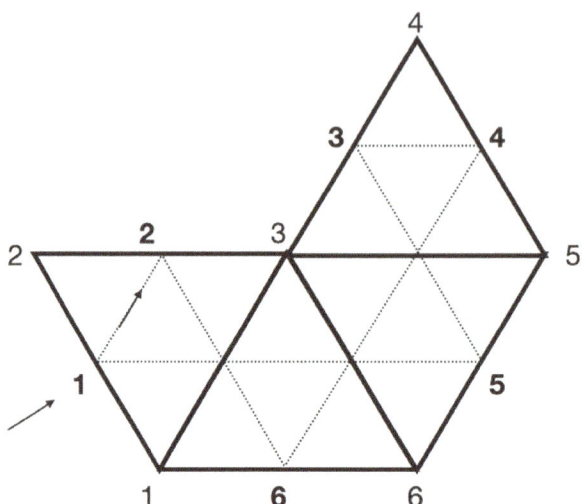

FIG. 24.7 Labeling the vertices and the edges of the hexa-hexaflexagon map. The vertices are labeled with regular-bold numbers, while the edges are bold.

Travel along the Tukey triangle network in the direction of the arrow starting from the number 1, and write down the numbers you find, alternating whether you put the number in the top row or bottom row. Using our example, we write the number 1 in the top row, the number 2 in the bottom row, 6 in the top row, 4 in the bottom row, and so on. When we return to the first number, add a vertical line and continue for a bit (we'll see why later on). Here is what the number rows look like for our example:

```
1       6       3       |  1       6    ...
    2       4       5    |      2        ...
```

Now write a plus sign (+) at the Tukey triangle network vertex number 1. Travel the Tukey triangle network, counting the number of (unlabeled) vertices that you cross until you get to the next vertex. If the number of vertices is even, write the same sign you just used at the new vertex, otherwise write its opposite sign, swapping + and −. In our example, we cross no vertices between vertex number 1 and vertex number 2. Since 0 is even, we write + next to vertex number 2. Between vertex 2 and vertex 6, we cross one vertex; 1 is an odd number, so we label vertex 6 with a minus sign (−). Continue until the whole network is traversed, as shown in figure 24.8.

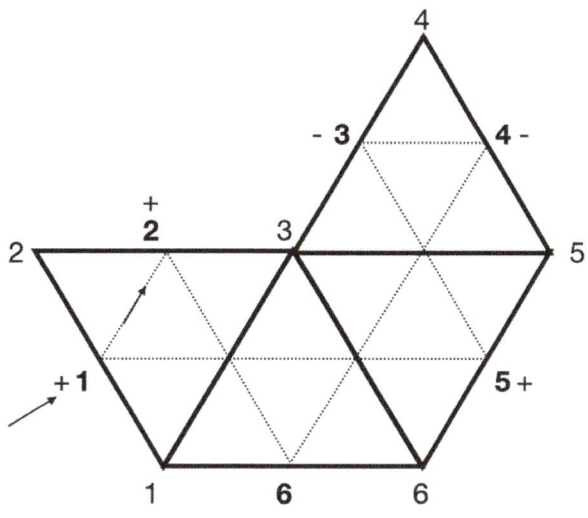

FIG. 24.8 Using + and − to label the vertices of the hexa-hexaflexagon map.

We write the +'s and −'s accordingly under the number rows, as follows:

```
1       6       3       |   1       6    ...
    2       4       5   |       2        ...
+   +   −   −   −   +   |   +   +   −    ...
```

Next add 1 to each number and write the sum, or 1 if the sum is $n + 1$, in the empty space above or below the number. This gives us the following:

```
1   3   6   5   3   6   |   1   3   6   ...
2   2   1   4   4   5   |   2   2   1   ...
+   +   −   −   −   +   |   +   +   −   ...
```

Now you have all you need to create the template. Note that the full template needs three sectors, so we repeat the pattern three times. Therefore, here is full number row series for our example:

```
1   3   6   5   3   6   |   1   3   6   5   3   6   |   1   3   6   5   3   6
2   2   1   4   4   5   |   2   2   1   4   4   5   |   2   2   1   4   4   5
+   +   −   −   −   +   |   +   +   −   −   −   +   |   +   +   −   −   −   +
```

The structure of each sector is given by the last row and the full template by the whole series. The numbers that label the top faces of the leaves in the template are in the top row with the labels for the other side in the corresponding position in the second row. Figure 24.9 shows the individual sectors and how they fit together to make the full template, and the full, labeled template appears in figure 24.10. Large labels indicate the numbers to be written on the top side of the leaf. Small labels indicate the numbers to be written on the bottom side of the leaf.

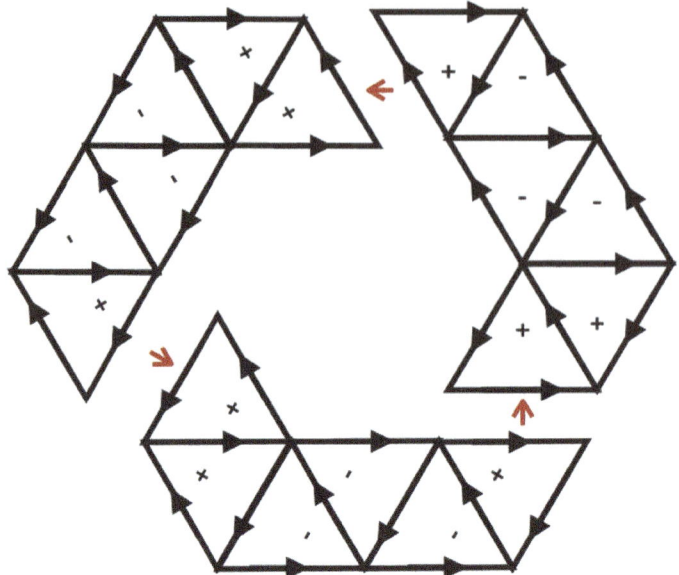

FIG. 24.9 Fitting together the sectors of the hexa-hexaflexagon map.

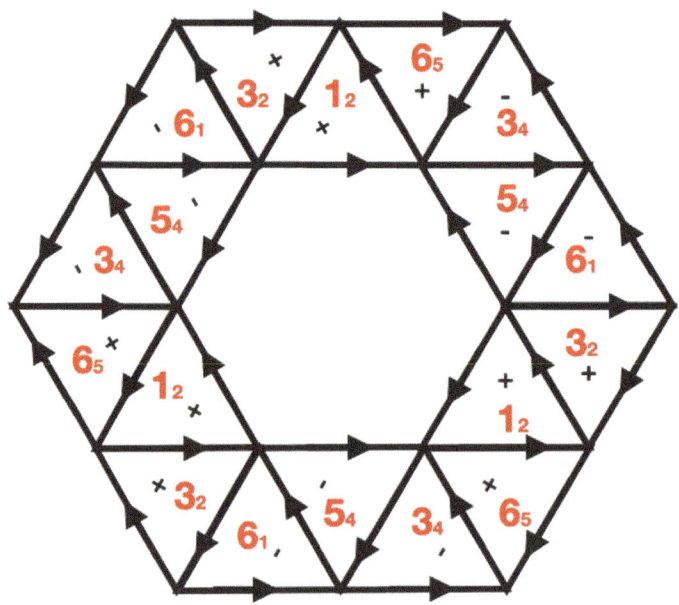

FIG. 24.10 The full, labeled hexa-hexaflexagon map. Small labels indicate the numbers written on the bottom side of the leaf.

There are a few notes we ought to mention about Conrad and Hartline's method. First, the reasoning behind this plus-minus labeling has to do with the directions in which arrows are arranged within an infinite Tukey triangle grid. However, going into the exact details of this is beyond the scope of this book. Second, in an earlier publication, Conrad suggested a slightly different scheme using R and L (right and left) symbols instead of pluses and minuses (Conrad, 1960). Again, those who wish to delve more deeply into these methods should refer to the actual publications.

Third, the connection between the numbering scheme on the maps and why it translates over to a numbering scheme on the template – where each number denotes a pinch face – might not be obvious at first glance. Here's a quick explanation. We begin by interpreting the map in a slightly different way than before. Instead of each vertex in the map denoting one pinch state (for example: a leaf with 1 on the top and 2 and the bottom) and its neighbor denoting the next pinch state where the bottom leaf becomes the new top leaf and a new face is exposed as the bottom leaf (in our example, the next vertex would reflect the state 2 on top and 3 on bottom), the vertices are interpreted to represent only the top leaf of each state. In this case, the line joining two adjacent vertices can be interpreted as the state itself. For example, the path between vertices 1 and 2 represents the state with face 1 on top and face 2 on bottom. Flexing between two states is represented by connecting the midpoints of adjacent lines.

The complete network of lines joining the midpoints of the paths in the map reflects the Tuckerman Traverse between pinch face states. Wherever the line is straight, a pinch flex is performed at the same hinge as the last pinch flex. When the line changes direction at the edges of the map, the flexagon must be rotated one hinge to continue flexing to a new face. The number of points where the line turns is the same as the number of lines and equal to the number of pinch faces of the flexagon. Therefore, sequentially numbering the turning points alternates between numbering the tops and bottoms of the leaf-faces in a sector. Readers should refer to the original paper for further details (Conrad and Hartline, 1962).

A final point we want to make concerns pat notation. Conrad and Hartline employed a pat notation originally devised by Oakley and Wisner (1957). They noticed that there can be at most only one complete thumbhole in any given pat, where the thumbhole is defined as a pocket that you can put your thumb in one side and it will come out the other. Another way of putting this is that all the leaves on one side of the thumbhole are only connected to each other, except for the hinge that connects the two sets of leaves. The only pat that doesn't have a thumbhole is a single leaf pat. The position of the thumbhole is denoted in their notation by a comma. By convention, the portion of a template to be folded into a pat is labeled consecutively with the numbers 1, 2, 3, … and then folded into a pat, with the numbers reading from top to bottom. For example, 324,1 represents a pat with four leaves where the three top leaves are 3 over 2 over 4, then a thumbhole, and finally a leaf labeled 1 at the bottom of the pat. By convention, the leaves below the thumbhole have lower numbers, while those above have higher numbers.

Conrad and Hartline's pat notation requires more effort to interpret. There is no distinction between the two sides of the leaf and it is difficult to infer where the hinges are. Compare their notation from the previous example – 324,1 – to ours – [[[3,–2],–4],1]. Ours makes it clear where the leaves are hinged and when a leaf-face is face up or face down.

Summary

Conrad and Hartline's approach to creating new flexagon templates was to make a diagram similar to the pinch state diagram and derive the template from that. They used their methods to make important combinatorial calculations on the number of hexaflexagon templates that can be created. They also expanded their methods to other kinds of flexagons. All this can be found in detail in their seminal paper.

Chapter 25

Les Pook's Flexagon Theory

Les Pook wrote two classic books on flexagons. His second book, *Serious Fun with Flexagons* (Pook, 2009), is a comprehensive catalog of many kinds of flexagons, their templates, and some of their interesting properties.

Flexagons as polygon rings

Les Pook's approach is systematic, focusing on the relationships between a flexagon's pats. Very broadly speaking, his definition of a flexagon (in its main state) is a polygon ring where each polygon is a stack of hinged leaves and the polygons are hinged either along their edges – edge flexagons – or at their vertices – point flexagons. The use of the word *ring* may seem a bit misleading, since when you fold up a tri-hexaflexagon for example, it doesn't have a "hole" in the middle. But mathematically it is considered a ring since the pats are connected around a ring centered in the middle of the flexagon. We've also seen flexagons that do have a hole in the middle, such as the pentagon hexaflexagon in figure 7.14.

One of the ways Pook classifies flexagons is by examining the angles between pats, using the following definitions:

- The **hinge angle** is the angle between two hinges, as shown in figure 25.1.
- In a **slant ring**, the sum of the center hinge angles is less than 360°, what this book calls a cup.
- In a **skew ring**, the sum of the center hinge angles is greater than 360°, what this book calls a mountain-valley state.

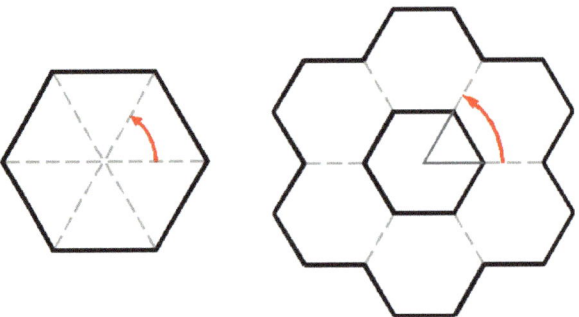

FIG. 25.1 The hinge angle on a regular triangle hexaflexagon and a regular hexagon hexaflexagon.

DOI: 10.1201/9781003433538-28

There are different kinds of polygon rings depending on the number and shape of the polygons that make them. Next, we introduce some of his classifications.

Regular even edge polygon rings

Regular even edge polygon rings consist of an even number of identical, convex, regular polygons hinged at their edges, with every polygon at an equal distance from the center of the ring and with all hinge angles the same. (Recall that a regular polygon is a polygon whose angles are all the same and whose sides are all equal – for example, an equilateral triangle, a square, or a regular pentagon.) See figure 25.2 for examples of regular even edge polygon rings. Note that the word *regular* in "regular even edge polygon rings" refers to the overall shape of the ring, not the individual polygons that make it (although they too are regular polygons). All sectors consist of two pats and are completely identical. For example, a hexaflexagon has three identical sectors while a tetraflexagon has two identical sectors.

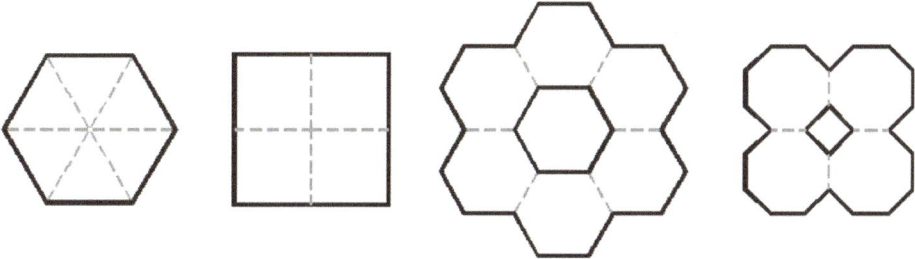

FIG. 25.2 Examples of regular even edge polygon rings.

Regular odd edge polygon rings

Regular odd edge polygon rings are the same as regular even edge polygon rings, except that there are an odd number of polygons in the ring. Figure 25.3 shows examples of regular odd edge polygon rings. Every pat is its own sector, and all sectors are identical.

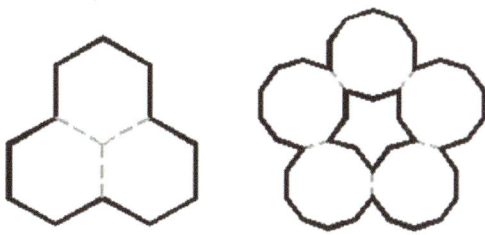

FIG. 25.3 Examples of regular odd edge polygon rings.

Note that, in this book, flexagons that are regular even or odd edge polygon rings are referred to as isoflexagons, and the sectors don't have to be identical. But since Pook focuses on the pinch flex, all sectors are identical.

Compound polygon rings

Compound polygon rings consist of an even number of identical, convex, regular polygons hinged at their edges, where alternate polygons are an equal distance from the center of the ring and alternate hinge angles are the same. See figure 25.4 for examples of compound polygon rings.

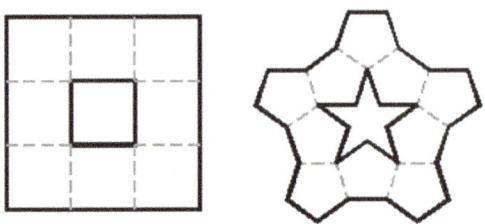

FIG. 25.4 Examples of compound polygon rings.

Irregular polygon rings:

Rings that are neither regular nor compound are **irregular polygon rings**. The flexagons in figure 25.5 are irregular rings because not all the polygons are at an equal distance from the center of the ring, nor do they alternate.

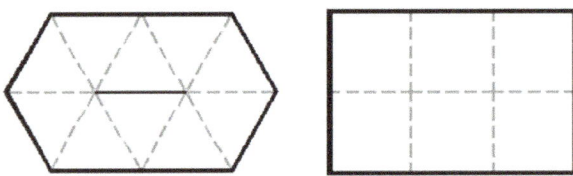

FIG. 25.5 Examples of irregular polygon rings.

Note that, in this book, flexagons that are compound or irregular polygon rings in their main state are referred to as non-isoflexagons.

Rings of irregular polygons

Not to be confused with irregular polygon rings (referring to the overall shape of the flexagon), **rings of irregular polygons** are rings made of irregular polygons, such as non-equilateral triangles or rectangles. Figure 25.6 contains several examples.

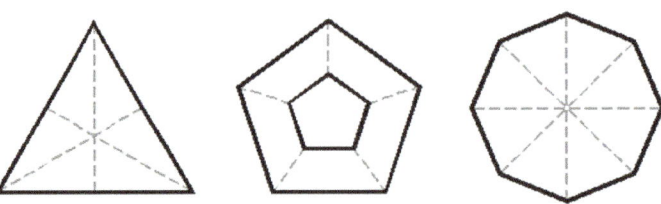

FIG. 25.6 Examples of rings made of irregular polygons.

Vertex polygon rings

Rings of polygons hinged at their vertices instead of their edges are known as **vertex polygon rings**. There are regular, compound, irregular polygon, and irregular vertex rings. Since it's practically impossible to create a model with joined vertices that don't tear, we add width to each vertex at the joint. Flexagons that are vertex polygon rings are usually called point flexagons, which is the term we use in this book. We showed a template for a point flexagon in figure 21.5.

Fundamental flexagon nets

All flexagons are created from templates, which Pook calls **nets**. Flexagons can be constructed by taping together multiple copies of the net of a sector. He calls the basic building blocks **fundamental nets** and the resulting flexagons **fundamental flexagons**, where the flexagon is the archetype flexagon of a family with similar characteristics. The other members of the family are called **complex flexagons**. To be more precise, the defining characteristic of whether a flexagon is fundamental or complex is the leaf structure within pats. Fundamental flexagons always have pats that are either single leaves or fan-folded leaves, while complex flexagons have pats in some main positions that include reverse-folded leaves. See figures 7.3, 7.9, 7.10, and 7.12 for examples of fan-folded leaves. Figures 8.1 and 8.3 are examples of reverse-folded leaves.

For example, the tri-hexaflexagon is a fundamental flexagon constructed from a fundamental net, but the hexa-hexaflexagon is not. The tri-hexaflexagon is the archetype flexagon of the hexaflexagon family and only has single leaf or fan-folded leaf pats. The hexa-hexaflexagon is a complex flexagon and in some main positions has pats where the leaves are reverse-folded. Complex flexagons are really just a combination of fundamental flexagons. They can be made in at least two ways: either by taping together copies of fundamental flexagon nets or by linking fundamental flexagons. The hexa-hexaflexagon, for example, can be made by linking tri-hexaflexagons together. **Linking** is a procedure used to combine two flexagons to get another one. This is done by removing alternate pats in the main position of two flexagons and then superimposing the flexagons and taping the pats together. To make an analogy with numbers, it might be useful to think of fundamental flexagons as prime flexagons, complex flexagons as composite flexagons, and linking as multiplication.

Pook defines a few families of fundamental nets, where each family differs by a few parameters. These nets can broadly be divided into first-order fundamental nets, second-order fundamental nets, and vertex nets (for point flexagons). We'll restrict our discussion to first- and second-order fundamental nets.

First-order fundamental nets and flexagons

First-order fundamental nets are used to build flexagons that in their main position look like regular or irregular edge polygon rings with an even number of polygons or compound edge rings, as we saw in figures 25.2 and 25.4.

We focus on the "mainstream" flexagons constructed with these nets, those Pook calls first-order fundamental flexagons. First-order fundamental flexagons and their nets have the following properties:

- The hinge angles are always the same, but the direction in which they're connected alternates, so the net looks like either a straight or a zigzag strip.
- They have at least two identical sectors made from regular convex polygons.
- Each sector has exactly two pats.

- The number of sectors is half the number of polygons in the flexagon's main position.
- In the principal main position, the number of leaves in each sector is equal to the number of edges of each leaf. For example, a flexagon made of pentagons has five leaves in each sector.

Numbering the nets is important. First-order fundamental nets are numbered in the following way. The leaves in the net are numbered on top and bottom. Each sector has the same leaf numbering, although there is a difference between nets that are made of sectors with an even number of leaves and those made of an odd number, in that the top/bottom numbering is reversed on alternate sectors in odd nets. The numbers never exceed the number of leaves in the sector.

- For nets made of sectors with an even number of leaves:
    - *Top leaves:* 1, 3, 3, 5, 5, ..., repeating the sequence for each sector consecutively.
    - *Bottom leaves:* 2, 2, 4, 4, 6, 6, ..., repeating the sequence for each sector consecutively.
- For nets made of sectors with an odd number of leaves:
    - *Top leaves:* First sector: 1, 3, 5, 5, ...; second sector: 2, 2, 4, 4, 6, 6, ...; repeat the sequences for the remaining pairs of sectors.
    - *Bottom leaves:* First sector: 2, 2, 4, 4, 6, 6, ...; second sector: 1, 3, 3, 5, 5, ...; repeat the sequences for the remaining pairs of sectors.

The three-sector first-order fundamental triangle net in figure 25.7 is an example of a fundamental net with an odd number of sectors. The first sector is numbered 1/2, 3/2, 3/1. The second is 2/1, 2/3, 1/3, which is just the same numbering as the first sector top/bottom reversed. The third is numbered 1/2, 3/2, 3/1.

The two-sector first-order fundamental square net in figure 25.7 is an example of a fundamental net with an even number of sectors. Using the top/bottom numbering style we have used in this book, each sector is numbered 1/2, 3/2, 3/4, and 1/4.

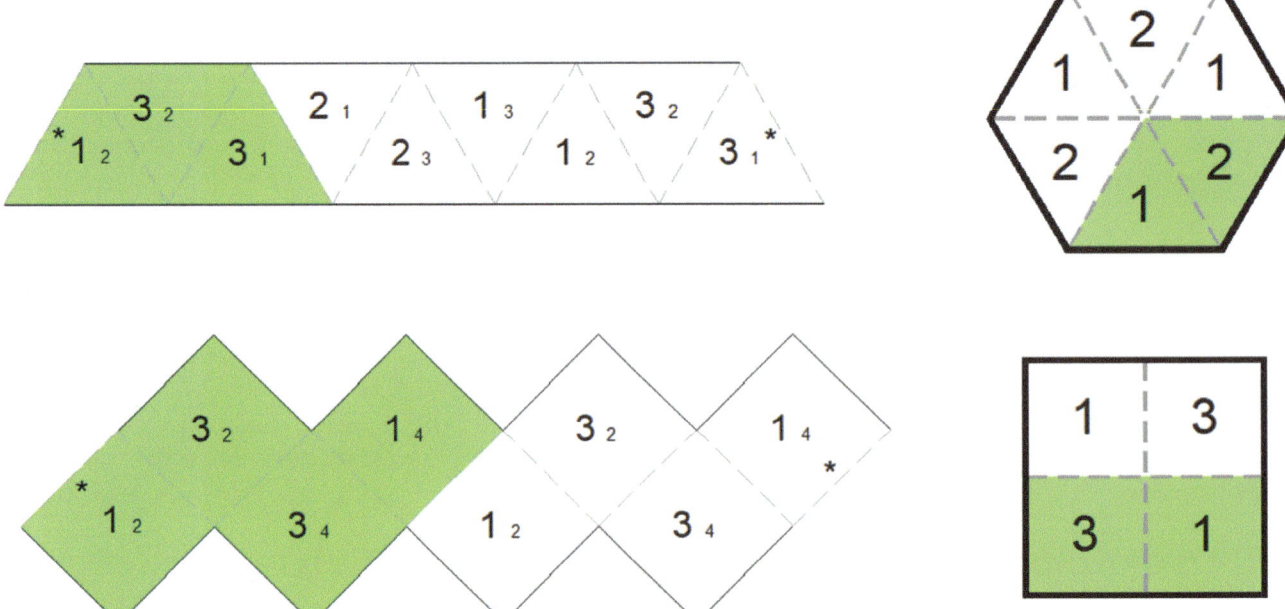

FIG. 25.7 Examples of first-order fundamental nets on the left, with the resulting flexagon on the right. The numbers on the flexagon indicate the number of leaves in the pat. Green represents a single sector.

To assemble a net into its corresponding fundamental flexagon, fold identically numbered leaves together starting from the highest number and working down until the numbers 1 and 2 are reached. When assembled, the topology of the flexagon is a simple, twisted band.

Translating first-order fundamental flexagons into the concepts in this book, you can think of them as flexagons created from the generating sequence (P>)n, where n is the number of edges in the leaves and P is the generalized pinch flex (sometimes called the book flex). For example, use the generating sequence (P>)3 for triangular leaves and (P>)4 for square leaves.

Second-order fundamental nets and flexagons

Second-order fundamental nets are used to build flexagons that in their main position look like regular edge polygon rings with an odd number of polygons, such as the ones shown in figure 25.3. The nets and their corresponding flexagons have the following properties:

- The hinge angles are always the same. However, in these nets pairs of hinge angles alternate in sign giving the net a more curvy or wavy look.
- They have at least three identical sectors made from regular convex polygons.
- Each sector makes up exactly one pat in the flexagon main position.
- In the principal main position, the number of leaves in each pat is exactly two.
- The number of sectors is the same as the number of polygons in the flexagon's main position.

Numbering of second-order fundamental nets is different from first-order nets, because each sector is made of exactly two leaves, each with a top and bottom face, so only four numbers are needed.

1. The top faces of the leaves in the first sector of the net (the first two leaves) are sequentially numbered 1 and 2, and their bottom faces are numbered 3 and 4, respectively.
2. The top faces of the leaves in the second sector are sequentially numbered 3 and 4, and their bottom faces are numbered 1 and 2, respectively.
3. And so on, alternating 1 and 2 and then 3 and 4, for all pairs of sectors.

To assemble a net into its corresponding fundamental flexagon, fold each leaf numbered 3 to its neighboring leaf numbered 4 from the same sector. When assembled, the topology of the flexagon is a simple, twisted band. See figure 25.8 for an example.

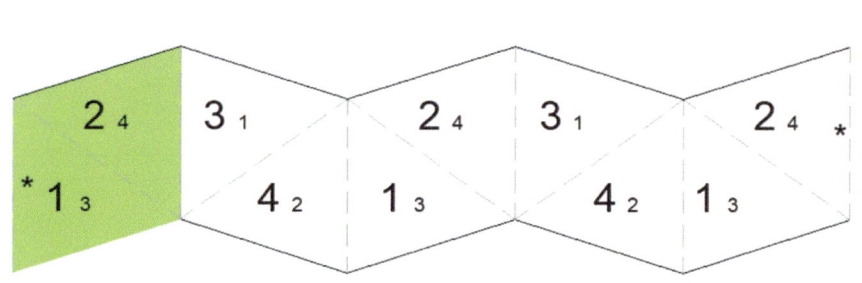 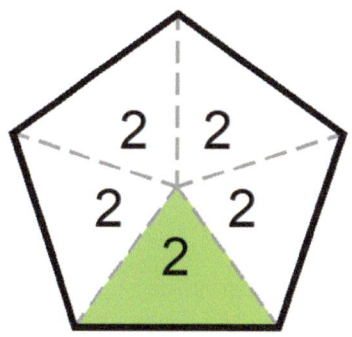

FIG. 25.8 Example of a second-order fundamental net on the left, with the resulting flexagon on the right. The numbers on the flexagon indicate the number of leaves in the pat. Green represents a single sector.

For those interested, there are many more properties of fundamental flexagons in Les Pook's books where most of the flexagons that can be made from these nets are tabulated and discussed.

Now that we know about fundamental flexagon nets, we are in an excellent position to label flexagons using Les Pook's flexagon symbols. These symbols are useful because they identify flexagons at a glance and contain important information about their structure and dynamics. We need one more ingredient, though, before we get to the flexagon symbol – the Schläfli symbol.

The Schläfli symbol

Ludwig Schläfli was a nineteenth-century Swiss mathematician who specialized in geometry, especially known for his influential work on geometry in high dimensions. Schläfli developed what is known as the **Schläfli symbol** for polygons, polyhedra, and polytopes in higher dimensions.

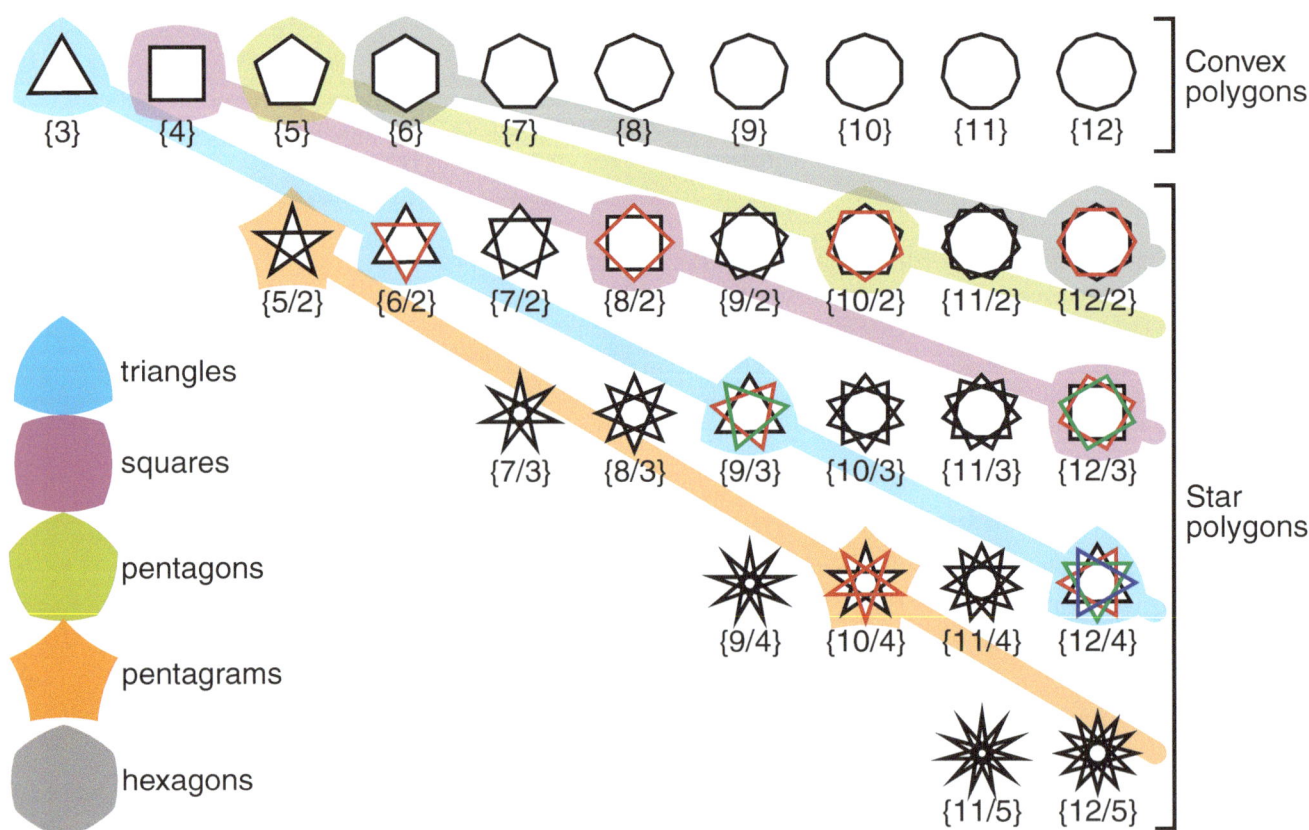

FIG 25.9 Regular convex and star polygons from 3 to 12 vertices labeled with their Schläfli symbols. (Image credit: CMG Lee – Own work, CC BY-SA 3.0.)

You can see the Schläfli symbols for some regular polygons in figure 25.9. The Schläfli symbol for a regular polygon is just the number of vertices (or edges) written inside curly brackets: {3} for a regular triangle, {4} for a square, {5} for a regular pentagon, {6} for a regular hexagon, etc.

The Schläfli symbol for a star polygon (a nonconvex regular polygon) is a fraction written inside curly brackets. The numerator is the number of vertices and the denominator is the number of vertices skipped when drawing each edge of the star, plus 1. For example, when drawing a line from point to point on the pentagram (five-pointed star) we skip over one point of the star, so its Schläfli symbol is {5/2} (remember to add 1 to the denominator). The Schläfli symbol for a nonagram (nine-pointed star) is {9/3} because two points are skipped.

Flexagon symbols

Pook created the **flexagon symbol** to characterize fundamental flexagons. It consists of three components: the number of sectors, the Schläfli symbol of the polygon that makes up the leaves of the flexagon, and what's called the flexagon net symbol.

For example, a hexaflexagon has three sectors, so the first component is 3. And the leaves are triangles with Schläfli symbol {3}, so the second component is also 3.

The third component is the **flexagon net symbol**, which is the Schläfli symbol of a polygon that describes the dynamics of the flexagon, corresponding to the flexagon's state diagram describing flexes between states, typically pinch flexes. This polygon is known as the **associated polygon**, not to be confused with the polygon that is the shape of the individual leaves. Here is how to find all the possible associated polygons of a given leaf:

1. Draw a leaf – a triangle for flexagons made from triangles, a square for flexagons made from squares, a pentagon for flexagons made from pentagons, etc.
2. Choose any edge as a starting point, and draw a straight line between its center and the center of the edge adjacent to it.
3. Do this until you return to the edge where you started
4. The inscribed polygon that is formed is the same shape as the leaf, and is called the trivial associated polygon.
5. For pentagons, heptagons, and higher, there is at least one other associated polygon. To find it, repeat the process, but this time, draw the lines between every other edge. If all edges are connected, this inscribed polygon is another associated polygon (it will be a star polygon).
6. If possible, repeat the process, but this time, draw the lines between every third edge. If all edges are connected, this is another associated polygon.
7. Continue until no further skips are available.

In figure 25.9, the polygons with only black edges are all the possible associated polygons corresponding to the regular polygons up to 12. Those that have two or more colors (e.g., {6/2}, {8/2}, etc.) are not associated polygons.

You can create a flexagon net once you know the associated polygon. To create a net for a first-order fundamental flexagon, you do the following:

1. Choose the number of sectors you want. Edge flexagons must have at least two sectors, while point flexagons can have one. The number of leaves in each sector must equal the number of edges of the leaf.
2. For each sector draw the leaves in a zigzag strip, connecting polygons so that the shortest distance between every two hinges is equal to the denominator of the Schläfli symbol for the associate polygon minus 1 – i.e., a (shortest) distance of no edges for a whole number Schläfli symbol, for example one edge for those whose Schläfli symbol denominator is 2 etc. – and making sure the hinge angles are the same and the sign of the hinge angle alternates. In figure 25.10, you can see two examples of such nets.

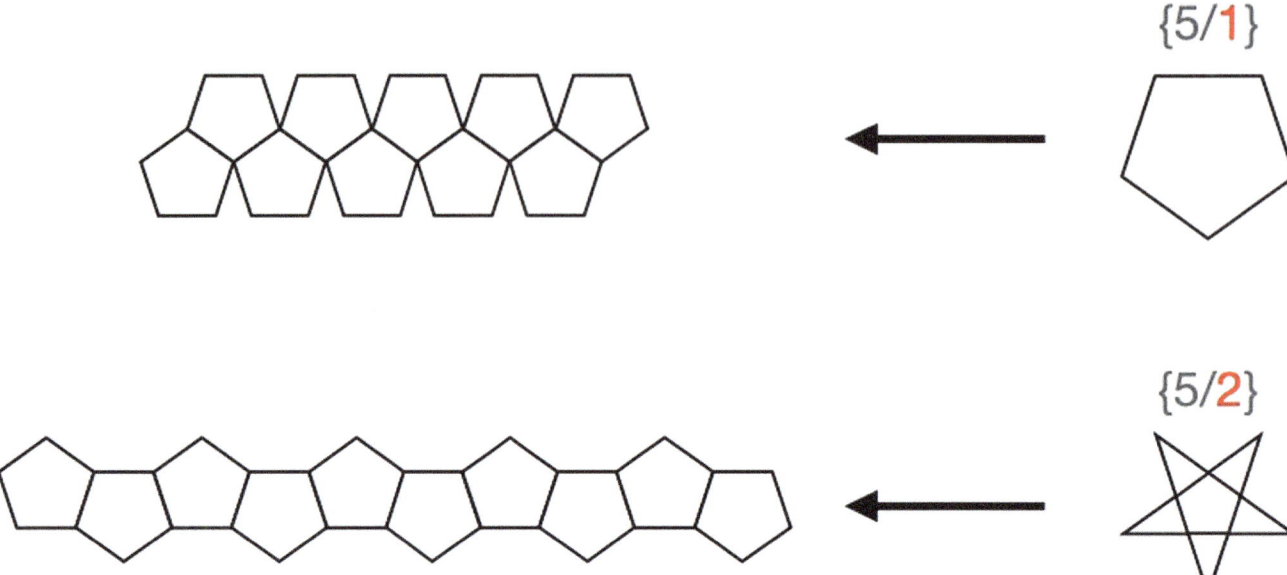

FIG. 25.10 Examples of associated polygons and the resulting nets.

For second-order fundamental nets the same technique is used, except that this time the net will be wavy because you need to make sure that the signs on pairs of hinge angles alternate.

Interestingly, if you inscribe the associated polygon within its corresponding leaf, you get a flexagon pinch diagram or map that describes the main pinch flex cycles between states, with the following properties:

- The edges of the circumscribing polygon are the main states of the flexagon.
- The vertices of the circumscribing polygon are the intermediate states of the flexagon (e.g., the propeller-looking state when you pinch flex a tri-hexaflexagon).
- The edges of the inscribed polygon – the associated polygon – show the flex connections between the states.

Now that we know the language, we can put the pieces together and construct the flexagon symbol. It looks like this (the Schläfli curly brackets are omitted):

component 1<component 2, component 3>

More specifically,

number of sectors<Schläfli symbol of leaf, Schläfli symbol of associated polygon>

For example, the flexagon symbol for the first-order fundamental tri-hexaflexagon is 3<3,3>. The square tetra-tetraflexagon's symbol is 2<4,4>.

Sector symbols

Pook also created a notation for the sector of a flexagon. The notation starts off the same way as the flexagon symbol, except that the first component – the number of sectors – is omitted. Two additional numbers are added to the component 2 and component 3 numbers. These are the number of leaves in each pat of the sector from the highest to the lowest. Note that for some flexagons the sector symbol is dependent on the state of the flexagons because the sectors in different states can have a different number of leaves in each pat. The sector symbol for the tri-hexaflexagon is <3,3,2,1>.

Summary

Les Pook's theory starts with first- and second-order fundamental nets as the basic building blocks for assembling more complex flexagons. He defines flexagon symbols and sector symbols as short-hand for classifying flexagons. This provides useful tools for expanding the universe of flexagons we can explore.

Chapter 26

Topology

So far, our approach to flexagons has been to study the different components that are involved: the structure of the pats, the templates, and, of course, the flexes. In this chapter we take more of a bird's-eye view of flexagons: what is their structure when we abstract away the specifics? In mathematics, we call this **topology**.

Topology is a field in mathematics that characterizes objects based on their properties rather than their exact shapes. In particular, topologists tend to imagine objects being made of stretchy and bendy materials, and then look at properties that stay the same even as they're stretched, like how many holes they have or how many knots there are.

Imagine you have a lump of clay. You punch a hole in it and mold it into the shape of a donut. By stretching the clay you can mold it into a coffee cup without punching another hole in it and without destroying the existing hole. Mathematicians describe the coffee cup and the donut as being **topologically equivalent** because you can transform one into the other just by stretching and bending. If, however, you punch another hole into the lump of clay, you create a *different* set of topologically equivalent objects, those whose topology is like a donut with two holes. The frame of eyeglasses (without the glass) is topologically a donut with two holes. It's important to note that adding or removing holes to objects changes their topology, so they are no longer equivalent.

The same goes for knots. Take a long rectangular rubber sheet and join the ends together to make a simple loop. Then take a second long rubber sheet, tie a knot in it, and join the ends. These two loops are *not* topologically equivalent because you can't turn one into the other by bending or stretching. You have to open the knotted loop, untie the knot, and rejoin the ends to make it topologically equivalent to the simple loop. Or go the opposite direction: open the unknotted loop, tie a knot in it, and rejoin the ends to make it topologically equivalent to the loop with one knot.

Punching or filling in holes, tying or untying knots, slicing open or joining ends of string-like things are all actions that change the topology of objects.

Möbius strips

What's all this got to do with flexagons? By defining the topology of flexagons, we gain more insight into some of their interesting properties. One of the first things that comes to mind when studying the topology of flexagons is their possible connection to something called a **Möbius strip** (or Möbius band), which was discovered independently by the German mathematicians Johann Benedict Listing and August Ferdinand Möbius. We mentioned them a few times earlier in the book, but now we'll take a closer look.

Take a strip of paper and join its ends together to make a band. The band has two sides, an inside and an outside. Cutting along the strip's centerline creates two new bands, each half the width of the original, as shown in figure 26.1. To more easily cut the band, fold a small portion of it in half, and

DOI: 10.1201/9781003433538-29

cut the middle of the folded edge. Then insert the scissors into the created slit and continue cutting around the band.

FIG. 26.1 Cutting a two-sided band results in two new bands each half the width of the original.

Making a half-twist before joining the ends of a strip makes a Möbius strip, which is shown in figure 26.2. This band is not topologically equivalent to the two-sided band you made before. You cannot take the Möbius band and make it into a two-sided band without slicing it open and undoing the half twist. Nor can you add a half-twist to the two-sided band to make a Möbius band unless you first slice it open.

FIG. 26.2 The Möbius strip.

The Möbius strip has some very interesting and unusual properties. First and foremost, in contrast to the regular band, it only has one side and one edge. This might seem impossible but, if you think about it for a moment, you see that this is what the half-twist does. You can check this for yourself by taking a pen or a marker and drawing a line along the center of the strip without lifting the pen off the paper until it returns to the point where you began. When you look at the resulting line, you'll see that it appears along the whole strip. Since you didn't lift your pen, this means that the strip has just one side. If you instead draw a line along the outside of a regular untwisted strip, you'll see that only the outside of the strip has a line drawn on it, but the inside does not. This means that the regular strip has two sides. Similarly, you can see that the Möbius strip has just one edge if you follow the edge all the way around with your finger.

Another interesting property of the Möbius strip is revealed when you cut the strip along its centerline. As we previously saw, if you do this to the regular strip, you get two strips, each the same length as the initial strip and half of its width. But if you cut the Möbius strip along its centerline, you get one long strip, double the length of the original strip and half its width. The new strip has four half-twists in it, as shown in figure 26.3. (Since we're only interested in the overall topology, we won't consider the lengths and widths after this.)

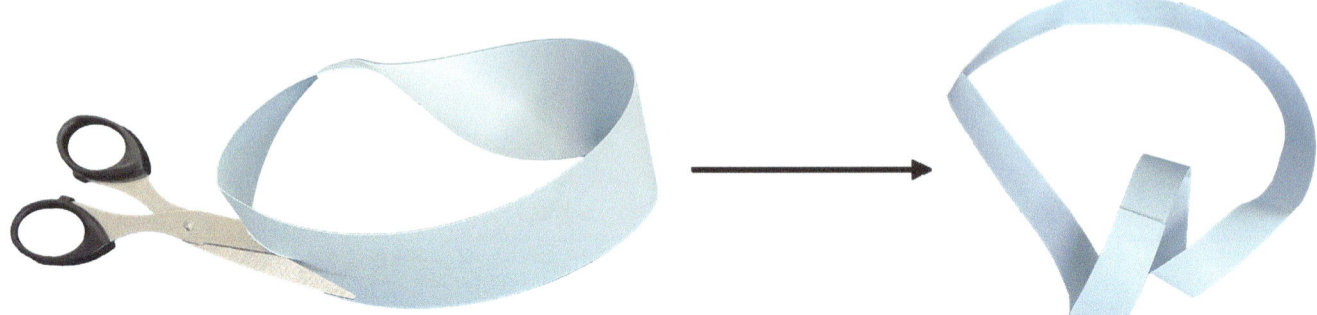

FIG. 26.3 Cutting the Möbius strip along its centerline results in a four-half-twisted strip.

There are many more things to explore with Möbius strips, like trisecting them instead of bisecting, or making multi-layered Möbius strips by laying a few strips of paper on top of each other, half-twisting them all together, and then joining their edges. However, this is beyond the scope of this chapter, so we leave it to you to experiment. But there's one more thing we need to see before returning to flexagons.

Half-twists

Once you get into Möbius strips, it's interesting to experiment with other numbers of half-twists. This raises many questions. What happens if you make two or three or four or more half-twists in a strip of paper and then join the ends? How many sides and edges do the resulting strips have? What happens if you bisect them along their centerlines? How many different ways can you fold a strip of paper that has been half-twisted any number of times? What is the topology of the resulting strips?

It turns out that strips with an odd number of half-twists have one side and one edge, and those with an even number of half-twists have two sides and two edges. You'll find in the literature several names that are used for multiple-half-twisted strips: Möbius strips (or bands), Möbius-like strips, paradromic rings, and others. For clarity, we'll just call them *n*-half-twisted strips, where *n* can be 0, 1, 2, etc. (Note that the term *Möbius strip* is reserved just for the one-half-twisted strip, and paradromic rings have other subtleties that are not needed in this book.)

It's important to note that even though we refer to them as *n*-half-twisted strips, they don't have the same topology when they have a different number of half-twists.

We saw that cutting a regular strip along its centerline creates two similar strips, while cutting a Möbius strip along its centerline creates a single, one-sided, four-half-twisted strip. Now try cutting the two- and three-half-twisted strips along their centerlines. Most people are surprised by the result. The two-half-twisted strip, when cut along its centerline, yields two two-half-twisted-strips that are linked together! When the three-half-twisted strip is cut along its centerline, you get an eight-half-twisted strip but with a knot in it! See figure 26.4.

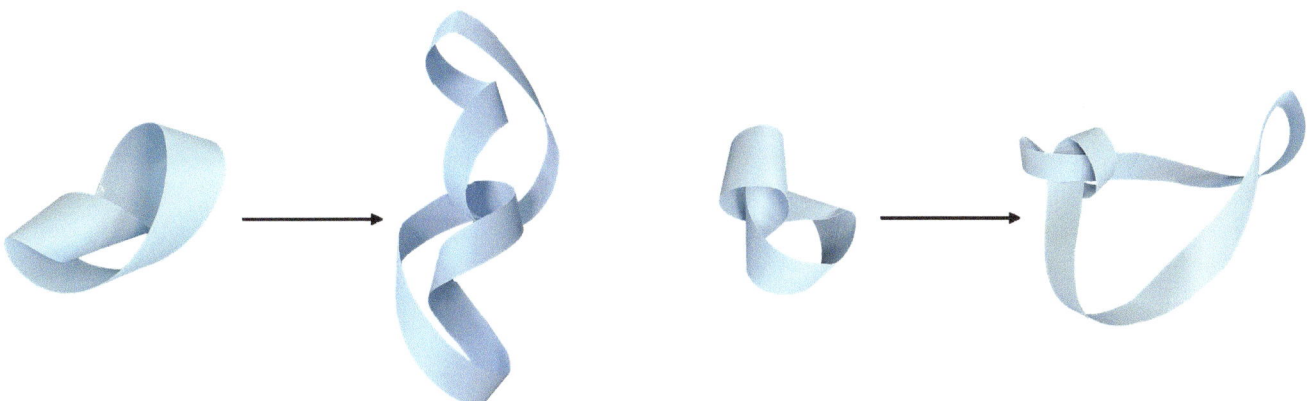

FIG. 26.4 Results of cutting the two-half-twisted strip (left) and the three-half-twisted strip (right) along their centerlines.

To generalize this, when you cut an *n*-half-twisted strip along its centerline, the following happens:

- If *n* is odd, you get a (2*n* + 2)-half-twisted strip with a knot in it.
- If *n* is even, there will be two bands linked together *n*/2 times, each of which has *n* half-twists.

Mathematicians found that the overall shape you get when you bisect a half-twisted strip along its centerline – not including half-twists – can be guessed even before cutting, just by observing the edge of the strip. For example, if you follow along the edge of the Möbius strip with your finger, you'll find that your finger traces the topological equivalent of a circle, which is what you get when you cut the Möbius strip down its centerline. Similarly, the boundary of the two-half-twisted strip is what's known as the Hopf link, the three-half-twisted strip is the trefoil knot, the four-half-twisted strip is Solomon's knot, and so on, in one-to-one correspondence with the topology of the same bands cut through their centers.

Flexagons and half-twists

Now we return to flexagons. It turns out that all edge flexagons are *n*-half-twisted strips. We introduced this idea in chapter 20, but now we'll look at it from a different perspective. Take a good look at the tri-hexaflexagon. It is equivalent (isomorphic) to the three-half-twisted strip, meaning that the tri-hexaflexagon is really a three-half-twisted strip in disguise. You can see this in a number of ways. One way is to take a strip of paper, make three half twists in it, join the edges, and flatten it down. Imagine that the three segments of the flattened down strip extend into the hole in the middle, meeting up at the center. What you get is the tri-hexaflexagon shown in figure 26.5.

FIG. 26.5 Flattening a three-half-twisted strip and interpolating to the center shows the correspondence between the strip and a tri-hexaflexagon.

Another way to see this is to cut a tri-hexaflexagon along its centerline. This is a bit tricky because the leaves are triangular, so you have to make sure you always cut along an imaginary curve, leaf by leaf, through the entire flexagon, always cutting through adjacent hinges and not through any of the other edges. To do this, start from a hinge on the outside of the flexagon, and make a cut through the hinge and through the center of the leaves, traversing all the leaves until you return to the initial cutting point. You must always follow a continuous path, meaning that when you reach the entrance to a pocket (see Fig. 26.6, left), you continue along the "band" underneath the upper leaf. Make sure you never cut the outer edges of the original template. You should get a knotted eight-half-twisted strip, same as if you were cutting through a three-half-twisted-strip (see figure 26.6, right).

 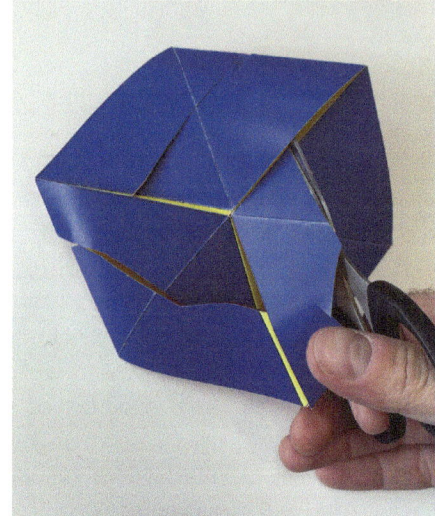 

FIG. 26.6 Cutting a tri-hexaflexagon along its centerline.

All *basic flexagons* are *n*-half-twisted (Möbius-like) strips.

How many half-twists does a given flexagon have? There are a few ways to calculate this, each with its advantages and disadvantages. We denote the number of half-twists in a flexagon with the letter *h*. Conrad and Hartline suggest the following formula: $h = 3N - 6$ where N is the number of pinch faces. This is easy to calculate, but only works for hexaflexagons with generating sequences that only consist of pinch flexes.

Les Pook provides formulas for *h* according to the different types of flexagons that he defines – fundamental first-order, second-order, etc. They work for all leaf shapes, but you need to first disassemble complex flexagons into fundamental flexagons and then into flexagon sectors. Moreover, Pook's formulas only work for those flexagons that can be disassembled into first- or second-order fundamental flexagons. Pook did not define specific expressions for other flexagon types such as compound or irregular flexagons, so the number of half-twists (which he calls the **torsion** of the flexagon) is derived by examination.

According to Pook:

- The number of half-twists in each sector of a first-order fundamental even edge flexagon is $s - 2C$, where s is the Schläfli symbol of the leaf and C the denominator of the net symbol (the denominator of the Schläfli symbol of the associated polygon).
- The number of half-twists in each sector of a second-order fundamental odd edge flexagon is 1.

Given this:

- The number of half-twists in each fundamental flexagon is the number of half-twists in each sector multiplied by the number of sectors.
- The number of half-twists in complex flexagons is the sum of the number of half-twists in its constituent flexagons.

A rigorous and quick way to find the number of half-twists in a given flexagon that works for all triangle flexagons is based on pat notation and atomic theory. It is simply the number of pairs of brackets in the pat notation of the flexagon. It is also equal to the number of Ur or Ul atomic flexes needed to unfold the flexagon completely, which is also half the number of folding labels larger than two (see chapter 27). Note that, at the time of this writing, atomic theory has only been developed for flexagons with triangular leaves.

It's important to note that the total number of half-twists and knots in a flexagon is a topological invariant. (We'll discuss knotted flexagons in chapter 31.) This means that the total number of half-twists and knots cannot change without taking it apart, regardless of what flex you make on the flexagon. Therefore, when a flex starts and ends with the same overall flexagon shape, it removes one or more half-twists in some locations of the folded template and redoes them at another location. You might think of it as "shuffling" the half-twists along the strip from one location to another. When we look at one face of a flexagon, which we call a *surface* of the flexagon, we are ignoring the internals of how it's folded, so the number of half-twists on this face can change after flexing. Since the total number of half-flexes in a flexagon remains constant, this local change in the number of half-twists will be compensated by a change at another location within the flexagon – within the folded parts of the template. Therefore, the total number of half-twists is equal to the number of half-twists in the flexagon surface plus the number of half-twists in the folded parts hidden within the flexagon.

Summary

By viewing flexagons through the lens of topology – the mathematics of stretching and bending – we learn more about their properties. We saw how a flexagon is folded into a continuous band with a fixed number of half-twists and possibly knots. These topological properties can't be changed without taking the flexagon apart and refolding it.

Chapter 27

Templates and Labels

This book is filled with labeled templates you can fold. How did we come up with the templates and how did we pick labels for them? And how can you design your own?

Creating new templates

We have discussed many techniques for creating new flexagons. Here is a list of strategies and the chapters that describe each one:

- Vary the properties of an existing template.
  - Double the length of a basic pinch flex template to double the number of faces. (Chapter 3)
  - Paste leaves together to reduce the number of leaves. This can be applied to an entire face or to any pair of leaves that fold together. (Chapter 3)
  - If the template repeats a pattern, change the number of times it repeats. (Chapter 4)
  - Change the shape of the leaves while preserving other properties of the template. (Chapter 5)
  - "Slit" one leaf to turn it into two leaves. (Chapter 24)
- Describe the desired behavior or structure of the flexagon, and use it to derive a template.
  - Describe the behavior you want a flexagon to have using a generating sequence. (Chapter 18)
  - Describe the internal structure you want a flexagon to have using pat notation. (Chapter 19)
  - Turn a generating sequence or pat notation into a template using atomic flex theory. (Chapter 20)
  - Use flexagon maps and Tukey triangles to derive a template based on the pinch flex. (Chapter 24)
- Use an *associated polygon* to create a fundamental net with a given leaf shape and number of pats. (Chapter 25)
- Connect pats from different flexagons to build up a more complex flexagon. (Chapter 25)

Labeling flexagons

Each leaf-face can be individually labeled with numbers, letters, colors, or other decorations. Labels can serve multiple purposes. They can tell you how to fold a template ("fold adjacent matched numbers together"). They can help you keep track of where you are as you flex ("green is on the front and red on the back"). Or you may choose to decorate the leaves in ways that make for interesting patterns, puzzles, or stories, which will be the focus of Part Five.

DOI: 10.1201/9781003433538-30

Throughout this book, most templates are labeled with numbers and sometimes colors. They're chosen with the dual purpose of helping you fold the templates and track where you are as you do a series of flexes. But how do we decide which labels to use?

First off, it's important to understand that *there is no single best way to pick labels* because there's such a huge variety of flexagons and flexes as well as different goals for the labels. Below we list some different strategies.

Label pinch faces

The flexagons in Part One generally use a different label for each pinch face, i.e., a set of leaf-faces that travel together when using the pinch flex. This technique is especially elegant for these flexagons because the labels provide guides for both folding the template and keeping track of faces as you pinch-flex.

There are different ways to choose what number you assign to a face. You might want to number them in the order you find them as you pinch-flex. Or use the lowest numbers for faces that appear most often in the Tuckerman Traverse. Or use Conrad and Hartline's Tukey triangle technique.

Note that this approach doesn't work for the flexagons described in chapter 6, since entire faces don't travel together, instead getting mixed up after every flex. Notice how the minimal enneaflexagon for the P333 flex (shown in figure 6.2) has nine 1's and 2's, but only six 3's and 4's. This is because the P333 flex only changes six of the nine pats.

Nor does this technique work in general for flexagons created from a generating sequence that's not all pinch flexes, such as the (V>)6 flexagon in figure 10.4, or on flexagons that can't support the pinch flex at all, such as the enneaflexagon.

If your goal is to be able to track individual states, only labeling pinch faces isn't sufficient for being able to distinguish between all of them. For example, face 1 from state 1/2 of the penta-hexaflexagon in figure 3.3 supports the pyramid shuffle at three different hinges, leading to three states where the individual leaves are arranged differently even though the states look identical without additional labels.

Advantages: The same labeling works for both folding and flexing. It's very elegant when you're only using the pinch flex.

Disadvantages: This often doesn't work for flexagons that are designed around flexes other than the pinch flex. Also, it doesn't allow you to distinguish some states.

Assign every leaf a unique label

In chapter 19, we assigned every leaf a unique leaf ID based on the order the leaves occur in the template, with the positive number on one face and the negative number on the other face. This allows us to precisely track where every leaf is as we perform any flex.

But this approach isn't perfect, either. It's harder to figure out which state you're in when you have to check the leaf IDs on every leaf-face. And leaf IDs don't tell you how to fold a template, so you'll need some additional folding instructions.

Advantages: This can be used to uniquely identify any state.

Disadvantages: It can be unwieldy because every leaf has a different ID, and requires additional instructions for folding.

Add labels as you flex

Label the two initial faces, then create new labels as you find new faces or unlabeled leaves during flexing. This is sufficient for labeling the P333 enneaflexagon, for example.

Advantages: This is easy to apply to new templates. Often, the resulting labels can also tell you how to fold the template.

Disadvantages: Sometimes ends up being somewhat arbitrary. This labeling isn't guaranteed to provide enough information to fold the template. It doesn't allow you to distinguish some states.

Use tree labeling

The **tree labeling** technique is used most often throughout this book, since it can be applied to any flexagon and provides similar results to pinch flex labeling. The basic idea is to view the nested pat structure as a tree in order to provide a consistent way to label leaves as you open up multiple layers of pats.

Consider the pat shown in figure 27.1. It has four leaves, where the top face is 1 and the bottom is 2. You open the pat to reveal two leaves labeled 3. When you open up the top subpat, you reveal two leaves labeled 4. When you open the bottom subpat, you find two leaves labeled 5. If you were to unfold this pat, you'd get a strip of triangles labeled 3/4, 1/4, 2/5, 3/5.

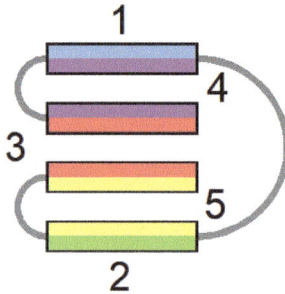

FIG. 27.1 Assigning labels to leaf-faces in a folded pat of four leaves, corresponding to [[- -][- -]].

You can continue unfolding subpats one layer at a time as you find them, numbering the newly revealed leaves with the next available number, starting at the top of the pat and working on down. But we need a couple more details to make this technique work nicely across a variety of flexagons.

First, we only use a number if there are leaves to assign them to. For example, if we only had one leaf in the top subpat instead of two (so the pat corresponds to [- [- -]]), we would instead wait to assign 4 till we had leaves to assign it to, as shown in figure 27.2. In practice, we also look across multiple pats to see if any of them have a top subpat that opens up, in which case we would assign it label 4. This strategy helps keep down the number of different labels and colors we need when the structure of the pats is lopsided, such as [- [- [- -]]].

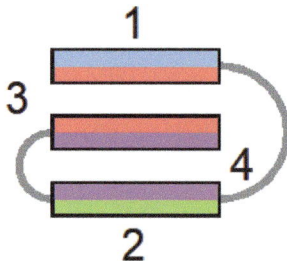

FIG. 27.2 Assigning labels to leaf-faces in a folded pat of three leaves, corresponding to [- [- -]].

Second, since we want tree labeling to give a similar result as pinch flex labeling, we apply this technique to sectors rather than individual pats. For example, consider a hexaflexagon with three sectors and two pats per sector. If we number the pats around the hexaflexagon 1 through 6, we consider the even-numbered pats as part of one group and the odd-numbered pats as part of another group. Then we apply this technique to each group, looking across the entire group to determine which labels to use. If you try it out on one of the hexa-hexaflexagons, you should end up with the same pattern of numbers as the templates in chapter 3, though the actual numbers might differ.

Advantages: This mimics what you get from labeling pinch faces, while working with any pat structure or generating sequence. Thus, it provides information for folding the template as well as useful patterns for tracking faces.

Disadvantages: It doesn't allow you to distinguish some states.

Summary

We now have a good selection of methods for creating new flexagon templates and assigning useful labels. We use these techniques to create the new flexagons we explore in Part Four.

PART FOUR

Exploring Flexagons

In Parts One and Two, we focused on individual flexes, trying them out on different types of flexagons. Now we'll do the opposite: focus on individual types of flexagons, exploring them through the flexes they support. We'll use many of the tools we developed in Part Three, such as state diagrams and flex sequences.

We chose the flexagons in the first two chapters of this section – the *square silver octaflexagon* and *hexagonal bronze dodecaflexagon* – because their rich symmetry leads to lots of interesting flexes and morphs. Our next choice, the *hexagonal silver dodecaflexagon*, has less symmetry since the pats don't all meet in the middle, which offers possibilities for new types of flexes and provides an example of a matryoshka, one flexagon nested inside another. The *silver bracelet* demonstrates that flexagons can be three-dimensional bands, possibly even twisted and knotted. The final example, the *octagonal ring 14-flexagon*, shows how you can use the resources in this book to create and explore a large variety of other flexagons. We finish by learning how to use the Flexagon Inspector tool to create and explore custom flexagons.

DOI: 10.1201/9781003433538-31

Chapter 28

Square Silver Octaflexagon

This chapter focuses on the square silver octaflexagon, as shown in figure 28.1, which demonstrates many flexes we've already seen plus some new ones. We saw this flexagon when investigating P, F, T, S, St, Fm, and S3. Now we'll try a template that supports all those flexes plus additional flexes.

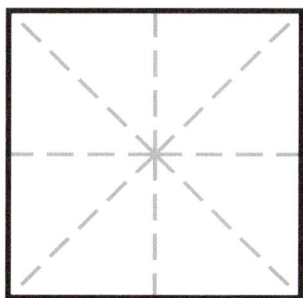

FIG. 28.1 Main position of the square silver octaflexagon.

Six faces

To fold the template in figure 28.2, first tape together the edges marked with "a". To connect the final edges together after folding, it's simplest to put tape on the edge of the leaf marked 3/4 before folding, attaching it to the leaf marked 2/5 after folding it into a square.

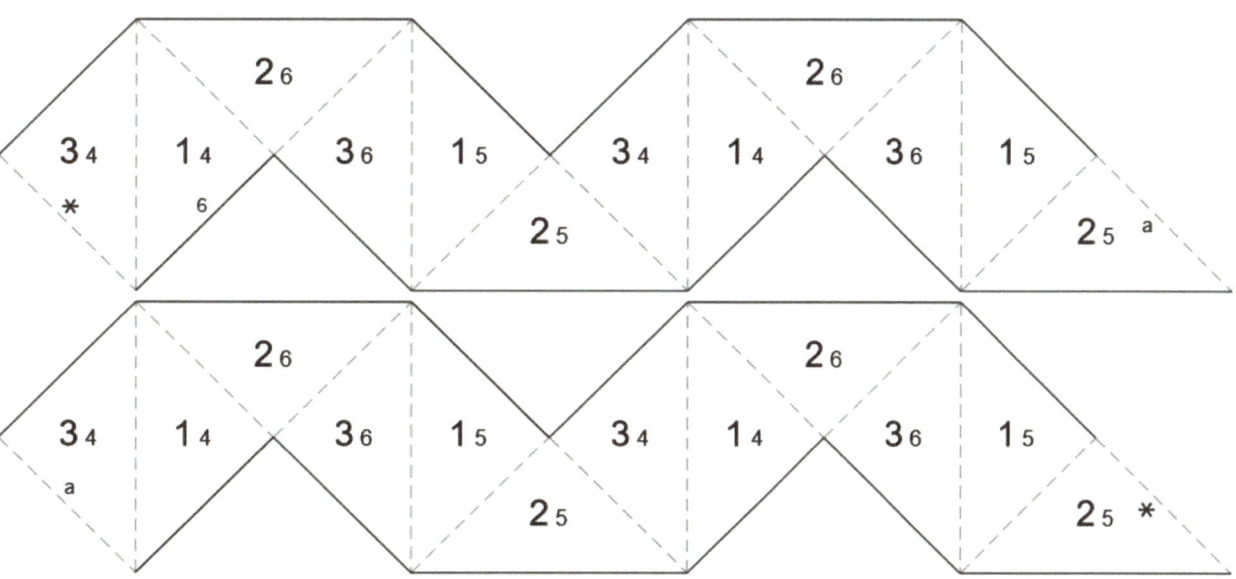

FIG. 28.2 Template for a square silver hexa-octaflexagon.

DOI: 10.1201/9781003433538-32

From face 1 of state 1/2, you can do the pinch (P), flip (F), v-flex (V), silver tetra (St), and Möbius flip (Fm) flexes. From face 2, you can do a pinch or a tuck (T). The labels in figure 28.2 show the flexes that can be performed at the various hinges of state 1/2. This includes the twist (Tw), which we'll discuss later in this chapter. Recall from chapter 11 that, on an octaflexagon, T2 refers to doing the tuck flex by opening the hinge opposite the reference hinge. Doing T2 from an edge hinge, which you can do from face 2, is fairly easy, while doing T2 from a corner hinge, such as on face 1, is difficult.

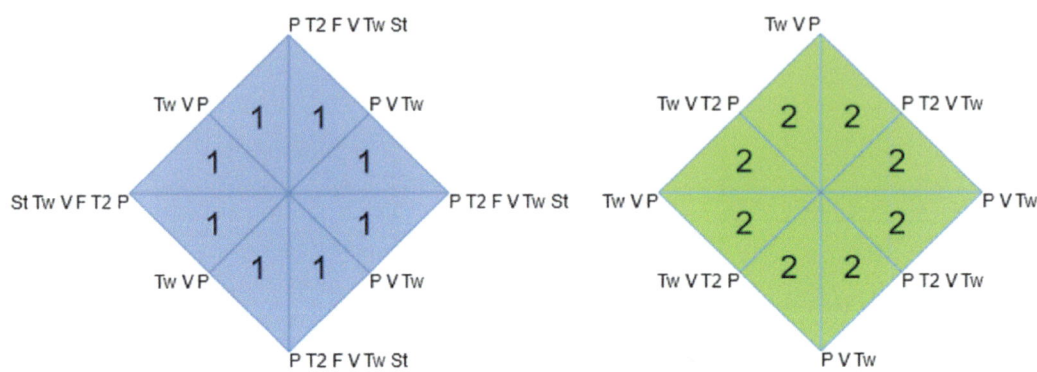

FIG. 28.3 Flexes available from state 1/2. Flexes are listed next to the reference hinge where you can perform them.

If you pinch-flex an edge hinge from face 1, it will take you to state 5/1. From face 1 of this state, you can do a tuck flex or pyramid shuffle (S). From face 5, you can do a silver tetra flex (St). You can pinch-flex between four flat states if you start from face 1 and continue to pinch-flex at an edge hinge, visiting 1/5, 2/1, 3/2, and 6/3. Figure 28.4 shows the flexes that can be performed at the various hinges of state 1/5.

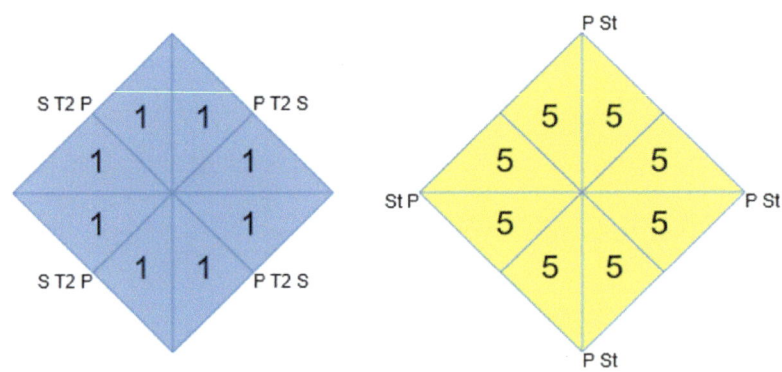

FIG. 28.4 Flexes that you can perform from various hinges from state 1/5.

From face 1 in state 1/2, if you do two pinch flexes starting at the corner of the square, you travel through the mountain-valley state 3/1 and continue on to the flat state 4/3. From here, pinching an edge hinge from face 4 takes you to flat state 1/4. Then from face 1, two pinch flexes at the corner of the square take you to 2/3.

Figure 28.5 shows the pinch state diagram for this hexa-octaflexagon with some added routes you can take using the flex sequence (S<<)4, a series of pyramid shuffles. Each state is marked with the leaf angle that will be at the center of the flexagon. States marked with 45° are flat and states marked with 90° are in a mountain-valley configuration.

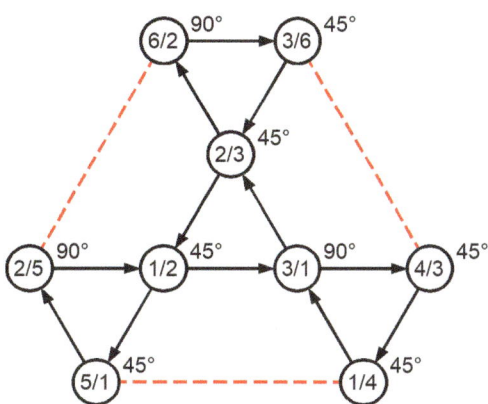

FIG. 28.5 Pinch state diagram for the silver hexa-octaflexagon with a few pyramid shuffle routes added. States are marked with the leaf angle at the center. The dashed red lines show where you can travel between states using the flex sequence (S<<)4.

You may wish to refresh your memory on how to pinch-flex around the faces of this flexagon by referring back to chapter 5. Figure 5.13 shows one way to cycle between three states, going from one flat state, to a second flat state, to a mountain-valley state, and back to the original flat state. For example, it can take you from 1/2 to 5/1 to 2/5 to 1/2, which is flex sequence (P>)3. Figure 5.17 shows the reverse pass-through sequence, which is just a fancy-looking way to do two pinch flexes in a row, represented as the flex sequence PP. This takes you from a flat state, through a mountain-valley state, on to a second flat state. For example, it can take you from 1/2 to 3/1 to 4/3. Or you can pinch-flex between a series of flat states: 3/6 to 2/3 to 1/2 to 5/1.

The dashed red lines in figure 28.5 show where you can use a sequence of pyramid shuffles in place of a sequence of pinch flexes. For example, from face 1 of state 5/1, you can get to state 1/4 by either doing the sequence (P^>)3 or (S<<)4. Since the sequence of pinch flexes travel through state 3/1, which isn't flat, this pinch sequence can be tricky to carry out. But since the pyramid shuffle is a local flex, the flexagon is flat both before and after each flex in the sequence. As a result, you may find the pyramid shuffle sequence easier than the pinch flex sequence. However, doing the pyramid shuffle from face 2 in state 2/5 is very tricky given how steep all the mountains are.

States 1/2 and 2/3 are especially rich in opportunities to experiment with flexes. Like the pyramid shuffle sequence, you should be able to find other ways of traveling between the single-colored states of the pinch state diagram besides the pinch flex. Here are some experiments to try:

- Instead of doing a single silver tetra, try doing two silver tetras at the same time, but from opposite corners.
- Instead of a single tuck, try doing two at once from opposite edges. From the new mixed-color state, see if you can pinch-flex.
- After doing a flip flex, see if you can figure out which hinge now supports a pyramid shuffle.
- From face 1 of state 1/2, see if you can do a Möbius flip. A well-placed pyramid shuffle and tuck flex can then restore state 1/2.
- Can you find any new flexes?

Twist flex

As you play with this flexagon, one flex you may discover is the ***twist flex*** (Tw). See figure 28.6 for the flex diagram. Figure 28.7 shows the template for the minimal octaflexagon for the twist.

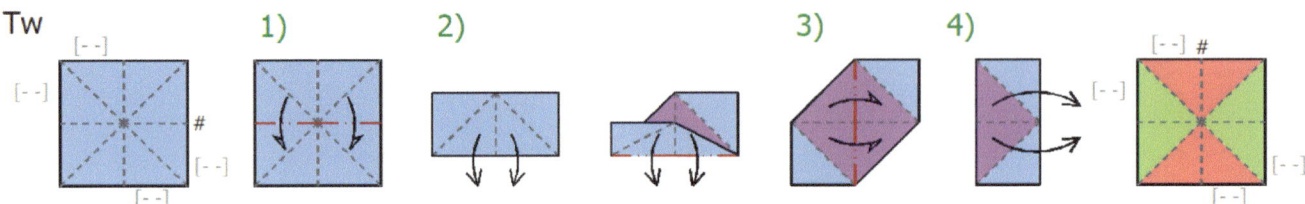

FIG. 28.6 How to do the twist flex (Tw).

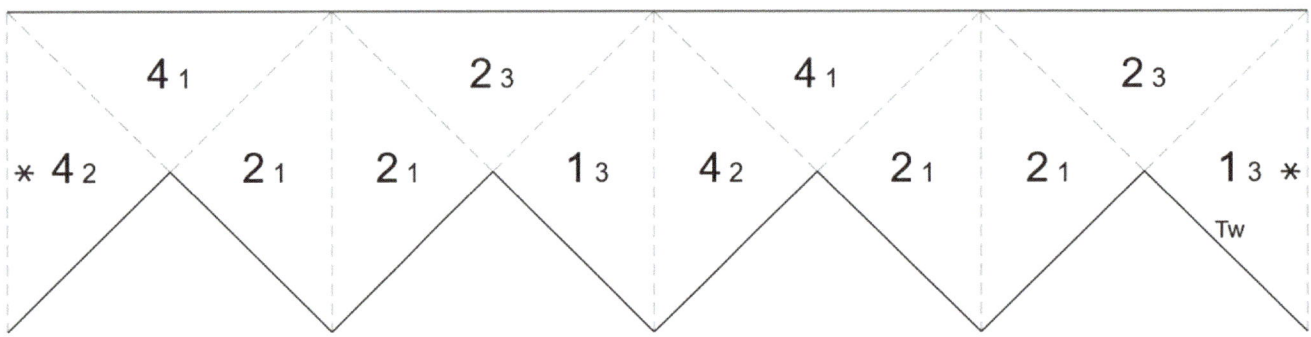

FIG. 28.7 Template for the minimal square silver octaflexagon for the twist flex.

Note that figure 28.6 shows the reference hinge for the twist flex at an edge hinge. But if you look back at figure 28.3, it says that you can do a twist flex at an edge hinge or a corner hinge. While the pat structure at a corner hinge supports the twist *in theory*, you'll find that it doesn't work very well *in practice* because you would have to turn a narrow cup inside out. Similarly, you will find that you can do a flip flex from a corner hinge, but not an edge hinge.

Summary

The square silver octaflexagon is a fun flexagon to use because it supports a wide variety of flexes we've previously seen, while offering many new possibilities because of its elegant symmetry.

You can easily mix up the faces by doing combinations of various flexes. Once the faces are mixed, can you figure out how to restore them to solid colors? Of course, if you get completely lost, you can always find a taped hinge, undo the tape, and refold the flexagon. But can you figure out how to flex your way back to solid faces?

Chapter 29

Hexagonal Bronze Dodecaflexagon

We first saw the hexagonal bronze dodecaflexagon (figure 29.1) in chapter 6. Now we'll explore it in more detail, demonstrating many of the flexes we've already seen plus some new ones. We'll take advantage of its flexibility to see more examples of morphs. Be aware that this flexagon can be trickier to manage, but also has lots of secrets hidden inside.

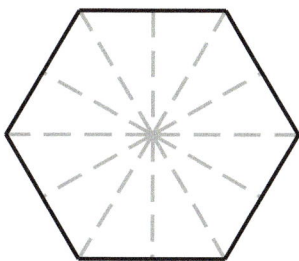

FIG. 29.1 Main position of the hexagonal bronze dodecaflexagon.

Six faces

We'll start with a bronze hexa-dodecaflexagon that follows a similar pattern to the square silver hexa-octaflexagon from the last chapter. See figure 29.2 for its template.

DOI: 10.1201/9781003433538-33

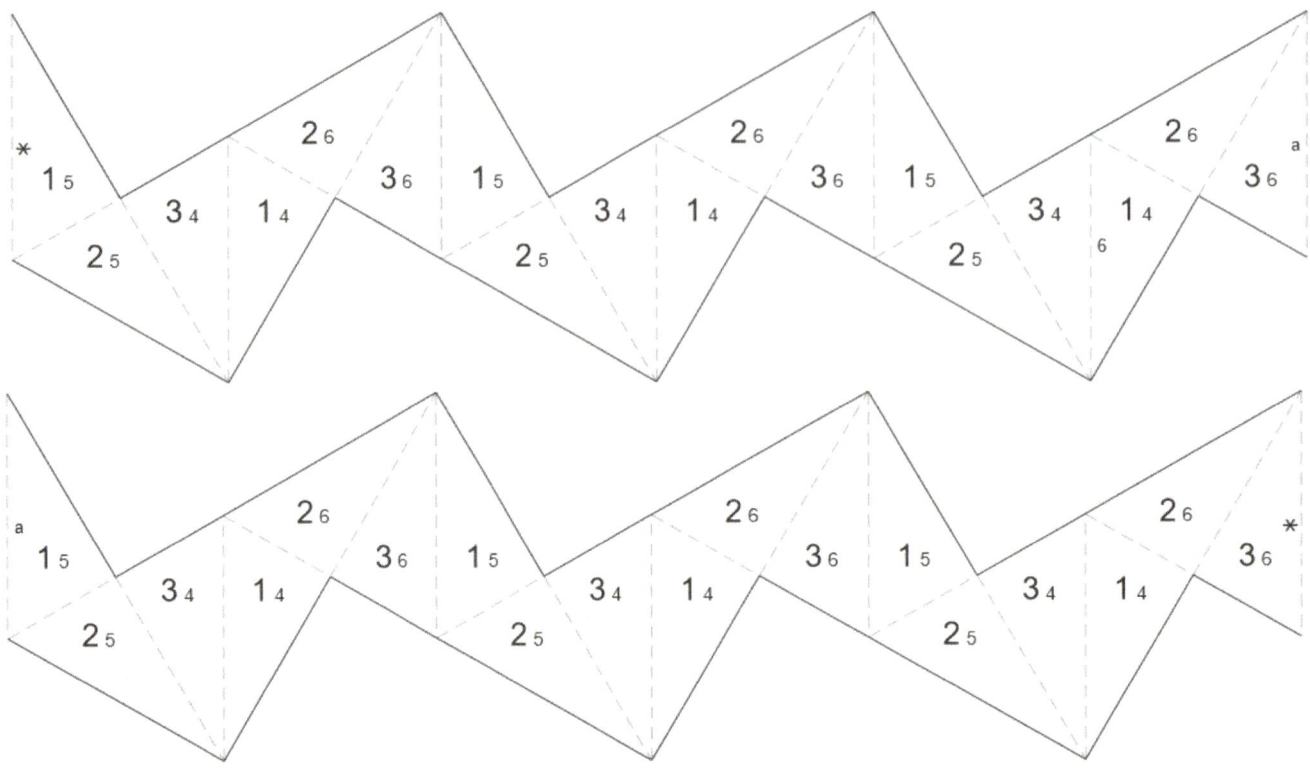

FIG. 29.2 Template for a hexagonal bronze hexa-dodecaflexagon.

Since this template follows the same pattern as the silver hexa-octaflexagon we previously explored, you can do pinch flexes (P), tucks (T), pyramid shuffles (S), flips (F), silver tetras (St), etc. from similar hinges. And the pinch state diagram looks very similar, as shown in figure 29.3, with shortcuts through it in the same places. Recall that, for the pinch flex, you pinch *every other hinge* all the way around the flexagon, which means you end up pinching *six hinges* on the dodecaflexagon. If we want to be explicit, we can refer to the pinch flex on the dodecaflexagon as the P222222. Also recall that the pinch flex often takes you to a non-flat state and you may sometimes need to reverse the mountains and valleys before your next pinch flex. And, as we saw in chapter 6, you can also choose to skip additional hinges when you pinch. For example, figure 6.8 shows the P444 flex, where you pinch every fourth hinge.

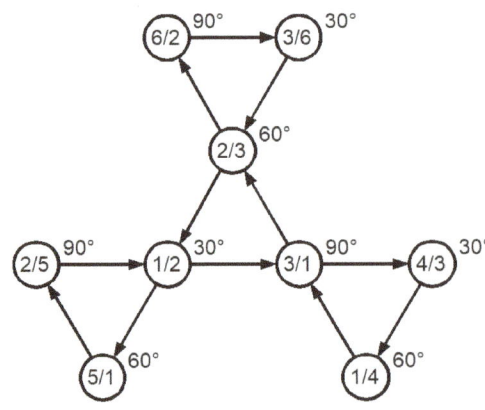

FIG. 29.3 Pinch state diagram for the bronze hexa-dodecaflexagon. On the dodecaflexagon, the pinch flex can be explicitly referred to as the P222222.

One big difference between the silver octaflexagon and bronze dodecaflexagon is that the latter is flat in fewer states. Since two of the three angles in a silver triangle are 45°, two of the three ways you can reflect eight of them around a common center are flat. But all three angles of a bronze triangle are different, so it'll only lie flat if you use four 90° angles, six 60° angles, or twelve 30° angles. Looking at the pinch state diagram, you can see that states 1/2, 3/6, and 4/3 are flat, while the other six states are not. So, while you can do the various flexes in similar places as you could on the silver octaflexagon, some of them may be harder to perform because the state may not be flat.

One implication of this is that, in general, the pinch flex (pinching every other hinge) is often difficult on the bronze dodecaflexagon because twelve 60° or 90° angles is fairly unwieldy. It can be much easier to use an alternate route between flat states rather than pinch-flexing. Many of the flexes described in the first two parts of the book work well on the bronze dodecaflexagon, especially the local flexes, so look to those chapters if you need a refresher on any of the flexes discussed next.

Here are various flex combinations you can try:

- From face 1 in state 1/2, do a silver tetra flex (St) every fourth hinge to get to 3/6, sequence (St>>>>)3.
- Alternately, do a tuck (T) every other edge-hinge to go from 2/1 to 6/3, sequence (T>>)6.
- From 3/6, six pyramid shuffles (S) take you to 4/3, sequence (S<<)6.
- From face 3 of 4/3, six T's at the corner hinges take you back to 2/1, sequence (T>>)6.
- From face 1 of 1/2, you can do P444 twice in a row, sequence (P444)2.
- From 1/2, you can use three flips (F) and three S's to go to 4/3, sequence (F>>>>)3 < (S>>>>)3.

Or try different pinch variations such as the P444 at either corner hinges or edge hinges, the P66 (pinching at two opposite hinges that are six hinges apart), or some other pinch variant. You'll find all sorts of different shapes and arrangements. Feel free to take the flexagon apart and refold it if you get too lost in your experimenting.

New flex variants

The silver tetra flex is especially interesting on a bronze dodecaflexagon because you can sometimes follow the same general strategy of folds, but do them in the middle of the flexagon instead of along an edge. Figure 29.4 shows the flex diagram for one silver tetra variant performed in the middle of the flexagon. Surprisingly, it changes two pairs of nonadjacent pats, leaving the pats between them unchanged.

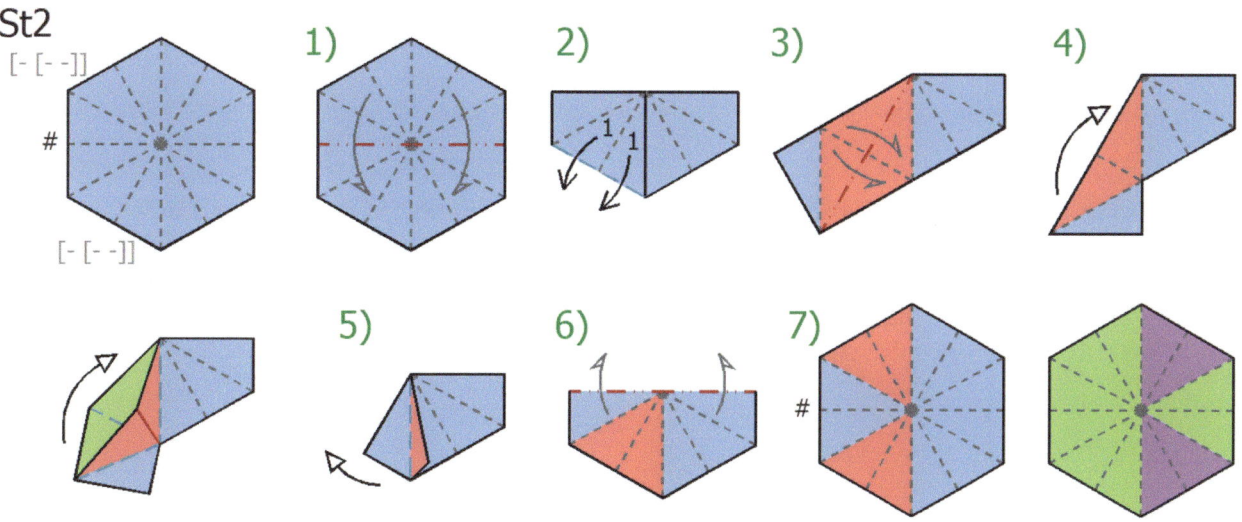

FIG. 29.4 Flex diagram for the silver tetra 2 flex, St2.

As we noted in chapter 18, the silver tetra flex is similar to doing two tuck flexes because St = >Tf'<<Tf'>. Similarly, this variant, called **silver tetra 2**, is also equal to two tuck flexes: St2 = >>Tf'<< <<Tf'>>. Can you find another silver tetra variant?

In a flip flex, you flip two pats from the front to the back of the flexagon. But you could choose to flip more than two pats. Figure 29.5 shows the flex diagram for the **flip 3 flex** (F3), where you flip three pats instead of two.

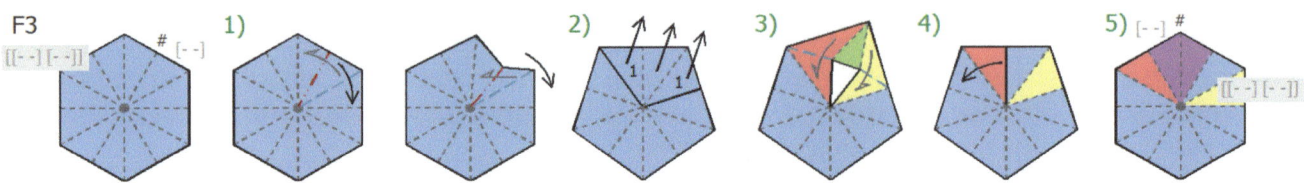

FIG. 29.5 Flex diagram for one variation of the flip flex, the flip 3 flex, F3.

If you use a corner of the hexagon as the reference hinge rather than an edge, you can sometimes flip four pats, called the **flip 4 flex** (F4). And if you don't do the final step in the flip where you open the flexagon back up, you can continue along with another flip variant, stringing together a whole series of different flip flexes. On this hexa-dodecaflexagon, there are many ways you can do a cycle of flip variants, starting and ending on face 1. One example is F4< (F3<)3 F. Can you find other flex cycles?

Morphs

The hexagonal bronze dodecaflexagon has two flat morphs, shown in figure 29.6. We saw the hexagon morph when we learned the P444 flex in chapter 6, so we call this shape its **pinch-morph**. You can also rearrange 12 bronze triangles into a rhombus, which we call its **rhombic-morph**.

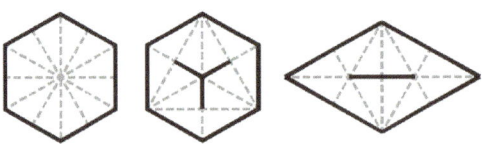

FIG. 29.6 The hexagonal bronze dodecaflexagon, its pinch-morph, and its rhombic-morph.

Let's take a closer look at the pinch-morph. Using the technique we learned in chapter 20 to describe the directions between pats, we can say that this morph has directions \//\\//\\//\\//\, which is the pattern \//\ repeated four times. Since \//\ is exactly the input to the inverse morph-kite flexes we saw in chapter 14, this means that we already know some flexes we can perform on this morph when it has the proper internal pat structure.

Let's see how we can do a flex on a pinch-morph using the hexagonal bronze hexa-dodecaflexa-gon from the beginning of this chapter. Restore it to state 1/2 if it's currently mixed up, and use one of the corners of face 1 for your reference hinge. Do a P444 to get to the pinch-morph shape with 2's and 3's on the front face. From here, you can do an inverse backflip flex, Bf', as shown in figure 29.7. If you then shift four hinges to the left, you can do another Bf', followed by another <<<<Bf'. From here, a second P444 takes you to state 3/6. Putting it all together, you can go from state 1/2 to 3/6 using the sequence P444 (Bf'<<<<)3 <<P444.

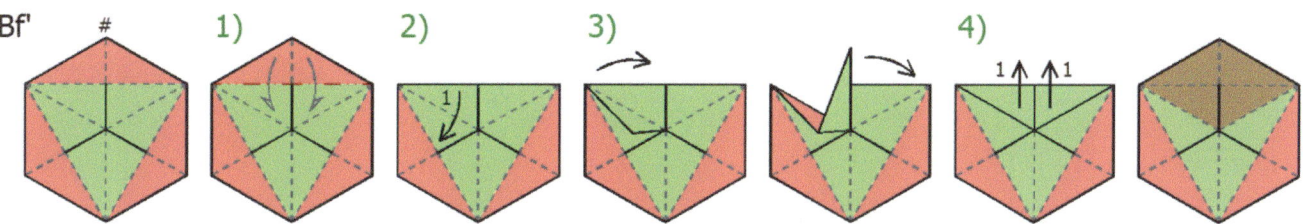

FIG. 29.7 Flex diagram for Bf' on the pinch-morph.

There are multiple ways to reach the rhombic-morph. One way is to start as if you're doing a silver tetra flex from both ends of the flexagon, but stop partway through and open it up instead of con-tinuing through. See figure 29.8 for the flex diagram of the **rhombic-morph flex**, Rhm. You can do this from state 1/2 on the same bronze hexa-dodecaflexagon we've been using, or you can make the minimal Rhm flexagon from the template in figure 29.9. Note that, as with most of the flex diagrams in the book, the colors and pat structure reflect how the flex behaves on the minimal flexagon.

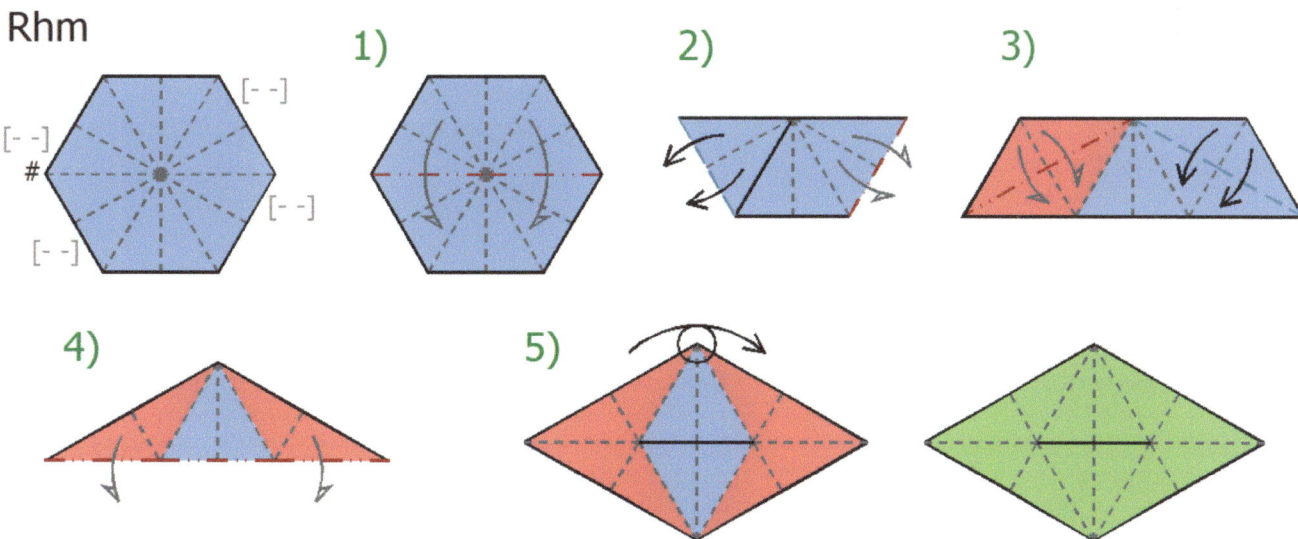

FIG. 29.8 Flex diagram for morphing from the hexagonal bronze dodecaflexagon to the rhombic bronze dodecafl-exagon using Rhm, the rhombic-morph flex.

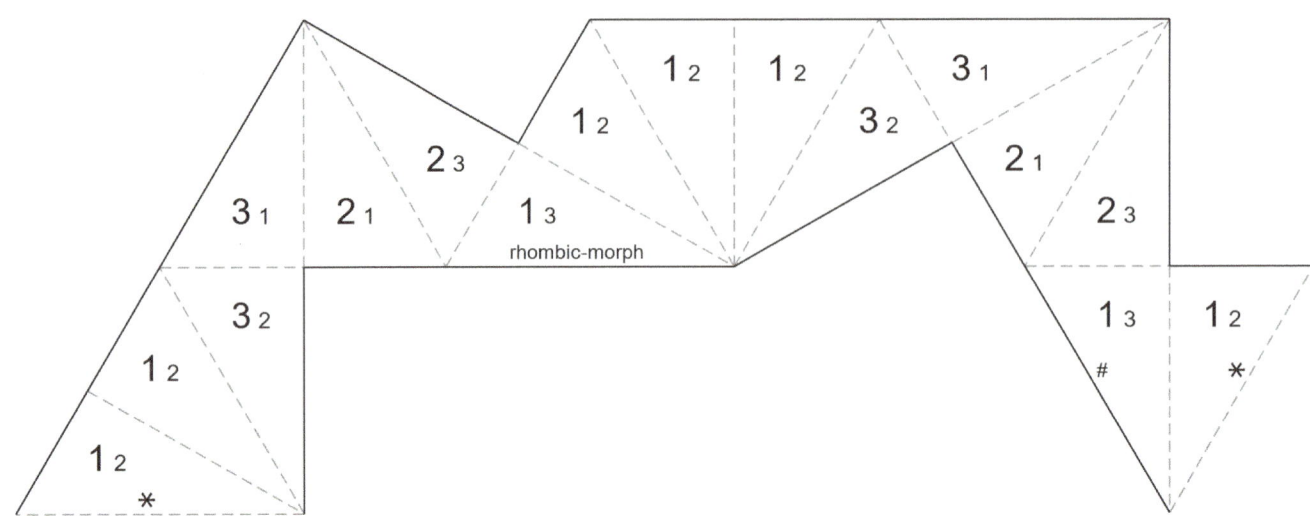

FIG. 29.9 Template for the minimal flexagon that supports morphing from the hexagonal bronze dodecaflexagon to the rhombic bronze dodecaflexagon.

This rhombic flexagon looks like two triangular bronze hexaflexagons connected together, and we've already seen flexes that work on that flexagon. In particular, we've used local flexes such as the flip, tuck, and pyramid shuffle. These also work on the rhombic flexagon when the underlying pat structure supports them.

For example, do a rhombic-morph flex from state 1/2 on our hexa-dodecaflexagon. From the face with a mix of 1's and 3's, pick as your reference hinge one of the sharp points with 3's on them. From here, you can do either a flip flex or a tuck flex.

Note that you need to be a lot more careful when flexing the rhombic dodecaflexagon because it's much less stable. Most flexagons we've explored have pats that all meet in the center, keeping them from moving around too much. Once the pats no longer all meet at a common point, they have a tendency to flex in unexpected places if not held down. In order to do a flip or tuck, you may want to use paper clips to help keep the rest of the flexagon from getting tangled, especially as you first try flexing the rhombic-morph.

Summary

The hexagonal bronze dodecaflexagon is a fun flexagon because it supports a large number of flexes, both ones we've previously seen and new ones that take advantage of having a dozen pats. And it supports multiple morphs, each with their own interesting set of flexes to explore.

Chapter 30

Hexagonal Silver Dodecaflexagon

In this chapter, we explore a flexagon where not all the leaves meet in the middle: the *hexagonal silver dodecaflexagon* in figure 30.1. This has left-to-right symmetry and top-to-bottom symmetry with two pointy ends and pat directions of //\//\//\//\. We'll try out some new flexes, reuse a flex we've seen before, and get our first example of a matryoshka, a flexagon nested inside a flexagon.

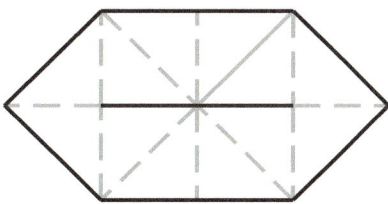

FIG. 30.1 Main position of the hexagonal silver dodecaflexagon.

Turn flex

The **turn flex** (Tu) is a very simple and elegant flex. You fold half of the flexagon backward and unfold a different section from the front to reveal previously hidden leaves. See figure 30.2 for the flex diagram for the turn flex and figure 30.3 for a template that supports multiple turn flexes. If you look carefully, you'll see that the flex changes which pats make up the two pointy ends of the hexagon.

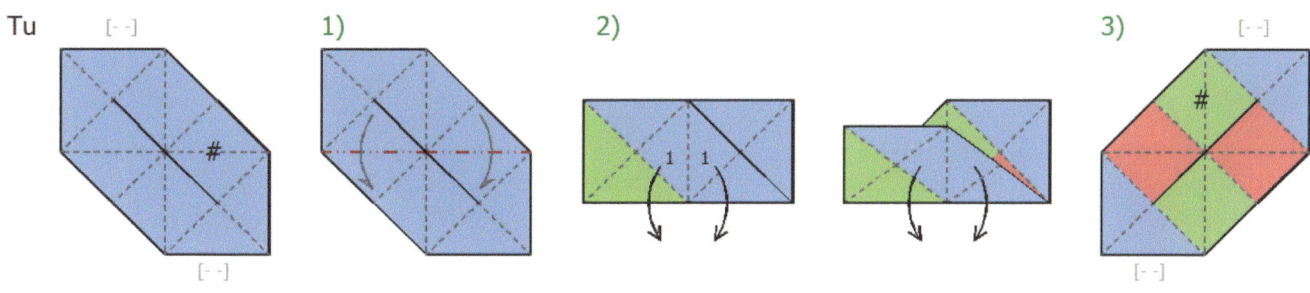

FIG. 30.2 Flex diagram for the turn flex, Tu.

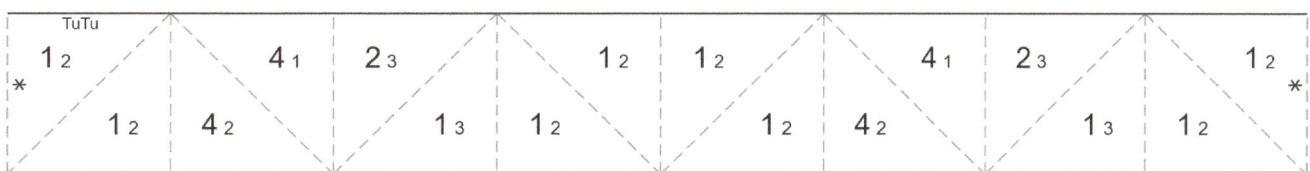

FIG. 30.3 Template for a hexagonal silver dodecaflexagon (generator: TuTu), which supports a cycle of four turn flexes.

DOI: 10.1201/9781003433538-34

This template supports a nice cycle of four turn flexes. After you do the first turn flex, rotate the flexagon 90° so that it's oriented the same way as the first picture in the flex diagram. This will allow you to do a second turn flex. If you repeat these steps four times, you'll end up back in the original state. The way this cycle mixes up the faces opens up some interesting possibilities for decorations.

Double slide flex

Another new flex you can do on this flexagon is the **double slide flex** (Ds). Figure 30.4 shows its flex diagram, and figure 30.5 contains a template for the minimal flexagon that supports it. As with the turn flex, this flex changes which pats make up the pointy ends.

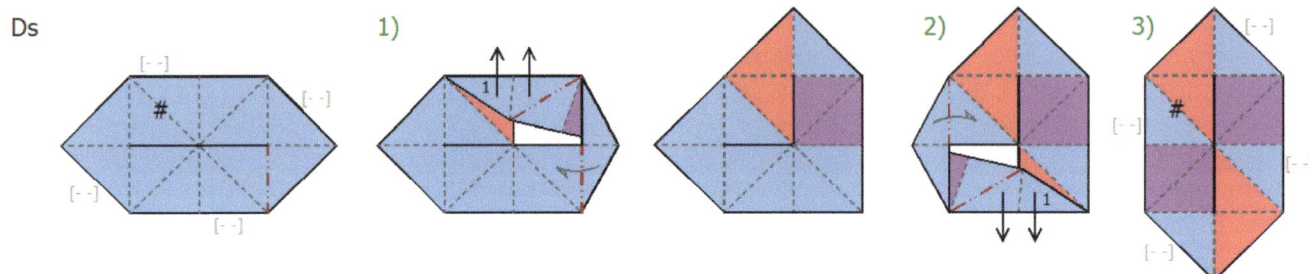

FIG. 30.4 Flex diagram for the double slide flex, Ds.

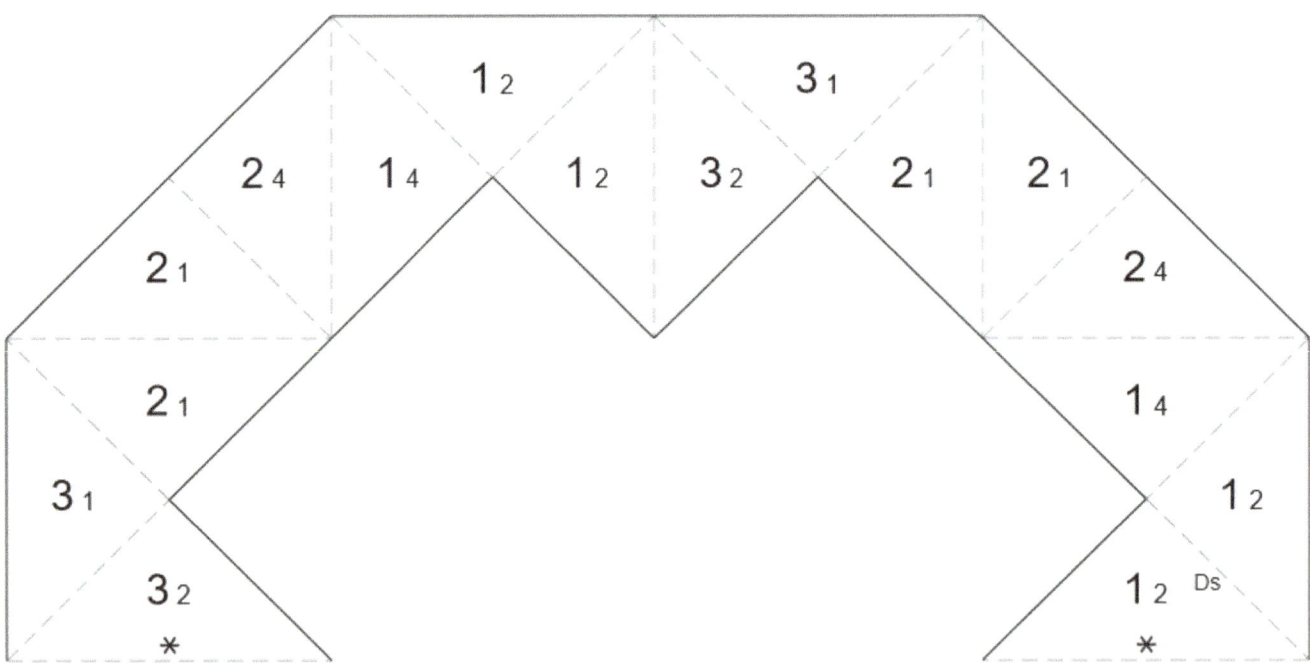

FIG. 30.5 Template for the minimal hexagonal silver dodecaflexagon supporting the double slide flex.

If you stop after step 1, you get the **single slide flex**, which is a morphing flex. It starts from a hexagonal dodecaflexagon and ends with a pentagonal dodecaflexagon.

Tuck flex

If you take a closer look at the directions between the pats, you'll find that they repeat the pattern //\ four times: //\//\//\//\. This means there are hinges where it supports the tuck flex, which requires

//, but not the pyramid shuffle, which requires ///, or the flip flex, which requires ////, or the pinch flex, which requires all /'s. However, the symmetry of its silver triangles combined with the fact that the pats don't all meet in the center opens up possibilities for new flexes.

Since the pat directions of our silver dodecaflexagon contain four pairs of adjacent /'s, there are four hinges where you might be able to do a tuck flex, as shown in figure 30.6. A template for a flexagon that supports all four tuck flexes can be found in figure 30.7. Its generating sequence is (Tf+>>>)4. Note that we use Tf instead of something like T1 or T2 because we don't need to add extra hinges across from the tuck flex.

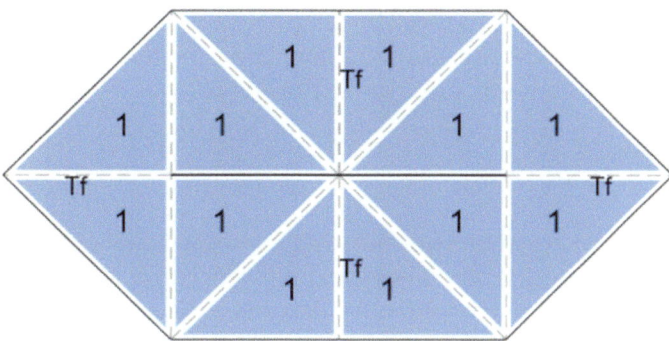

FIG. 30.6 The four hinges of a hexagonal silver dodecaflexagon where you can perform a tuck flex.

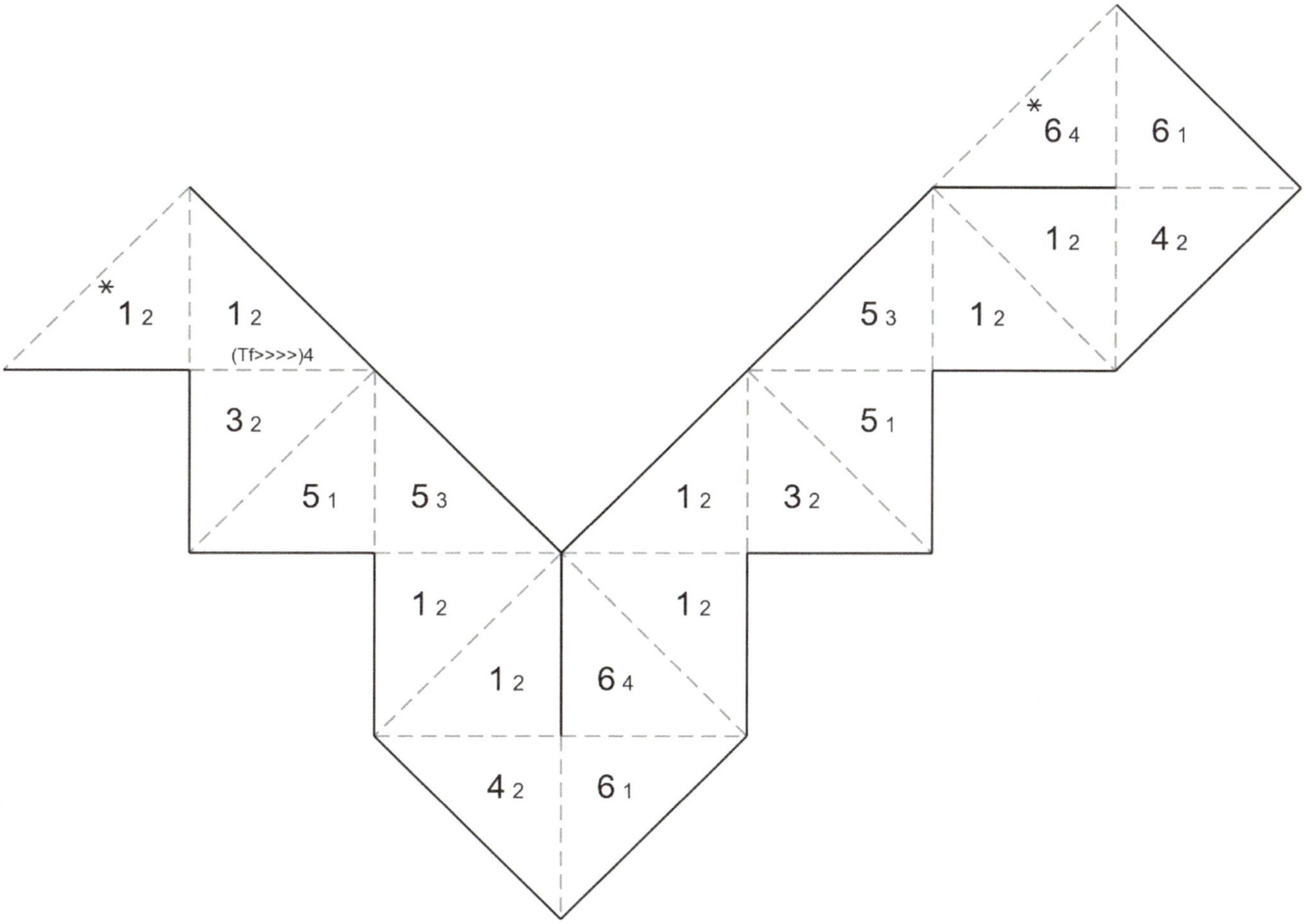

FIG. 30.7 Template for the silver dodecaflexagon (generator: (Tf>>>>)4).

If you're careful, it's also possible to do two opposite tucks at the same time, either the pair at the pointy corners or the pair along the inner edges. As described in chapter 6, these can be classified as double pinch flex variants at locations that are six hinges apart, so they're called the P66. An even trickier flex is to do all four tucks at the same time, which is the P3333 pinch variant:

(P66)2   =   Tf >>>>>> Tf>>>>>>
(P3333)2   =   Tf>>> Tf>>> Tf>>> Tf>>>

From face 1 of state 1/2, you can do a turn flex followed by a double slide, so this flexagon supports all the flexes we've discussed in this chapter. In fact, there are over 100 states you can reach with these flexes, so you can easily get lost as you mix up this deceptively simple flexagon. And there are additional shapes you can uncover with enough experimenting.

Matryoshka

The hexagonal silver dodecaflexagon has yet another new trick tucked up its sleeves: a matryoshka. Like a set of nested dolls, there are times when you can find smaller flexagons nested inside larger ones.

To try this out, use the silver dodecaflexagon (generator: (Tf>>>>)4) from figure 3.7 and refer to figure 30.8 for the flex diagram. Starting from face 1 in state 1/2, fold back both points to make a square that looks like a square silver octaflexagon. This is a matryoshka of the hexagonal silver dodecaflexagon. From here, you can do a pinch flex by pinching the edge hinges. When you open it back up, you'll again have a square silver octaflexagon, but now with two flaps sticking up, which you can fold down in either direction. Pats that stick out and can fold in either direction are called **toggles**.

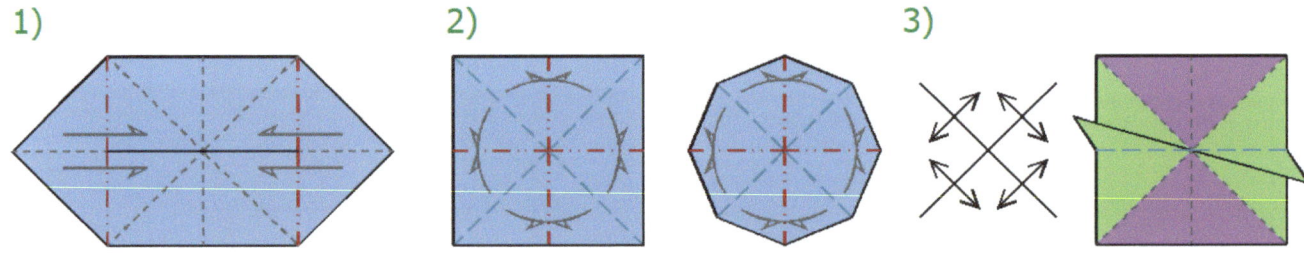

FIG. 30.8 Flex diagram for the matryoshka pinch flex.

When we defined *flexagon* and *flex*, we stated that they obey the atomic flex axioms, then said that matryoshkas seem to break those rules. But it turns out that matryoshkas still follow the rules when viewed through the right lens. If you interpret step 1 of the matryoshka pinch flex as taking you from a flexagon with 12 pats to a flexagon with only eight pats, you would conclude that some of those pats don't follow the axioms because of the way they're folded. But if you instead interpret this new arrangement as still having 12 pats, you can see that it's just that they no longer lie flat, instead winding around in three dimensions. The key here is that you view the flexagon band as a chain of pats that don't have to lie flat in two dimensions. This is what allows us to continue to use atomic flex theory for cases such as matryoshkas.

The technique of folding together adjacent pats to create a smaller flexagon can be done in other flexagons as well. For example, you could fold under six of the pats in the hexagonal bronze dodecaflexagon pinch-morph to make a triangular bronze hexaflexagon. However, folding together pats like this often blocks the freedom of the hinges, so it doesn't always create a workable flexagon. As you flex a matryoshka, you may find that sometimes the extra flaps are tucked away neatly inside, while other times you can freely toggle them back and forth. It's even possible to have matryoshkas nested inside matryoshkas. If you look through the pictures in appendix B, can you see any that might support multiple levels of matryoshkas?

Summary

The hexagonal silver dodecaflexagon has pats that don't all meet in the middle, which leads us to several new flexes. We also saw how to flex a matryoshka, a simpler flexagon nested inside a larger one. This is a fun flexagon to explore, with a nice combination of symmetry in a different arrangement than the flexagons we've seen previously.

Chapter 31

Silver Bracelet

We call the flexagons in this chapter **silver bracelets** because they are made from silver triangles and could be worn as bracelets. We've mostly focused on flexagons that lie flat, so the bracelets serve as an interesting example of how flexagons come in a wide variety of three-dimensional shapes as well. Figure 31.1 shows the main position of a silver bracelet with 16 pats.

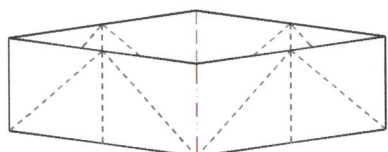

FIG. 31.1 Main position of the silver bracelet 16-flexagon.

The directions between pats in these bracelets all repeat the pattern //\\. When we use silver triangles, this pattern gives us a straight band. These directions allow for new types of flexes, like the transfer flexes, and a new strategy for using flexes such as the flip flex by wrapping it with a morphing flex.

But be warned that bracelets can be a bit harder to work with, since they don't lie flat. While trying to do a flex on one portion of the flexagon, it can be hard to keep the rest of the flexagon from rearranging itself at the same time. You may wish to use a paper clip or two to keep the other portions of the flexagon in place as you flex.

Transfer 2 flex

We'll start out with the **transfer 2 flex** (Tr2), so named because it seems to "transfer" a stack of pats between two adjacent pats. See figure 31.2 for the template for the minimal silver bracelet dodecaflexagon for Tr2. When you attach the first and last leaves after folding, there should be 12 1's around the outside of a three-dimensional loop that looks like a bracelet, with 2's on the inside. See figure 31.3 for the flex diagram for the Tr2.

DOI: 10.1201/9781003433538-35

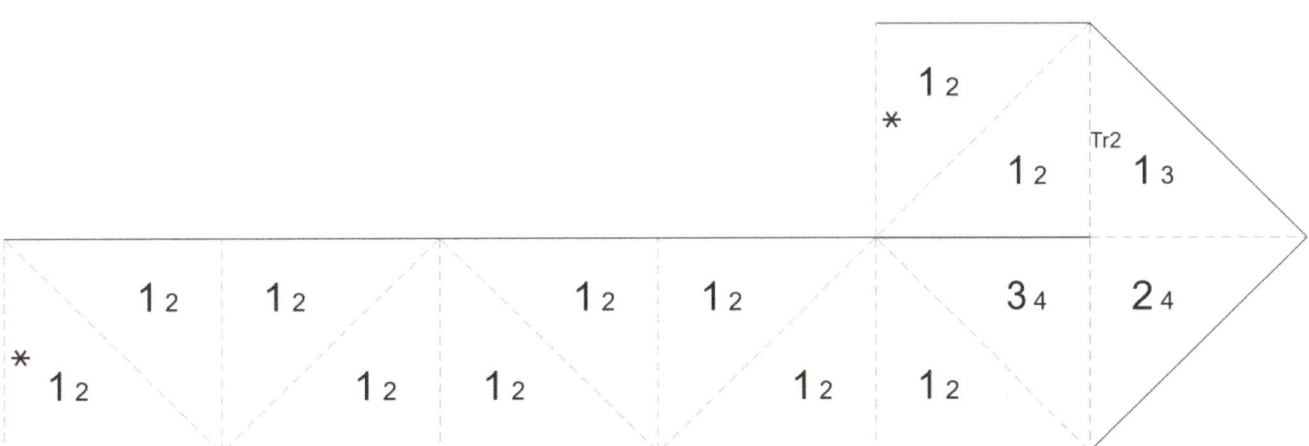

FIG. 31.2 Template for the minimal silver bracelet dodecaflexagon for the transfer 2 flex.

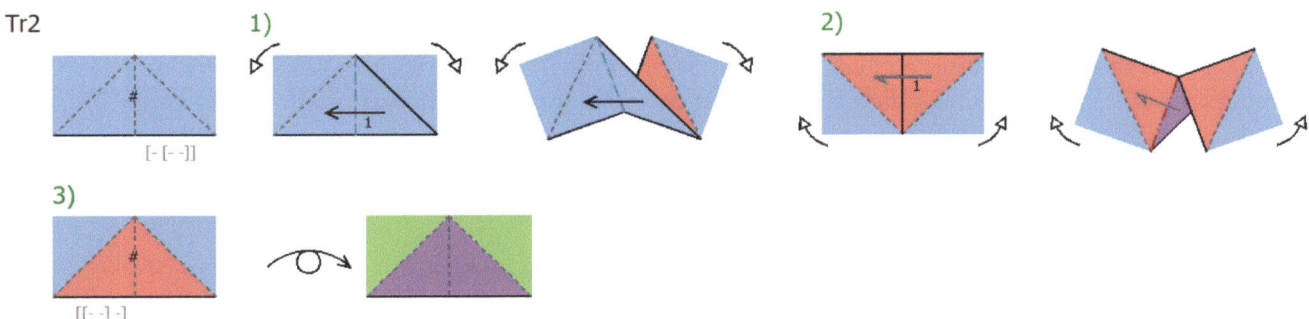

FIG. 31.3 Flex diagram for the transfer 2 flex, Tr2. Note that the bracelet extends to the left and right of the portion shown.

Perhaps you noticed that the transfer 2 flex has a similar feel to the tuck flex. If we write out the atomic pat notation for Tr2, we can see that it's really just the forced tuck in disguise:

$$
\begin{aligned}
\text{Tr2} \;=\;& a \, 1 \setminus \# \, [2,[4,-3]] \setminus b \rightarrow a \, [[-2,1],3] \setminus \# \, 4 \setminus b \\
=\;& {}^{\wedge}\!\!\sim \text{Tf}' \sim^{\wedge} \\
=\;& \text{Xl Xr}
\end{aligned}
$$

Next, let's try our hand at a silver bracelet tetra-16-flexagon, with four faces and 16 pats in the bracelet. See figure 31.4 for the template. As you start to fold it, you'll notice that the 4's occur four in a row. Fold the first and second adjacent 4's face-to-face then the third and fourth 4's face-to-face. This will give you four 3's in a row, where you again fold the first and second together then the third and fourth together. Arrange 1's on the outside of the bracelet when you connect the first and last leaves together, with 2's on the inside.

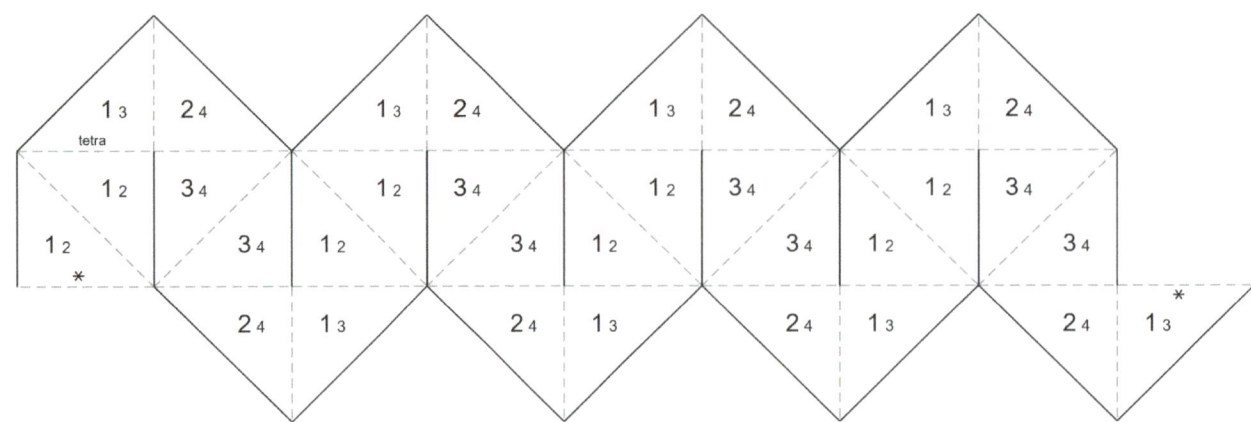

FIG. 31.4 Template for a silver bracelet tetra-16-flexagon.

On this bracelet, you can do a series of eight transfer 2 flexes all the way around it to reveal the two hidden faces, with 3's on the outside and 4's on the inside.

Unfortunately, doing a single transfer 2 on a bracelet flexagon can be tricky without accidentally changing other pats at the same time or having to use a paper clip to make them behave. But fortunately, in the same way a single silver tetra variation on the bronze dodecaflexagon can replace two tuck flexes, we can do a single flex that carries out two transfer 2 flexes at the same time. This flex is much easier to perform without accidentally affecting any other part of the bracelet. See figure 31.5 for the **double transfer 2 flex**. It's intriguing that you can seemingly tangle together opposite sides of the bracelet, yet it has the same effect as if you had done two independent transfer 2 flexes.

1. Pick an initial reference hinge such that the two adjacent hinges are pointing up as shown in the first image of figure 31.5. Align the reference hinge with the vertical hinge on the opposite side of the bracelet. Fold the bracelet flat.
2. For the next few steps, the flex diagram will just show the pats near the center hinge. Grab the left side of the flexagon and pull forward, folding down until the flexagon is flat. This will include the top two pats to the left of the center hinge and the top pat to the right of the center hinge.
3. Fold the left half of the flexagon backward, swinging across the center hinge until it's folded along the back of the right half. On the silver bracelet with 16 pats, the flexagon will now be a square.
4. Stick your finger into the center hinge to separate the front and back pats and open a pocket. Then bring corner A up to corner B, flattening the pocket.
5. Open a pocket at corner C, opening until the flexagon is flat.
6. Open up the center of the bracelet.

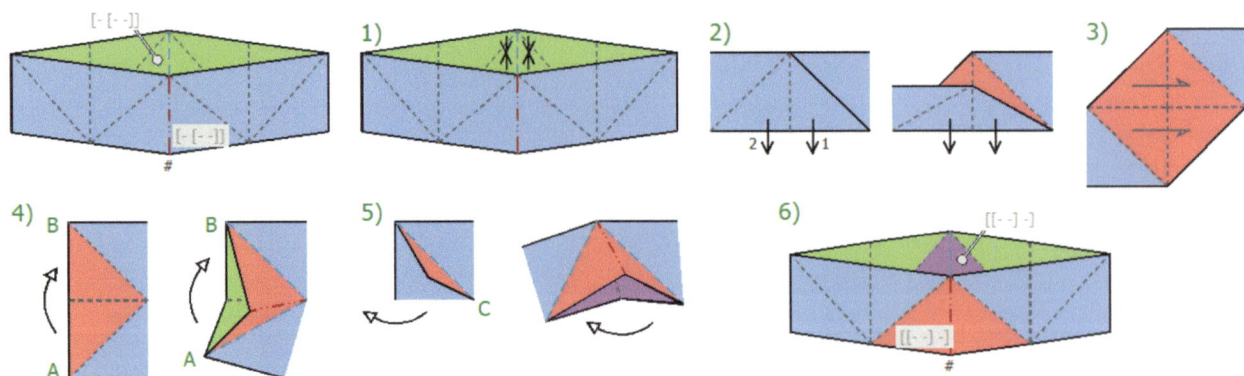

FIG. 31.5 Flex diagram for the double transfer 2 flex.

Rollup flex

Another interesting flex is the **_rollup flex_** (Ru) shown in figure 31.6. This will take you from state 1/2 to 2/3. A second rollup flex takes you to state 3/4. Thus two rollups have the same effect as a sequence of eight transfer 2 flexes.

1. Pick an initial reference hinge such that the two hinges adjacent to it are pointing down as shown in the first image of figure 31.6. Align the reference hinge with the vertical hinge on the opposite side of the bracelet.
2. Flatten the bracelet. Unroll by opening up the pats from the front starting from the right, then the middle, and finally the left side.
3. Roll the flexagon back up by folding the pats backward, starting from the upper right, then the middle, and finishing with the bottom.
4. Rotate the flexagon 90° to the right.
5. Open the entire shape from the bottom and flip around to the back.
6. Open up the flexagon to restore it to the bracelet shape.

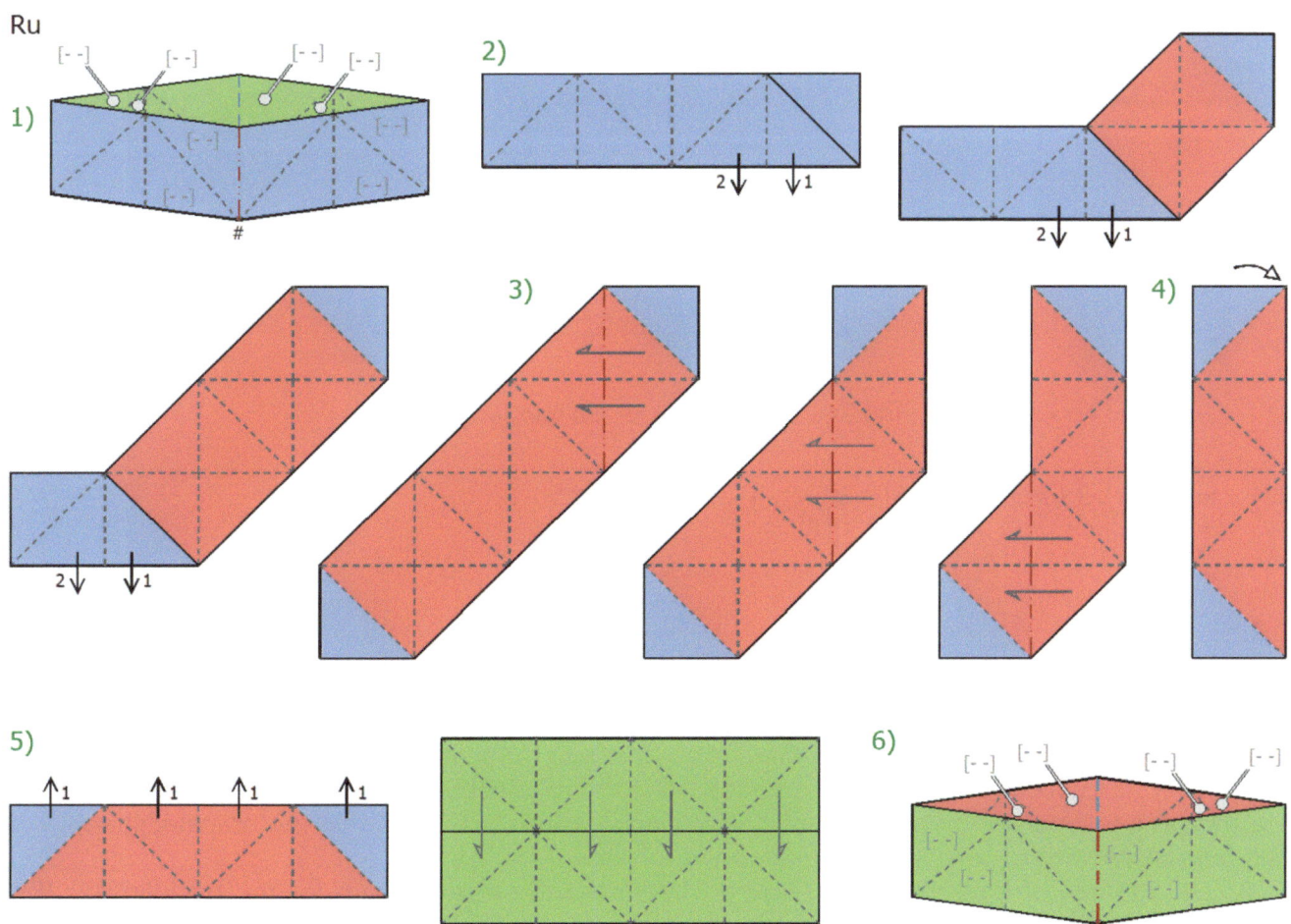

FIG. 31.6 Flex diagram for the rollup flex on a silver bracelet 16-flexagon.

Transfer 4 flex

Next, let's look at the **transfer 4 flex** (Tr4), named for the way it seems to transfer a stack of pats from one end of a set of four pats to the other. Figure 31.7 contains the minimal silver bracelet dodecaflexagon.

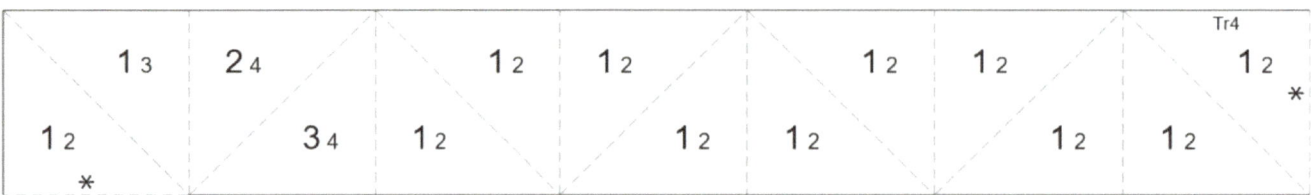

FIG. 31.7 Template for the minimal silver bracelet dodecaflexagon for the transfer 4 flex.

Follow along with figure 31.8 to perform the transfer 4 flex:

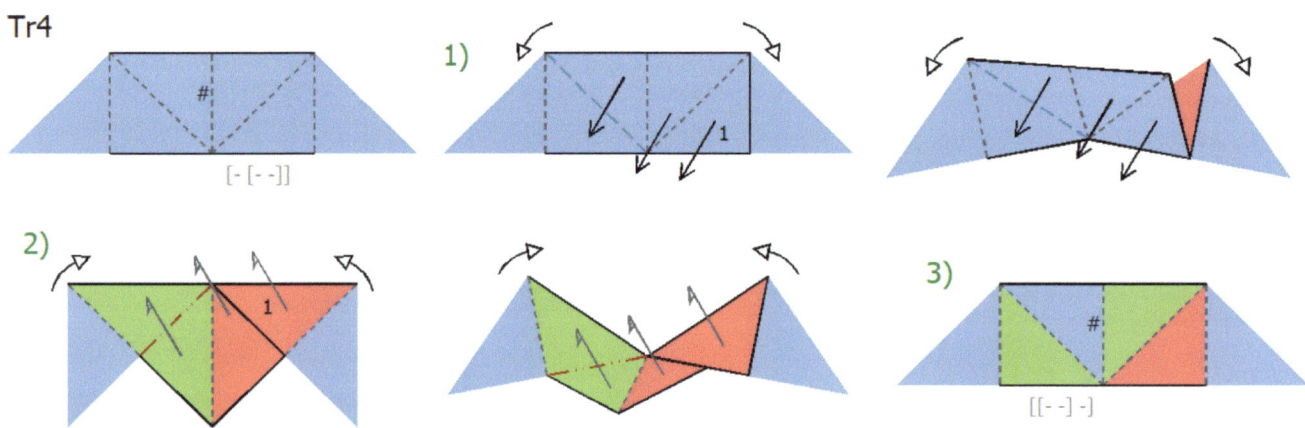

FIG. 31.8 Flex diagram for the transfer 4 flex.

$$\text{Tr4} \quad = \quad a \; 1 \setminus 2 / \# 3 / [4,[6,-5]] \setminus b \to a \; [[-2,1],3] \setminus 4 / \# 5 / 6 \setminus b$$
$$= \quad > \text{Ul} <<< \text{Ur'} >>> \text{Ur} <<< \text{Ul'} >>$$

Morph-kite flexes

Let's examine the pat directions to figure out other flexes we can do. We previously noted that these silver bracelets repeat the pattern //\\; for example, the silver bracelet dodecaflexagon has directions //\\//\\//\\. Recall that the morph-kite flexes start with //// and end with \//\. Since \//\ is a subset of the directions in our bracelets, this means that, with the necessary pat structure, we can do any of the morph-kite inverses, going from \//\ to ////. And if you want to preserve the bracelet shape, you can follow that with a morph-kite flex, taking you from //// back to \//\.

As an example of how this works, figure 31.9 shows the flex diagram for the backflip flex (Bf), which is equivalent to Mkf' Mkb. See the template in figure 31.10 for the minimal silver bracelet dodecaflexagon for the backflip.

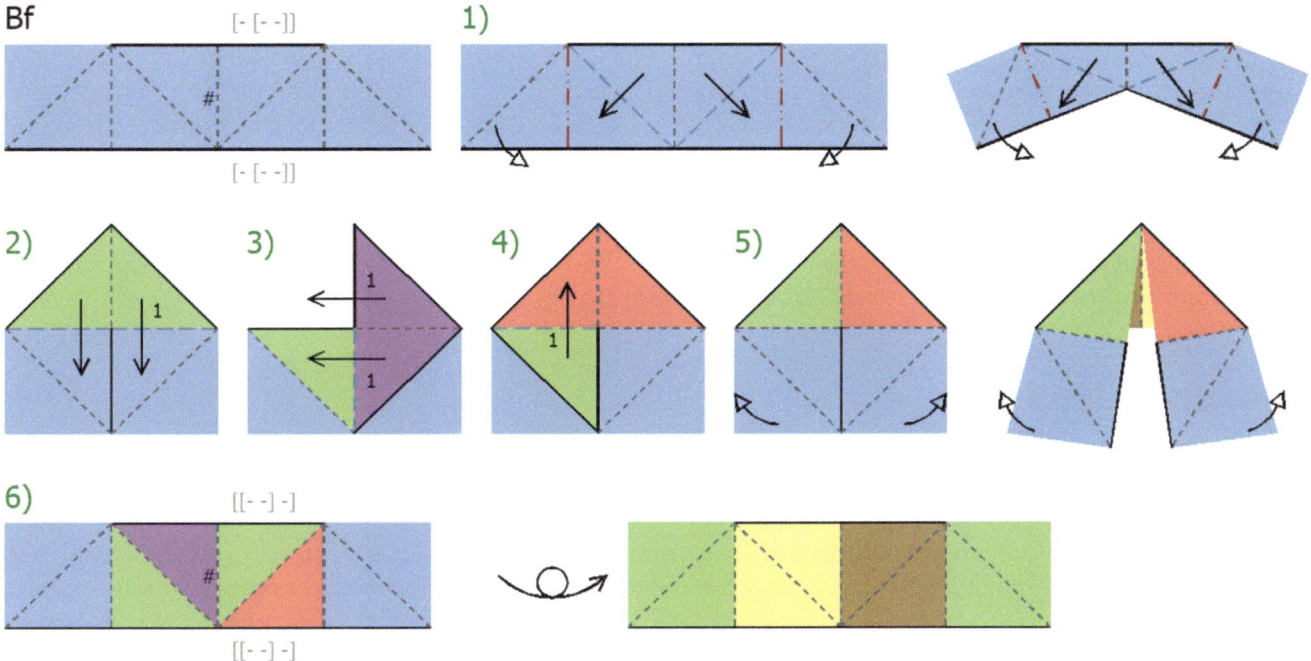

FIG. 31.9 Flex diagram for the backflip on a silver bracelet.

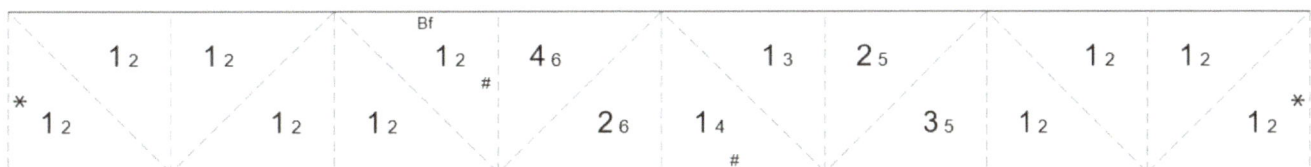

FIG. 31.10 Template for the minimal silver bracelet dodecaflexagon for the backflip.

Notice that the bracelet starts with adjacent pats that have structure [- [- -]], [- [- -]] and, after the backflip, ends with adjacent pats of [[- -] -], [[- -] -]. If you now do a ^ flex, rotating the bracelet left to right 180° so you're looking toward the inside of the bracelet rather than at the outside, you'll see the original structure of [- [- -]], [- [- -]]. That means you can now do a second backflip, but from the inside of the flexagon, which undoes the previous backflip. In other words, Bf' = ^ Bf ^.

But if you instead flip the flexagon top to bottom along an axis perpendicular to the reference hinge, which is flex ~, you can complete a cycle of eight backflips in a row before returning to your original state. In flex notation, this cycle is (Bf <<~)8. If you look at what the backflip does, you'll see that it shifts which pair of pats consist of three leaves over two pats to the left. The orientation of those two pats is flipped, which is why you need the ~ flex before doing the next backflip. Continuing this process all the way around the bracelet returns the pair of three-leaf pats to their starting position. The flex sequence is like a wave traveling around the bracelet.

It may seem surprising that such a simple flexagon supports a cycle. However, it turns out that this flexagon also supports a cycle of transfer flexes: (Tr4<<Tr2<<Tr4<<~)8. Furthermore, (Bf<<~)3 = Tr4<<Tr2<<Tr4<<~, so you can mix and match the two sequences to create a variety of different cycles.

231

Since the backflip simply moves the pair of three-leaf pats to the left, you can extend the length of the cycle by extending the length of the bracelet. For every additional four leaves you add to the template, you need two more backflips to complete the cycle. Or, if you want to throw in an extra twist, you can add two leaves instead of four so you only need one extra backflip to complete the cycle. This requires you to add a half-twist to the folded template before taping it, making the flexagon surface a Möbius strip with only one visible face.

With longer bracelets, you can add an arbitrary number of half-twists to the flexagon surface, though the only property that affects a sequence of transfer flexes is whether or not it has a whole number of twists. With a long enough bracelet, you could even add knots to the flexagon surface. Knots have no impact on local flexes, but may impact global flexes.

Wrapper flexes

At first glance, it may seem like you can't do a flip flex on a silver bracelet, since it requires four pats in a row that all go the same direction, ////, while the bracelets we've been looking at don't have that pattern, instead alternating between // and \\.

But if we first use the proper morphing flex to change the directions between pats, we can then perform the flip flex, finishing with the inverse morphing flex to restore the original directions and return the flexagon to the original bracelet shape. We refer to such a morphing flex as a **wrapper flex**, a general technique for adapting a flex to a particular flexagon.

Figure 31.11 contains a template for the minimal silver bracelet dodecaflexagon for the flex sequence WFW', where W is our chosen wrapper flex.

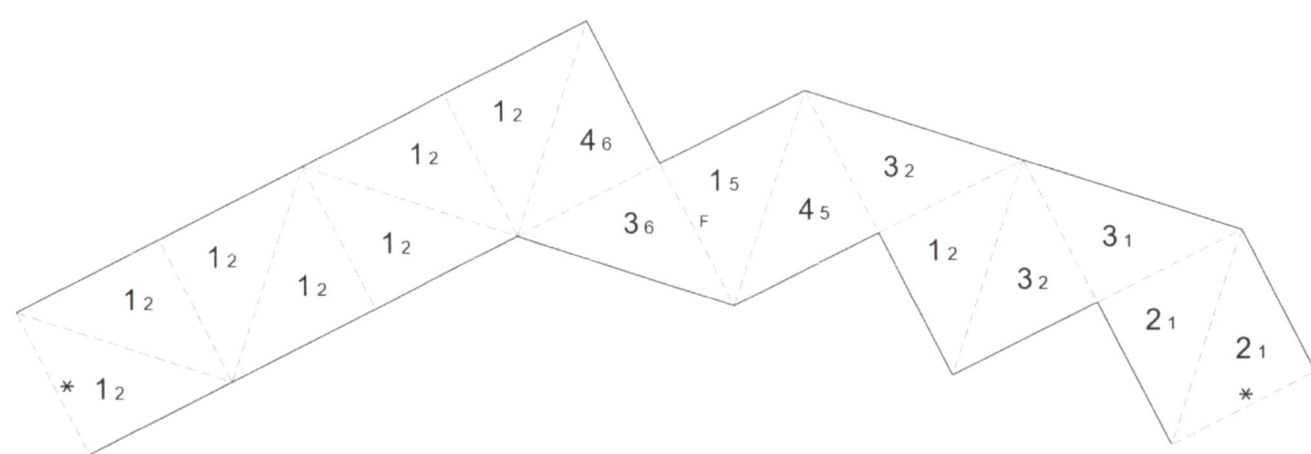

FIG. 31.11 Template for the minimal silver bracelet dodecaflexagon for a wrapper flex around the flip flex.

Use the directions below while following along with the flex diagram in figure 31.12 to do a wrapper flex around the flip flex. Start with the hinge marked F as your reference hinge, with the two adjacent hinges angling up.

1. Pull down on the left and right sides of the reference hinge, which should unfold the pat immediately to the left of the reference hinge. Fold the back half of the left pat against the back of the pat to its right.

2. Use the hinge in the upper right corner as the reference hinge for the flip flex. Start the flip by folding corner A backward up to corner B.
3. Continue with the flip flex, flipping over the top layer of the two triangles pointing down before opening it back up. Refer to figure 12.2 if you need a refresher on the details of the flip flex.
4. Lift up on the left and right sides of the hinge between the leaves labeled 4 (purple) and 5 (yellow), which should unfold the pat immediately to the right of this hinge. Fold the back half of the right pat against the back of the pat to its left.
5. Return to the bracelet position.

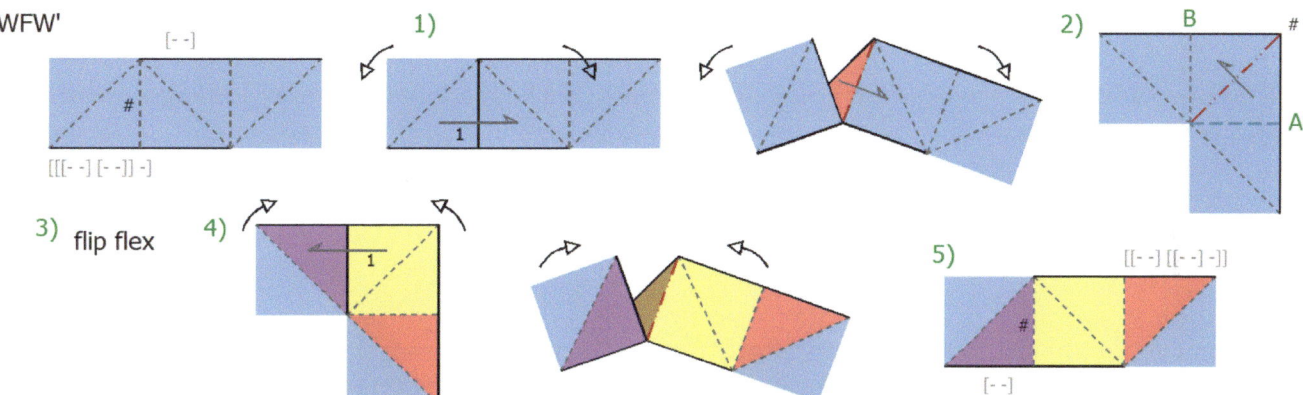

FIG. 31.12 Flex diagram for adding a wrapper flex around the flip flex. Steps 1 and 2 are the wrapper flex. Step 3 is the flip flex. Steps 4 and 5 do the inverse of the wrapper flex.

This wrapper flex has the following definition:

$$W \quad = \quad a\,[1,-2]\,\backslash\,\#\,-3\,\backslash\,-4\,/\,-5\,/\,b \to a\,1\,/\,[-3,2]\,/\,\#\,-4\,/\,-5\,/\,b$$
$$= \quad <Ul >Ur' >$$

You can see how this flex changes the directions \\// into ////, setting the stage for being able to do a flip flex. Other flexagons and flexes may use different wrapper flexes.

Summary

The silver bracelet represents a different arrangement than we've seen before, forming a loop in three dimensions. We learned the transfer 2, double transfer 2, rollup, and transfer 4 flexes. We've seen how to use the morph-kite flexes on bracelets. We explored a bracelet that supports cycles of backflips and transfer flexes, noting that we can extend it to arbitrary lengths, including a Möbius strip or knotted surface. And finally, we learned how a wrapper flex can temporarily morph a flexagon so that you can do flexes that aren't otherwise supported.

Chapter 32

Octagonal Ring 14-flexagon

We've provided lots of specific examples of flexagons and flexes, but there are far more out there to explore. In this chapter, we show how you can use this book's resources to explore other flexagons and flexes by walking through how to create a new flexagon.

Finding a flexagon and flexes

Take a look at appendix B, which contains several charts that categorize flexagons in different ways. In particular, look through the section titled "Directions between pats," which groups flexagons by the patterns of their pat directions. The sorting starts with the isoflexagons (all pats meet in the middle) followed by highly symmetric rings. The last couple rows have some less regular patterns. Let's pick the octagonal ring tetradecaflexagon, with pat directions of //\\//\//\\//\, as shown in figure 32.1. We shorten *tetradecaflexagon* to *14-flexagon* in this chapter, since Greek prefixes (like tetradeca) can get a bit unwieldy for larger numbers.

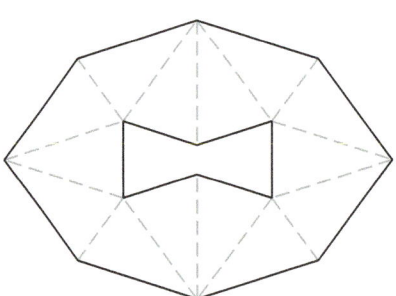

FIG. 32.1 Main position of the octagonal ring 14-flexagon.

Next, take a look through the flex compendium in appendix C to find some flexes that will work on our chosen flexagon. For simplicity, let's limit our search to flexes that support the directions in our flexagon without changing them, skipping over morphing flexes and wrapper flexes.

Find the table of atomic pat notation for local triangle flexes, which groups flexes by pat directions. This 14-flexagon's pat directions contain //, so it supports Tf. Likewise, it includes \\ so it supports Tr2, /\/ for Tr3 (which we'll see later in the chapter), and \//\ for Bf and Tr4. But it doesn't contain the longer sequences of /'s required by S, F, P, or many other flexes. And while it supports the input requirements for Lkk, that flex changes the pat directions, so we'll skip it.

DOI: 10.1201/9781003433538-36

See figure 32.2 for a picture of which hinges can theoretically support the local flexes we found. We can figure this out by comparing the flexagon's pat directions to the flex's requirements for pat directions. For example, the top two center pats have directions \\ and the input to Tr2 puts the reference hinge in the middle of those pats.

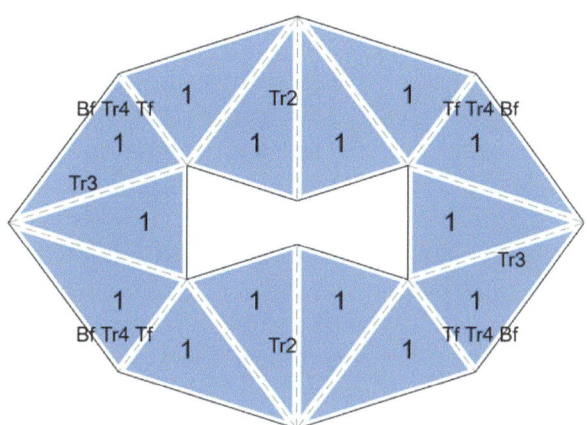

FIG. 32.2 A 14-flexagon showing the hinges that might support various local flexes.

We'll use this picture to help us figure out how to make useful generating sequences for our custom flexagon. For example, if we start with the reference hinge at the top middle hinge, we can use the generating sequence Tr2+. Or, if we want to be able to do the Bf (backflip flex), we need to first shift two hinges to the right, so we'd use the generating sequence >>Tr4+.

Note that the backflip flex is composed of morph-kite flexes, since it's equal to Mkf' Mkb. As we saw in chapter 14, there are lots of ways to combine morph-kite flexes as we did with the backflip, so there are many other flexes we could use on this flexagon. We'll use the backflip as a representative of the class of flexes that consist of an inverse morph-kite and a morph-kite.

If we look through the table of global triangle flexes in appendix C, we don't find any matches. However, the following section mentions that the requirement for a pinch flex variant is to have multiple pairs of /'s around the flexagon, while the directions between those pairs don't matter. We represent a hinge direction that can be either / or \ using a question mark. If we take //\\//\\//\ and replace the hinges between the // pairs with question marks, we get //??//?//??//?. This is the pinch variant P4343 or P3434, depending on which hinge you start with. It can be done at the same hinges where we can do the tuck flex. As with other pinch variations, a single P3434 changes the pat directions, while a double P3434 – denoted by P3434d – preserves the pat directions.

So, we've figured out that our 14-flexagon has the ability to do the non-morphing flexes Tf, Tr2, Tr3, Tr4, Bf, and P3434d. Now what?

Flexagon Inspector

We could use the flex definitions and atomic flex theory to figure out how to create templates that support these flexes. But an easier way is to use an online tool called Flexagon Inspector, a companion app to the book, to make the templates for us.

Go to https://loki3.github.io/flex/inspector.html from your favorite browser. The top of the page should look similar to figure 32.3, with possible variation depending on your system.

| clear | overall shape: | ▾ | leaf shape: | ▾ | pat count: 6 ▾ |

| generating sequence | | create pats | |

FIG. 32.3 Interface for creating a custom flexagon.

By default, it shows a hexa-hexaflexagon made from a straight template, but we want to tell it to display our octagonal ring 14-flexagon. Pick the list to the right of *pat count* and select 14. The flexagon below these controls should switch to now show an isosceles 14-flexagon. Next, pick the list to the right of *overall shape* and select *octagonal ring*. The display should now show the octagonal ring 14-flexagon plus useful details about it – flexes it supports, the leaf angles, the directions between the pats, and the number of leaves – as illustrated in figure 32.4. The numbers in square brackets represent the angles of the triangular leaf: in this case 72°, 72°, and 36°. The large image on the left is the front face of the flexagon, while the small image on the right is the back face, flipped left to right. The star in the top middle of the front face marks the current hinge. Each pat on the front face shows its visible leaf ID and internal pat structure, neither of which is interesting yet.

flexes: Tr2 Tf P3434d Bf Tr4 Tr3

octagonal ring tetradecaflexagon   [72, 72, 36]   V/V/\V/V/\

14 leaves

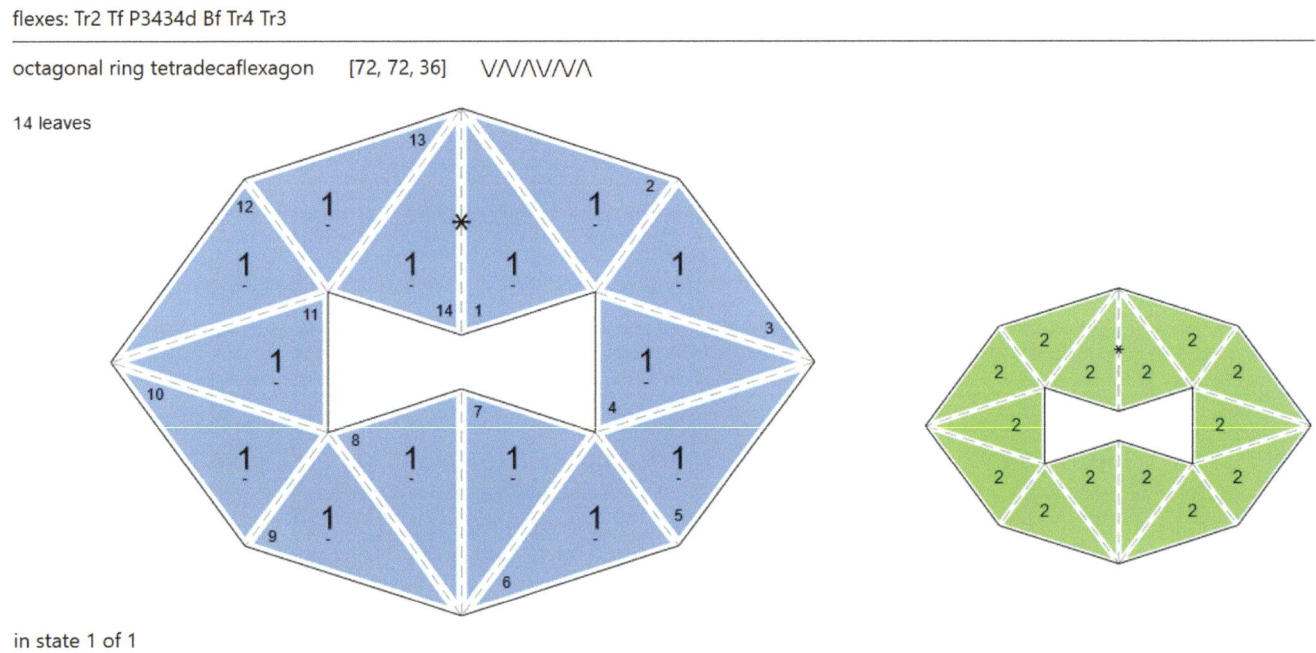

in state 1 of 1

FIG. 32.4 Resulting display after selecting pat count: 14 and overall shape: octagonal ring.

Notice that the display says this flexagon has 14 leaves. Since this is a 14-flexagon, it has 14 pats and hence a single leaf per pat, and we can't yet do any flexes on our new flexagon. Let's use a generating sequence to make a flexagon that we can flex.

Looking back at figure 32.2, we can see that the upper middle hinge could support the Tr2 flex. Type Tr2+ into the box to the right of the *generating sequence* button and press the Enter key. Recall that this says to create the pat structure needed to support the Tr2 without actually performing the flex. Several portions of the display should change as a result, as shown in figure 32.5:

- The name has been updated to *octagonal ring tetradecaflexagon (generator: Tr2+)*.
- The number of leaves is now 16.
- The front face of the flexagon now says Tr2 at the top hinge, indicating that you can perform that flex there.
- The pat to the right of Tr2 says [- [- -]], indicating that it now has three leaves in it.

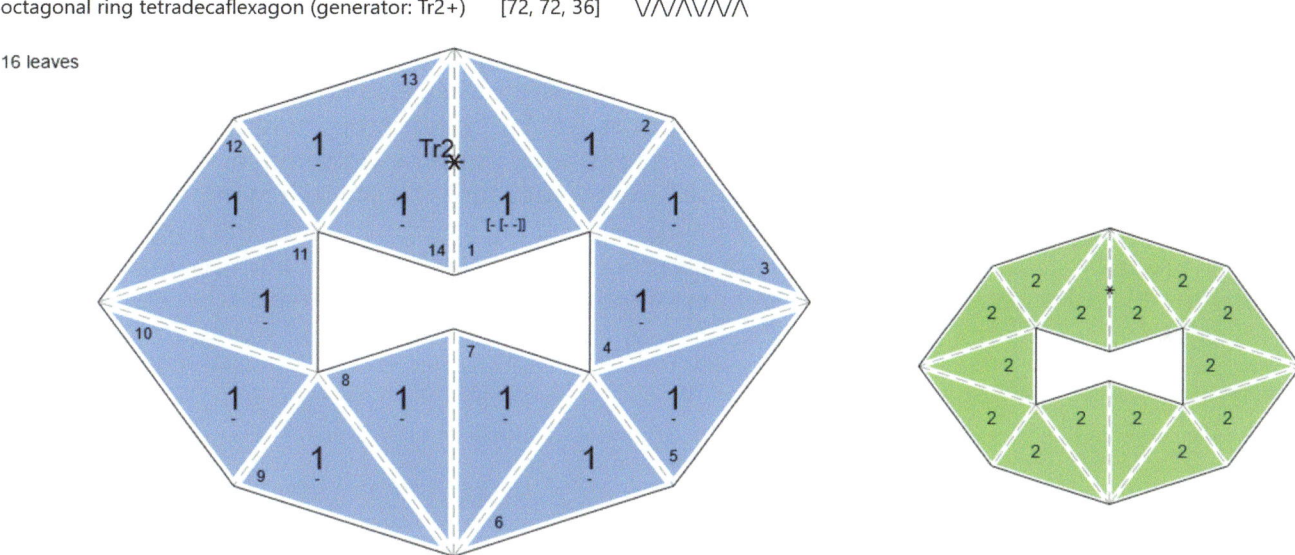

FIG. 32.5 Resulting display after entering generating sequence: Tr2+.

If you click on the Tr2 text at the top hinge, the app will perform the flex and show the results, as seen in figure 32.6.

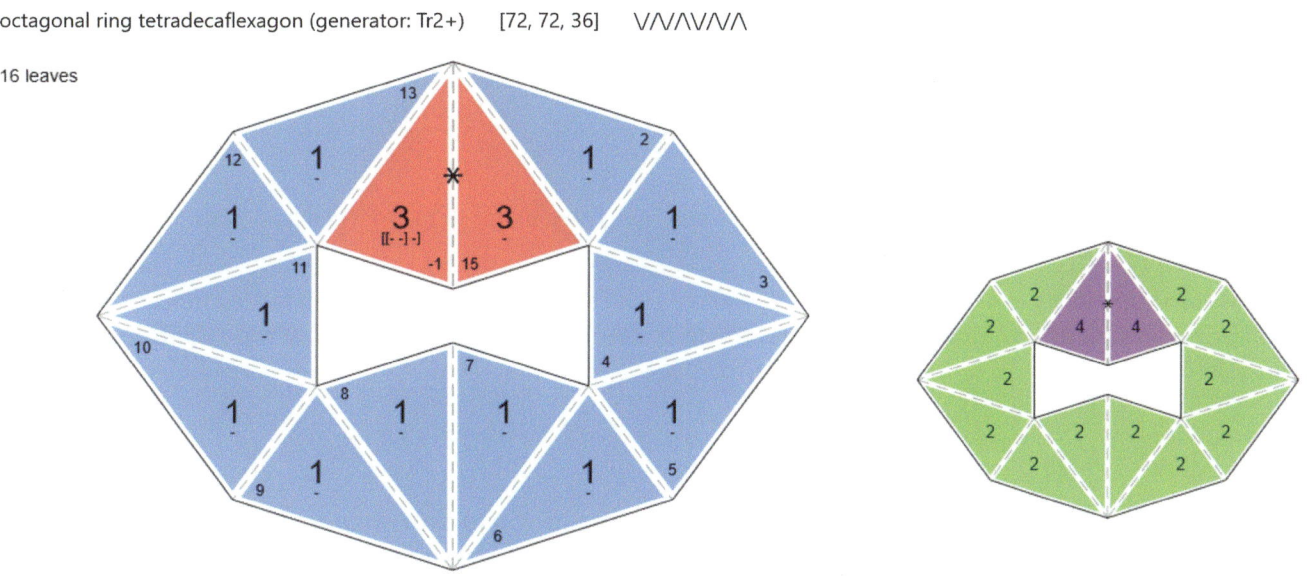

FIG. 32.6 Resulting display after clicking on Tr2 to apply the flex.

Note that it also changed the text in the lower left corner to reflect that the flexagon is in a new state. We have visited two different states, one before Tr2 and one after, and we're currently in the second state. Click on the smaller flexagon on the right to turn the flexagon over, making the green and purple face the front face. You should once again see Tr2 in the upper middle portion of the larger flexagon. Click on it to perform the Tr2 flex a second time, restoring the two faces to solid green and solid blue. Clicking on the smaller blue flexagon will turn it over again, returning you to the original position. Of course, this is a fairly uninteresting flexagon so far, since the only flex you can do is a Tr2 at a single hinge, which simply switches you back and forth between two states.

Let's come up with a generating sequence that adds the structure needed to do Tr2 and Tr4 at every hinge that could support them. Looking back at figure 32.2, we see that the top middle hinge can support Tr2. Two hinges to the right can support Tr4. Stepping three more hinges to the right can also support Tr4. And stepping two more hinges brings us to the bottom hinge, where we can repeat the same pattern on the other half of the flexagon because the shape is symmetric. Thus, the generating sequence (Tr2+ >> Tr4+ >>> Tr4+ >>)2 will add the structure needed to support Tr2 and Tr4 at all the appropriate hinges. Type in that generating sequence (replacing any sequence that was already there) and press Enter to create our new flexagon, as shown in figure 32.7.

octagonal ring tetradecaflexagon (generator: (Tr2+ >> Tr4+ >>> Tr4+ >>)2)   [72, 72, 36]   V/\V/\V/\

26 leaves

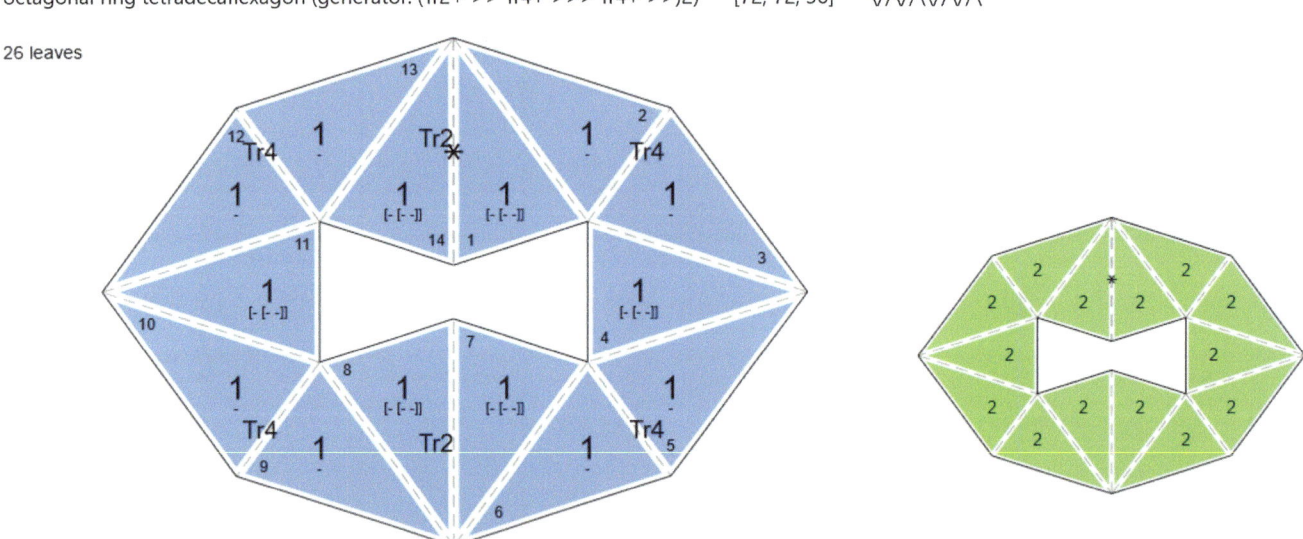

in state 1 of 1

FIG. 32.7 Resulting display after entering generating sequence: (Tr2+ >> Tr4+ >>> Tr4+ >>)2.

You can now click on any of the hinges that say Tr2 or Tr4 to perform the flex, clicking on the smaller flexagon to flip it over if you want to do a flex on the other face. How many different states can you find?

Custom templates

Now that we can create flexagons from a generating sequence and explore them in the Flexagon Inspector app, we would also like to create a template so we can make an actual flexagon.

Let's first try this with the minimal 14-flexagon for the Tf. Enter >>Tf+ as a generating sequence to create the flexagon. Below the flexagon, you will find the template. Depending on the size of your screen, you may need to scroll the page down to see it. See figure 32.8 for the resulting template.

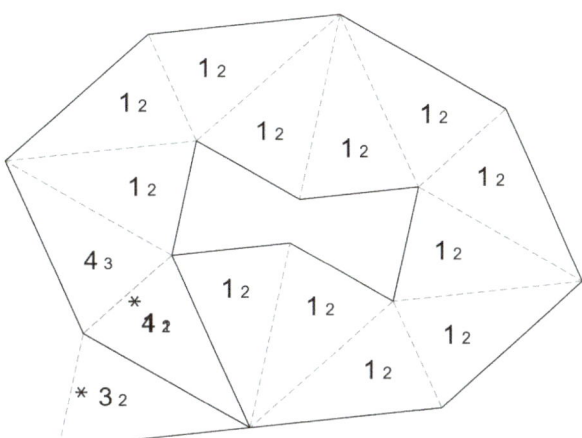

FIG. 32.8 Template for the octagonal ring 14-flexagon (generator: >>Tf+).

The first thing to notice about this template is that it overlaps itself, with one leaf seemingly connected to three other leaves and having a mix of numbers sitting on top of each other. Overlapping leaves is a common challenge when making templates, since they can wind all over the place. There are a few ways to address this.

Sometimes you can simply shift the leaf that the template begins with to avoid overlap. In the box next to the *apply flexes* button above the template, type > and press Enter. This shifts the current hinge one to the right, causing the template to be sliced in a different place, which avoids the overlap, as shown in figure 32.9. Sometimes you may need to try shifting multiple times, though this only works in some cases where there's minimal overlap.

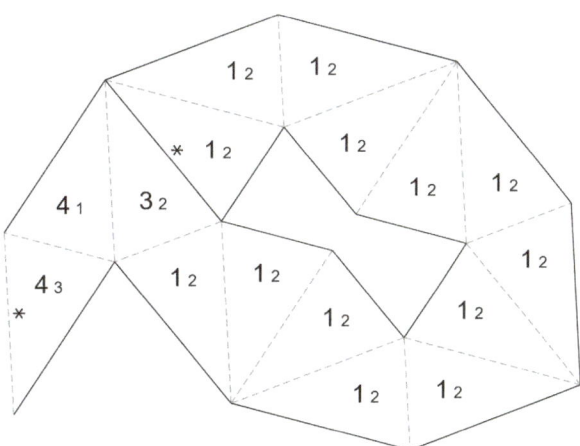

FIG. 32.9 Template for the octagonal ring 14-flexagon (generator: >>Tf+) after shifting one hinge.

But if we switch back to the generating sequence (Tr2+ >> Tr4+ >>> Tr4+ >>)2, we'll find that the template twists over itself so many times that it can't be fixed simply by changing the current hinge. In this case, we need to slice the template into multiple pieces.

Find the *leaves per slice* box right above the template area. This lets you tell it how to slice up the template by specifying how many leaves to put in each slice. If you type in 6 and press Enter, it will put six leaves in the first slice and the remaining leaves in the second slice right below. Six is the largest slice we can start with, since the first seven leaves overlap. Typing in 6,6 puts six leaves in the first slice, six in the second slice, and the remainder in the third slice. With some further experimenting, you'll find that 6,6,6,6 gives you a template sliced up into non-overlapping pieces.

These tools allow you to create custom flexagons and templates for any generating sequence.

Transfer 3

Earlier, we glossed over one of the flexes we found that can work on this 14-flexagon: the ***transfer 3 flex*** (Tr3). We saw both the transfer 2 and transfer 4 flexes in chapter 31, but the transfer 3 is new. Similar to the other transfer flexes, it gives the impression of transferring leaves from one pat to another, but this time across three pats. And, like the other transfer flexes, it pivots pats through the center of the ring of this 14-flexagon.

Use a generating sequence of <<<Tr3+>>> to make the minimal octagonal ring 14-flexagon for the Tr3 in the Flexagon Inspector. This will produce a non-overlapping template that supports the Tr3 at one hinge.

There are several clues you could use to try to figure out the Tr3 flex. Clicking on Tr3 in the Flexagon Inspector will show you what the flexagon looks like after the flex. The flex compendium in appendix C shows the atomic pat notation for its starting and ending state. And the table in the compendium that decomposes flexes into simpler flexes shows that it can be expressed as Ur << Ul' >> Ul << Ur' >. You might find it interesting to try to figure out how to do the flex on the minimal flexagon just given this information.

Or you can refer to figure 32.10 to see how it's done and compare these steps to the decomposition.

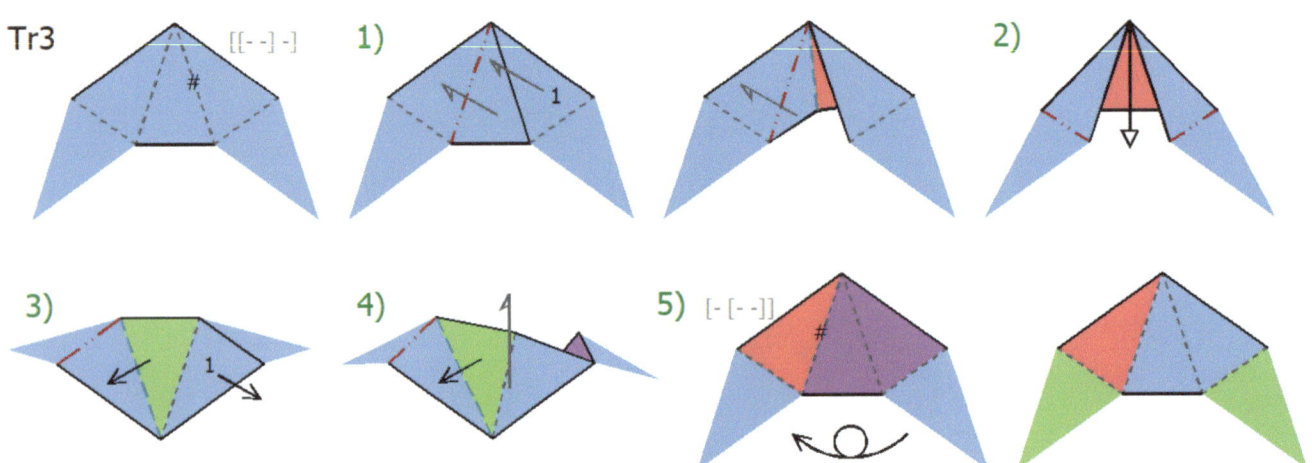

FIG. 32.10 Flex diagram for the transfer 3 flex, Tr3.

We could do more with this flexagon: create more complex templates, try different flex sequences, discover new flexes, add wrapper flexes to enable flexes such as the flip and pyramid shuffle, and try morphing flexes. But we'll leave that for you to explore.

Summary

In this chapter, we saw how to custom design a new flexagon. We first used the appendices to pick an interesting shape and figured out some flexes that work on that shape. We then used the Flexagon Inspector tool to build templates based on generating sequences. These techniques can be used to experiment with many other flexagons and flexes. The next chapter dives more deeply into using Flexagon Inspector to create and explore.

Chapter 33

Flexagon Inspector

There's an endless variety of flexagons and templates to explore. A book can only contain a tiny sampling, which is why we also included recipes for how to create new ones. And since it can be tedious to follow that recipe by hand, we've also provided a tool called Flexagon Inspector to help you create and explore your own custom triangle flexagons. We introduced it in the last chapter, but now we'll go into more detail about what you can do with it.

Flexagon Inspector can be used to generate templates for flexagons with custom behavior and to explore the states you can flex between. An interactive version is available at https://loki3.github.io/flex/inspector.html. It's built using a more general-purpose flexagon library called flexagonator. The full source code for flexagonator is available at https://github.com/loki3/flexagonator.

The Inspector consists of three sections: one for *creating*, one for *interacting*, and one for the *template*. To create a new flexagon, you choose the parts of its name that determine its outward appearance, then use a generating sequence or pat structure to determine its internal structure. You can interact with the resulting flexagon, seeing the results of performing flexes at different hinges. The final section shows the template for your flexagon, allowing you to slice it up if needed.

When you first open the Inspector, it shows a hexa-hexaflexagon made from a straight template. You can click on the flex shortcuts such as P or T to perform a flex at the associated hinge. Looking at the bottom of the page (scrolling down if needed), you will find the corresponding template.

Next let's look more closely at what you can do in each section of the Inspector.

Create

The options for creating a flexagon are at the top of the page. See figure 33.1 for what this may look like (the exact details depend on your system and may change if the Inspector app is updated).

| clear | overall shape: [▾] leaf shape: [▾] pat count: [6 ▾] |

| [generating sequence] [_____] [create pats] [_____] |

FIG. 33.1 Initial view of the creation portion of Flexagon Inspector.

The first step in designing a flexagon is to choose its outward appearance by picking the components of its name. Start by selecting the *clear* button to clear out any parts of the name already selected. Take some time to look through the options by clicking on the dropdown lists for *overall*

DOI: 10.1201/9781003433538-37

shape, *leaf shape*, and *pat count*. You will see many names you recognize from other parts of the book plus some new ones.

Note that these lists don't contain all possible flexagons. They only show triangle flexagons that generally lie flat in their main position and have a limited number of pats because these are easier to print out and flex.

When you pick an option in one of the lists, it will filter the other lists down to just the items that work with the selected option. For example, if you pick *pentagonal* for *overall shape*, it filters *leaf shape* to *isosceles*, *right*, and *silver*, while *pat count* is filtered down to *5* and *10*. The Inspector doesn't have any pentagonal flexagons with bronze leaves or seven pats, so those aren't listed as options.

Once you have selected a pat count, the named flexagon appears in the interactive section and you're ready to define its internal structure using either a generating sequence or pat structure.

To demonstrate how to use a generating sequence, let's first pick *overall shape: pentagonal* and *pat count: 10*. The top portion of the Inspector should look something like figure 33.2. You should see a list of some of the possible flexes that you could perform on the pentagonal decaflexagon. (Note that P334d and P55d are the double versions of the P334 and P55, respectively, equivalent to (P334)2 and (P55)2, which preserve the overall shape.) Also, note that its leaves are triangles with angles of 36°, 90°, and 54°, indicated by the numbers in square brackets.

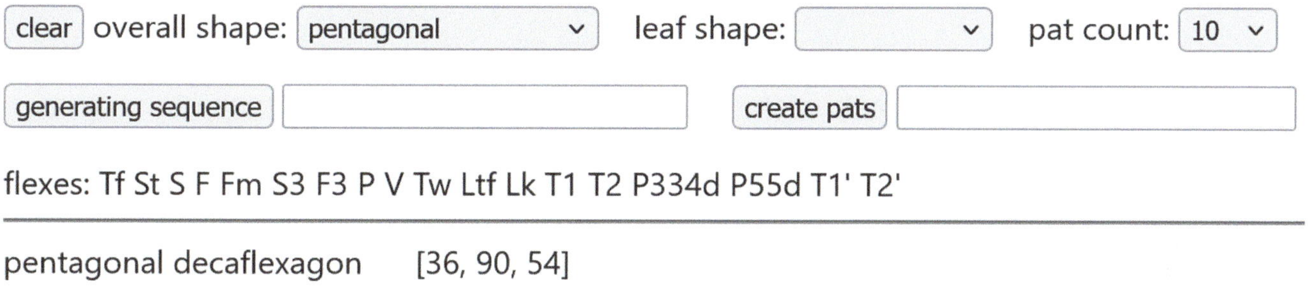

| clear | overall shape: | pentagonal ⌄ | leaf shape: | ⌄ | pat count: | 10 ⌄ |

| generating sequence | | create pats | |

flexes: Tf St S F Fm S3 F3 P V Tw Ltf Lk T1 T2 P334d P55d T1' T2'

---

pentagonal decaflexagon   [36, 90, 54]

FIG. 33.2 The top portion of Flexagon Inspector with the features for a pentagonal decaflexagon selected.

Let's say we want to create the minimal flexagon for the pyramid shuffle. Type S+ into the box next to the *generating sequence* button and press Enter. You should see the interactive flexagon change to indicate that you can now perform flexes at a couple different hinges. You can use any valid generating sequence to create the flexagon's structure, such as S*>>Tf*>>^St*. If you've mixed up the flexagon by selecting a series of flexes and want to restore it, you can click the *generating sequence* button to start over in its original state.

The second way to define the flexagon's internal structure is to use pat notation. You can type it into the box next to *create pats* and press Enter. If you want to reset your flexagon after mixing it up with various flexes, you can click on the *create pats* button to return it to the original state.

For example, if you want your flexagon to alternate between one and two leaves per pat, you can type in [[0,0],0, [0,0],0, [0,0],0, [0,0],0, [0,0],0] (where the 0's indicate that you don't care what ID gets assigned to each leaf). Of course, for such a simple repeating pattern, that's a lot to type in, so you can also just type in two pats – [[0,0],0] – and the Inspector will automatically repeat the pattern until it's defined all the pats. Note that you wrap the entire list of pats with square brackets, so [[0,0],0] means the two pats [0,0] and 0.

The Inspector also offers shortcuts for specifying pats with one, two, three, or four leaves, covering all the possible ways to make pats with up to four leaves:

1 = 0
12 = [0,[0,0]]
112 = [0,[0,[0,0]]]
12-1 = [[0,[0,0]],0]
4 = [[0,0],[0,0]]

2 = [0,0]
21 = [[0,0],0]
211 = [[[0,0],0],0]
1-21 = [0,[[0,0],0]]

This allows you to more succinctly specify alternating one and two leaf pats using 1,2,1,2,1,2,1,2,1,2 or just 1,2. Note that you leave off the square brackets when using these shortcuts.

Flexagon Inspector uses the tree labeling technique described in chapter 27 to assign numbers and colors to each leaf after you've specified a generating sequence or pat structure.

Interact

The middle section of Flexagon Inspector allows you to interact with the flexagon. It shows the outside of the flexagon with a summary of the internal pat structure, and allows you to apply various flexes. It also includes buttons for *undo* and *redo* followed by a history of the flexes you've performed.

To explore what the interactive section can do, create a *hexaflexagon (generator: (P*)3)* using the following steps: in the creator section, pick *clear*, select *pat count: 6*, and enter a generating sequence of (P*)3. After you do this, the interactive section should look similar to figure 33.3.

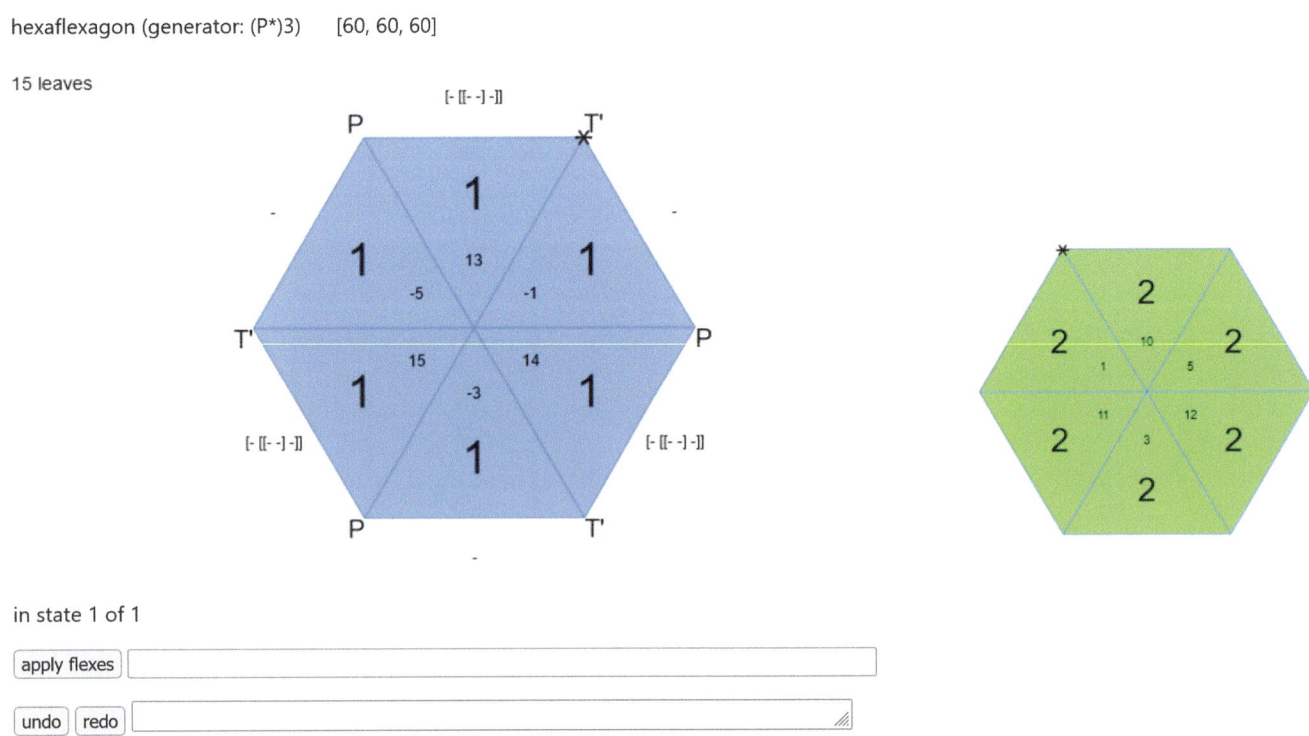

hexaflexagon (generator: (P*)3)   [60, 60, 60]

15 leaves

in state 1 of 1

apply flexes

undo   redo

FIG. 33.3 The interactive portion of the Inspector showing the hexaflexagon (generator: (P*)3).

In this view, you can see the name of the flexagon, the leaf angles, the number of leaves, the front and back of the flexagon, a state tracker, a place to enter flex sequences, and the flex history. The large hexagon is the front face and the small hexagon is the back face. Each face shows the label (large number), leaf ID (small number), and color of each visible leaf-face. Each hinge shows the flexes that can be performed at it (if any), with a star (*) marking the current reference hinge. Each pat shows pat notation for its internal structure. For the initial state of the hexaflexagon we're looking at, it shows that three pats have a single leaf (marked as -) and three pats have four leaves with more complex structure (marked as [- [[- -] -]]).

Click on any of the flexes to apply that flex at the given hinge. Click on the smaller hexagon representing the back face to turn over the flexagon so that it becomes the front face.

To walk through an example, start by clicking on the T′ at the upper right (where the asterisk is) to apply the inverse tuck flex. You should see two pats on the front and back faces change. Now click on the P on the right side of the flexagon to perform a pinch flex. The resulting view should look like figure 33.4.

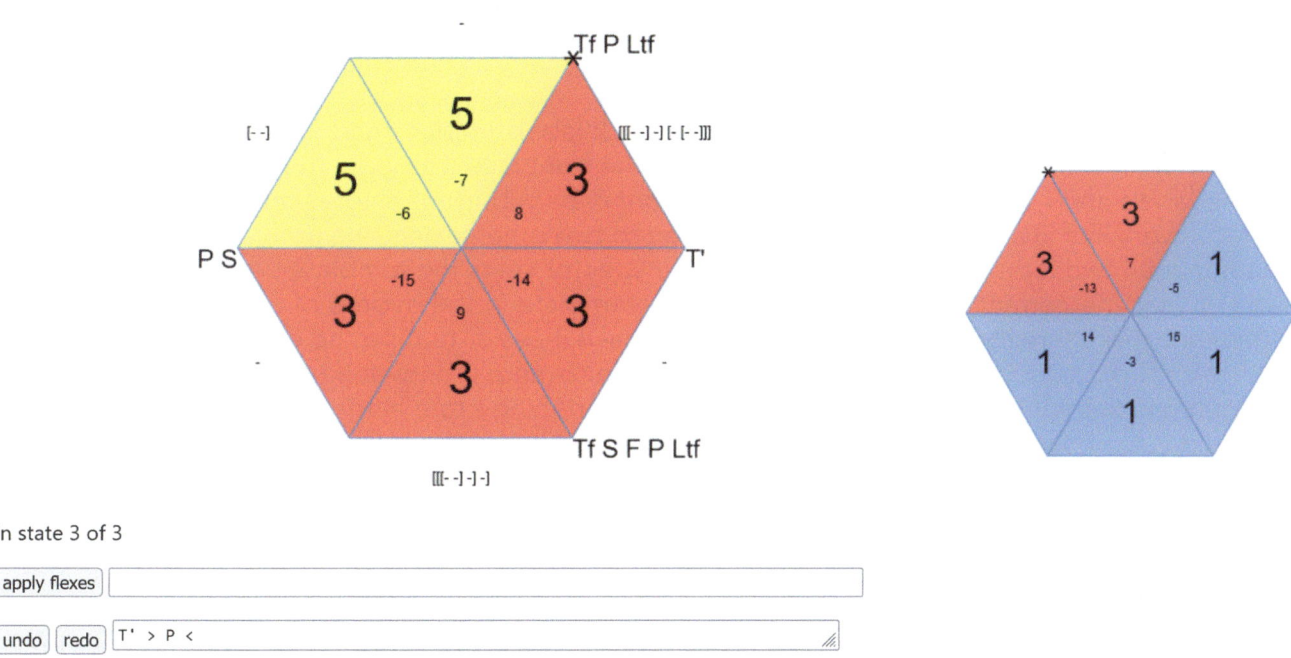

FIG. 33.4 The results after applying the flex sequence T′>P<.

Notice that the box at the bottom next to the *redo* button is keeping track of the flexes you've done. You can click on *undo* and *redo* to step back and forth through these flexes. Or, if you undo all the flexes you've done, you could type T′ >P< into the *apply flexes* box and press Enter to do those same flexes all at once.

Note that the Inspector only displays prime flexes at the hinges in order to keep the display from getting too messy, since sometimes a single hinge could support a dozen or more different flexes. This means that it may be possible to perform flexes that aren't listed at the hinges. In chapter 19, we learned that the relatively prime flexes on the hexaflexagon are F, Ltf, Ltb, Ltb′, P, S, T, T′, and V. The creation section at the top of the page lists additional flexes such as Fm and S3 that you may also be able to do.

The other useful piece of information is the state tracker. After you apply T′ >P<, it says *in state 3 of 3*. This is keeping track of how many unique states you've visited so far and which of those states you're currently in.

Feel free to mix up this flexagon as much as you want, since you can always undo your flexes or reset it by clicking on generating sequence again. How many different states can you find?

Template

The section at the bottom of the Inspector page shows the template for the current flexagon. The template follows the same conventions we've used throughout the book for labeling and assembling. You can choose whether to show the front and/or back face and to split the template up into pieces. The options interface looks something like figure 33.5.

show faces: front & back front back   leaves per slice [   ]

FIG. 33.5 Interface for controlling how the template is displayed.

By default, the template shows the front and back labels in black and white. If you click on the *front* button, it shows just the labels and colors for the front face. Likewise, the *back* button will show just the labels and colors of the back face. The *front & back* button restores the default view.

Since every leaf in a folded flexagon is attached to two other leaves, the unfolded template has to split between two leaves somewhere. Flexagon Inspector splits the template between the two pats that straddle the current hinge. It never splits the template in the middle of a pat, since this could sometimes make it really difficult to assemble. To make it easier to tape the ends together, you may want to shift the current hinge to a different hinge, for example between two single-leaf pats if the flexagon has them. Or you may want to shift the current hinge to a hinge that gets less use, so the wear and tear of repeated flexing doesn't ruin the taped hinge too quickly. This is as simple as typing > into the *apply flexes* box.

The one tricky aspect of templates is that sometimes they can twist around themselves, causing them to overlap. There are several possible ways to fix this: shift the current hinge, add additional pat structure, or slice the template into multiple pieces.

The previous chapter on the octagonal ring 14-flexagon gave an example of how you can shift the current hinge to see if that's sufficient to avoid overlap. Sometimes typing > and hitting Enter a few times can get rid of the overlap. Note that this only works some of the time, if the overlap isn't that severe.

A second possibility is to add more leaves to the template (or remove some). Perhaps adding another flex to your generating sequence or more nested structure to your pat notation will cause the template to twist in a different direction, avoiding the overlap. This may take some experimentation.

A third option is to slice the template into multiple pieces. You can do this by typing how many leaves you want in each slice in the *leaves per slice* box, with a comma separating each slice count, and hitting Enter. For example, typing in *4,2* puts four leaves in the first slice, two leaves in the second slice, and any remaining leaves in the third slice. Chapter 32 gave an example of how to do this.

Script

Flexagon Inspector is built on the flexagonator toolkit, which does more than what the Inspector interface shows. It can be told what to do through **script** commands. You can get access to the script commands by clicking on the *advanced options* checkbox located below the flex history box. This will reveal the *creation script* and *apply script* boxes.

Every time you create a new flexagon by changing the name or internal structure, the creation script is updated to reflect the current settings. You can make a copy of this script in order to save a particular flexagon or to use it with other tools that can be found with the flexagonator library such as the Flexagon Explorer, which can find every state of a flexagon.

The *apply script* box can be used to customize the behavior of the Inspector, such as adding new flex definitions or using a different strategy for labeling the leaves. As a simple example, let's see how we can change the flexes displayed on the flexagon in the interactive section. As we noted earlier, the Inspector defaults to just showing the prime flexes that can be performed on a hinge. You can see this list in the creation script under the *searchFlexes* command. If you instead wanted it to just show P and St (note that the silver tetra flex isn't prime), you could type the following script into the box under *apply script* then click the *apply script* button:

[{"searchFlexes": "P St"}]

Note that the Inspector will display an error message immediately under the box if there's a problem with the script. See the flexagonator documentation at https://github.com/loki3/flexagonator/tree/master/docs for further script commands.

"Solving" a flexagon

After applying a few flexes on even a relatively simple flexagon, it can be challenging to get back to the original state. It can end up feeling a lot like trying to solve a twisty puzzle such as a Rubik's Cube. But there are several key differences that make it a very different sort of puzzle to solve.

A Rubik's Cube has over 43 quintillion arrangements, while the flexagons in this book only have dozens to thousands to hundreds of thousands. That may sound like flexagons must be much easier to solve, but that's not the case.

One wrinkle to flexagons is that much of their state is hidden. You can see the outside leaf-faces, feel how thick the pats are, and probe some of the pat structure with your fingers, but you can't always see all the internal details. In contrast, you can always see every interesting part of a Rubik's Cube simply by rotating it in your hands.

A second wrinkle is that moves on a Rubik's Cube form a mathematical group, while flexes on a flexagon typically don't, as we saw in chapter 22. The practical impact of this is that you can always do the same repeatable sequences of moves on a Rubik's Cube to make predictable changes, but this strategy doesn't work on a flexagon, since the required pat structure isn't always present.

There's also the question of how you define *solved*. Is it when you get to one particular state? Or is it when you get back to any set of solid-colored faces? Another option would be to design your faces such that there are multiple objectives, with some more difficult than others.

So how do you solve a flexagon? You could make a diagram of all the states using the techniques described in chapter 17, then refer to your diagram to help you navigate from where you are to where you want to be. Of course, this is only practical for a small number of states.

But what do you do when you have a random, possibly very complex flexagon? Then it helps to have a more thorough understanding of what each flex does plus the ability to recognize when the current pat structure supports a given flex or flex sequence. A combination of practice and looking at the pat notation and flex diagrams for the various flexes can help. Knowing all the prime flexes for a given flexagon is essential for being able to solve from any state, but non-prime flexes can offer shortcuts.

For example, a couple interesting flex sequences that make localized changes are F>T^>>S<^, which changes the outside leaf-faces of a single pat, and T<T^<T>T<T^, which only changes a single outside leaf-face on a single pat. If you've worked with a Rubik's Cube, sequences that make such

small changes may sound wonderful. But of course, on a flexagon, those sequences can only be done if the directions between the pats match the flex requirements and the proper internal pat structure is in place because they actually rearrange more leaves than you can see from the outside.

Can you figure out strategies for how to apply flexes to restore a mixed-up hexaflexagon? How about other flexagons?

Explore

You now have tools you can use to set off on unguided exploration. Here are a few flexagons and generating sequences you can try with Flexagon Inspector if you want to experiment with more flexagons;

- The hexagonal regular decaflexagon:
    - The generating sequence Tf+>>>>>>S+F+ produces a flexagon that supports the tuck, pyramid shuffle, flip, and transfer 3 flexes.
    - Or learn the **three-and-open flex** (Tao) as shown in figure 33.6 and try the generating sequence <(Tao*)14. Can you figure out what's special about repeating this flex 14 times?
    - Make a hexagonal regular decaflexagon where every pat has two leaves. Can you discover new flexes?

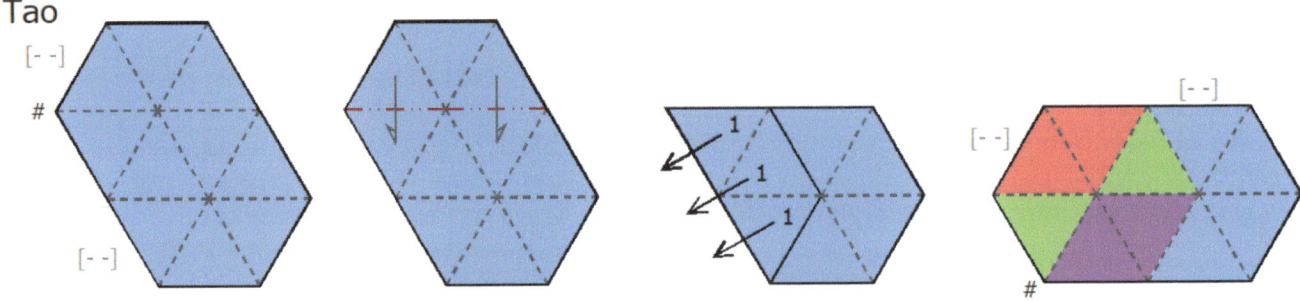

FIG. 33.6 Flex diagram for the three-and-open flex (Tao) on the hexagonal regular decaflexagon.

- The octagonal ring dodecaflexagon:
    - Experiment with different lengths of the sequences (Bf*<<<)n, (Bf*>>>)n, (Tr4*<<<)n, or (Tr4*>>>)n. How many times do you need to repeat the sequence before the original two faces are entirely replaced?
    - Flexagon Inspector points out that the turn flex (Tu) and double slide (Ds) are both possible flexes. We learned those flexes in chapter 30 when we looked at the hexagonal silver dodecaflexagon, which has the same number of pats and the same pat directions. Can you figure out which hinges to use in order to create valid generating sequences? Can you figure out how to perform those flexes on this different shape? Do you understand why the double slide works better than the turn flex on this flexagon?
- The bracelet regular dodecaflexagon:
    - Try the generating sequence (Tr3*<)2 to give you a flexagon that supports the cycle (Tr3<)15.

What other flexagons and flexes are interesting to experiment with?

Summary

After introducing the Flexagon Inspector app in the previous chapter, we explored it in more detail here. We saw how to create flexagons, define their internal structure, explore their states, and create templates. After covering the basics, we introduced the script tools that allow advanced customization, using the flexagonator toolbox. To cap off the chapter, we suggested some avenues for further exploration.

PART FIVE

Fun with Flexagons

In the final part of the book, we'll use what we've learned to be creative. A flexagon's ability to reveal and rearrange leaves provides a flexible canvas for drawing pictures, creating puzzles, and telling stories. The following chapters contain many examples that can serve as inspiration for your own experiments.

DOI: 10.1201/9781003433538-38

Chapter 34

Decorating Flexagons

Flexes can rotate and rearrange the leaves of a flexagon, providing lots of opportunity for adding designs that change in interesting ways as you flex between different states.

In the first chapter of the book, you learned how to make a tri-hexaflexagon and cycle between its three faces using the pinch flex. The chapter finished by having you draw patterns on the leaves to see how they change as you flex. This demonstrates that the leaves rotate as you pinch-flex, rearranging any pictures you draw.

Try labeling the corners of the leaves or drawing pictures on other flexagons we've explored in the book. You will find that global flexes often rotate the leaves, while local flexes swap which leaves are visible without changing the leaf rotation. There are lots of ways we can use this behavior to create fascinating designs.

Example: Comedy/tragedy mask

Take a look at figure 34.1 for an example of how we can take advantage of the way leaves rotate when you pinch-flex, transforming a face between a comedy mask and a tragedy mask.

FIG. 34.1 A flexagon that transforms between a comedy mask and a tragedy mask when you pinch-flex.

Figure 34.2 contains the front and back of a template for this tri-hexaflexagon. Paste the blank sides of the two strips together, then fold adjacent X's together before taping together the first and last edges. For added effect, you may also wish to cut out the eyes and mouth.

DOI: 10.1201/9781003433538-39

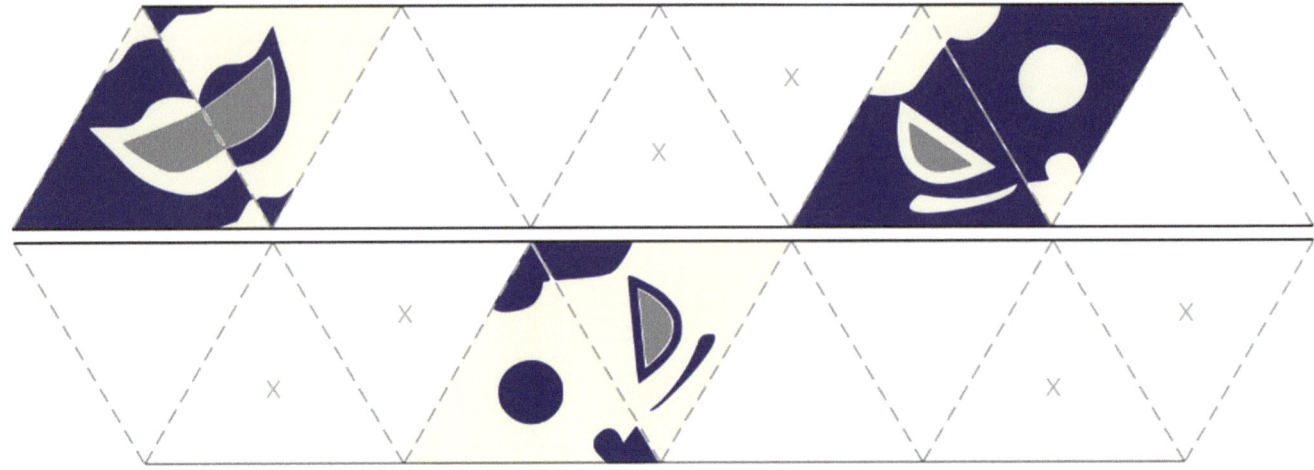

FIG. 34.2 Front and back of a template for the comedy/tragedy tri-hexaflexagon.

As you flex between the two arrangements of the mask, pay attention to how different parts of the face change. For example, while the mouth turns upside down, the noses for the two faces are made up of completely different shapes, including the eyebrows from one of the faces. Also notice how the blue and ivory background colors swap sides.

Pivoting sectors

The comedy/tragedy mask works because each of the three sectors pivots across a centerline and recombines on the back face in a different arrangement. There are other flexagons that share this ability to pivot symmetric sectors across a centerline, which can then be decorated in interesting ways. Figure 34.3 shows some examples of flexagons with this property. The related templates are figure 34.4 for the silver tri-octaflexagon, figure 34.5 for the star isosceles tri-decaflexagon, and figure 34.6 for the star isosceles tri-dodecaflexagon.

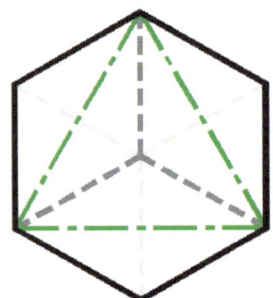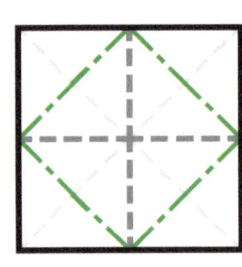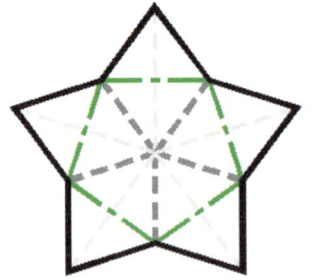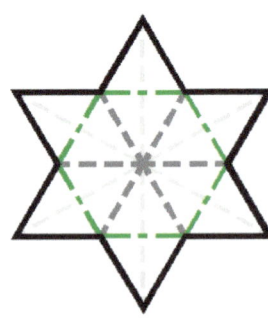

FIG. 34.3 Flexagons that can pivot an image across a centerline of a sector represented by the green line: the regular hexaflexagon, silver octaflexagon, star isosceles decaflexagon, and star isosceles dodecaflexagon.

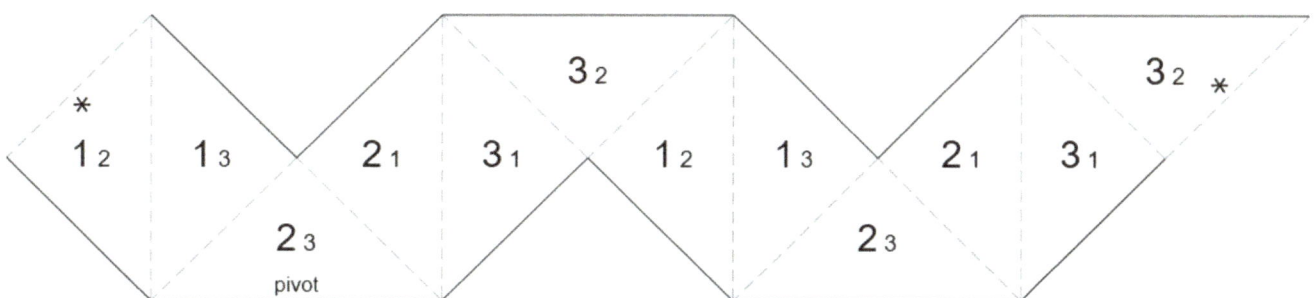

FIG. 34.4 Template for a silver tri-octaflexagon that keeps its shape and pivots sectors after a pinch flex.

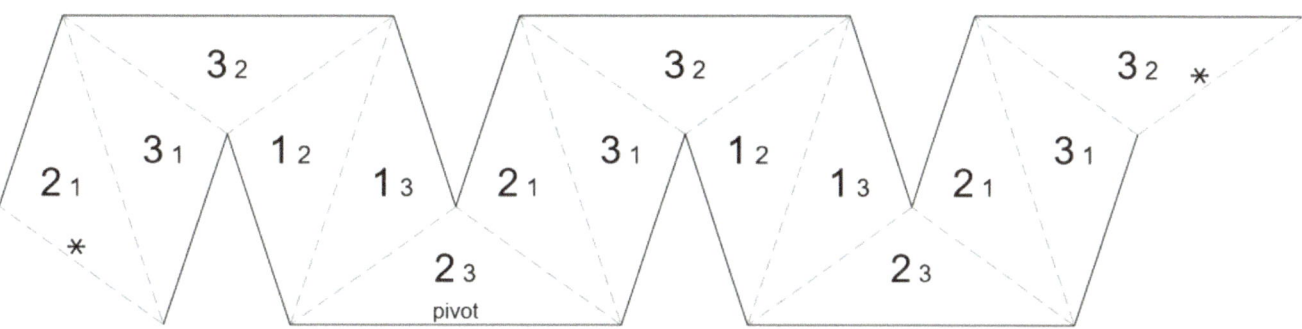

FIG. 34.5 Template for a star isosceles tri-decaflexagon that keeps its shape and pivots sectors after a pinch flex.

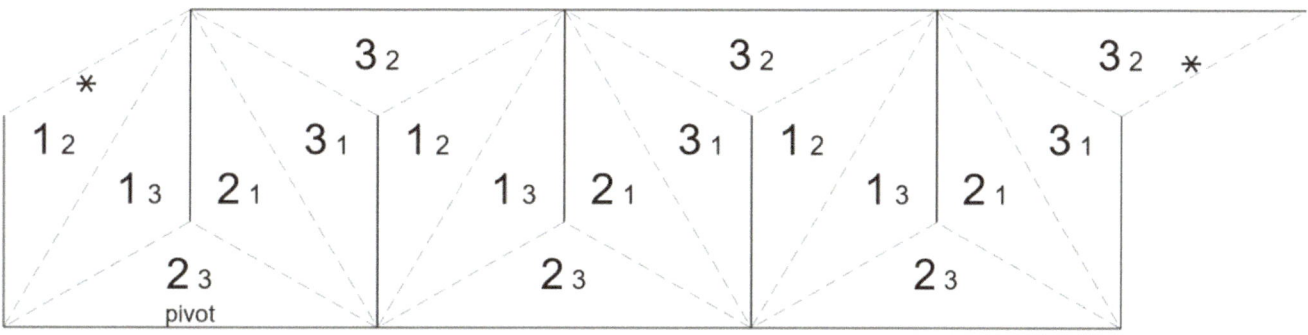

FIG. 34.6 Template for a star isosceles tri-dodecaflexagon that keeps its shape and pivots sectors after a pinch flex.

Example: The globe

Sometimes the easiest way to add artwork to a flexagon is to cut up a picture and paste the pieces onto its leaves. We'll use that technique to put a map of the world onto a flexagon. In one position, the North Pole is in the center of the face; after using a pinch flex to pivot the sectors, the South Pole is in the center.

Print out the world map and template in figure 34.7. Slice the map into four squares by cutting along the vertical and horizontal black lines. Assemble the tri-octaflexagon from the template.

Carefully paste the four map squares onto the four squares of face 1 so they align as in the original map. Now when you do a pinch flex, the map squares pivot across the equator (thin red line) and the South Pole ends up in the middle of the face, giving you a new perspective on the world.

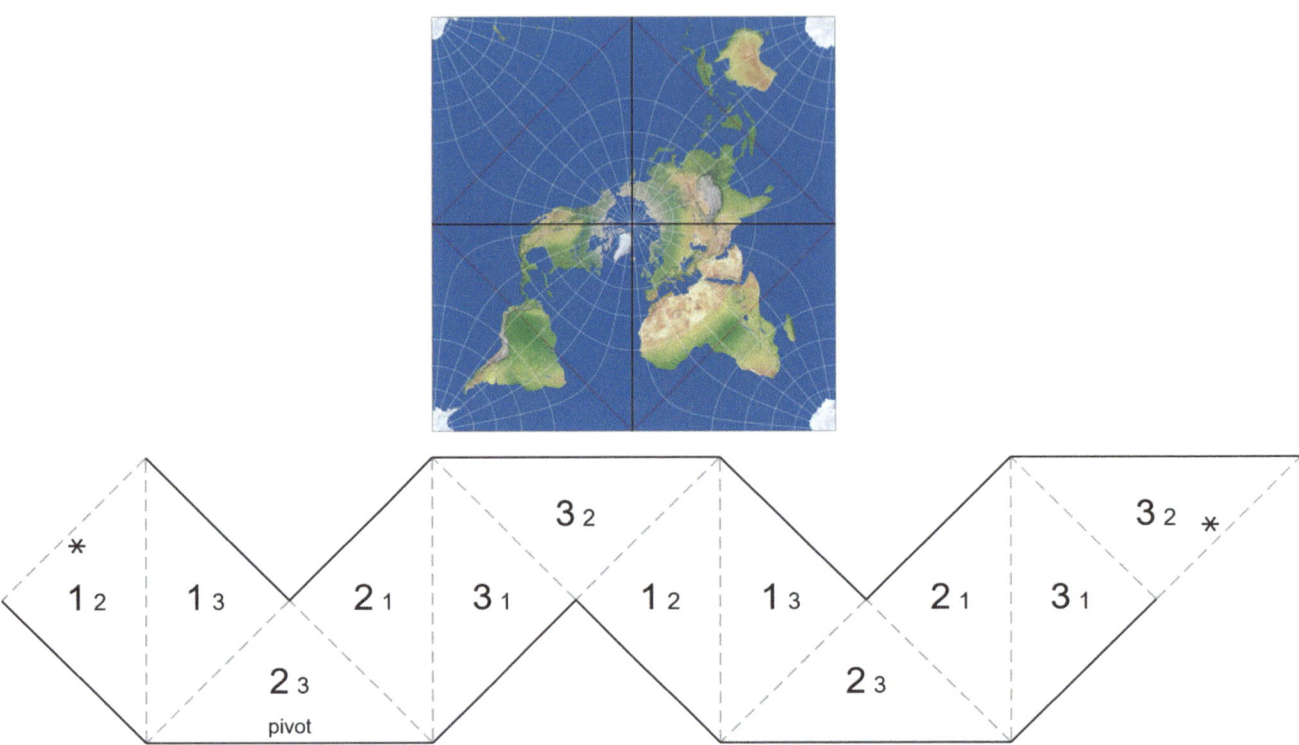

FIG. 34.7 A world map that can be cut up and pasted onto a silver octaflexagon. (The map image is by Tobias Jung and is licensed under a Creative Commons Attribution-ShareAlike 4.0 International License.)

This particular map projection is called the **Peirce quincuncial map projection**. Attempting to put the curved surface of the Earth onto a flat piece of paper always comes with some compromises, but this projection does a fairly reasonable job of keeping the continents in their recognizable shapes and relative sizes while fitting it all into a square. And the ability to pivot the map from pole to pole is an especially nice way to take advantage of a flexagon's behavior. You can even keep it partially flexed so that it stands up, looking a bit like a globe.

Example: Ela Schwartz's time-themed face

Ela Schwartz, who just happens to be a sister of one of the authors as well as a flexagon enthusiast, specializes in designing faces that take advantage of the way pats on the hexa-hexaflexagon (variation 6a) can rotate and rearrange. The time-themed face shown in figure 34.8 was painted on one of the three faces that appear more often during the Tuckerman Traverse. One way to see all the pictures is to do the traverse from one face, then turn it over and do it from the other face. The Kit Kat was painted in the clock arrangement, but startled Ela by putting its hands on its hips in one of the rotations!

FIG. 34.8 Different arrangements of a single decorated face of a hexaflexagon.

These decorations go further than the face pivoting technique of the mask and globe, revealing how the leaves can get more mixed up on more complex flexagons. We'll show more details on how this works in chapter 36.

Example: Audrey Nasar's decorated flexagons

Mathematician and artist Audrey Nasar, from New York's Fashion Institute of Technology (FIT), created some beautiful artwork on tetraflexagons that collectively tells a short story when flexed the right way by using a flexagon's ability to cycle back to the beginning. Figure 34.9 presents two such flexagons from her collection. They're drawn on the square tetra-tetraflexagon (generator: B>B) from figure 7.3, which supports the elegant (B>)4 cycle.

  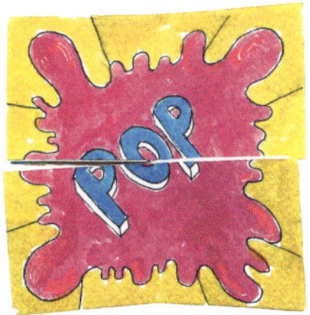 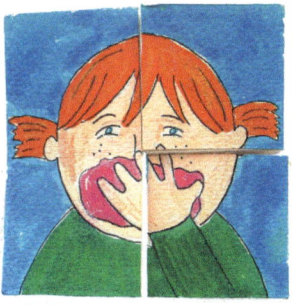

FIG. 34.9 Two of Audrey Nasar's decorated flexagons.

Local flexes

So far, our examples have leveraged how a global flex like the pinch flex rotates leaves. Another interesting technique is to instead take advantage of how a local flex like the tuck or pyramid shuffle only changes a few pats, while leaving others in place.

Here are a few ideas for how you can decorate a flexagon with local flexes in mind:

- **Replacement:** Draw pictures of animals where local flexes replace the head, legs, or tail. Or a picture of a person where you can replace their head, arms, and legs. Or a picture of a face and you can replace the hair, mouth, or ears.
- **Cut-away diagrams:** For example, the main face shows a person. One set of flexes shows you the skeleton, another the vascular system, and another the organs. Or the outside of a building, its internal floor plan, the electrical system, and so on. Or the internal workings of various mechanisms.
- **Fortune tellers:** The main face could be fancy and elaborate, with alchemical symbols and such. Different flexes take you to different types of predictions.
- **Alternate views:** Same view of a forest and meadow during different seasons. Or the same view but one's empty, one has people, one has animals, etc. A single flex could add snow to a portion of the scene or change the people into animals.
- **Words:** Each local flex replaces a subset of letters or words.

If you want to limit the flexes to only changing two visible leaves, you can use T, T', S, and S' at each vertex, giving you a total of eight hidden faces on a flexagon with an even number of pats (since T or S at the even vertices and T or S at the odd vertices gives you different faces). And you could do the same thing on the back face, giving you a total of 16 hidden faces.

Example: Tennis player

The example in figure 34.10 has been drawn on a pentagonal decaflexagon, where you start with an abstract, visible person and a hidden person dressed for sports. You can tuck-flex at any of the hinges in the middle of a pentagon edge to reveal a portion of the other version of the person. You can find a blank template in figure 34.11 so you can draw your own design.

FIG. 34.10 A decaflexagon with two faces that you can tuck-flex between, allowing you to replace portions of a picture.

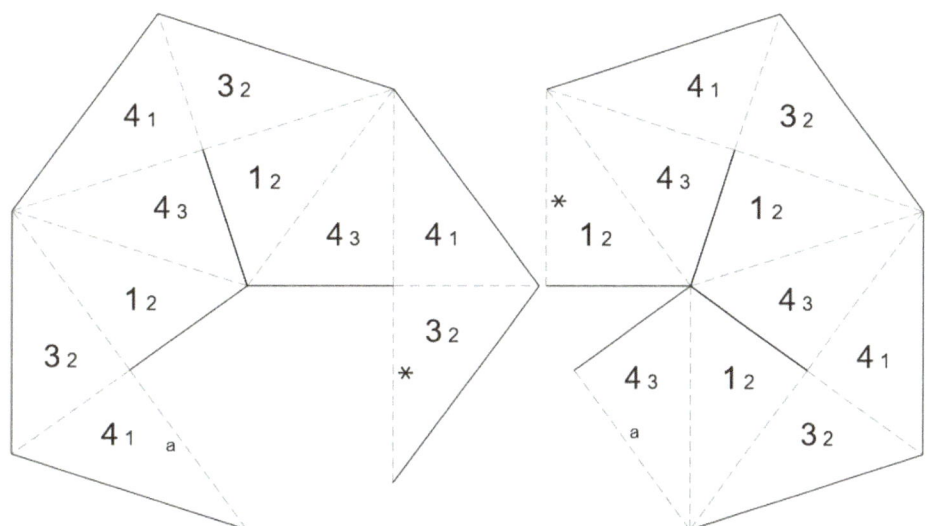

FIG. 34.11 A decaflexagon that supports a tuck flex at each edge hinge from face 1.

Summary

We've seen several examples of how to add interesting decorations to flexagons, taking advantage of their behavior: the comedy/tragedy mask on a hexaflexagon, the globe on an octaflexagon, the time-themed hexaflexagon that remixes pictures, telling a story on a series of panels on a tetraflexagon, and revealing portions of a drawing using local flexes on a decaflexagon.

Chapter 35

Flexagon Books

A flexagon's ability to walk through a series of faces opens up some intriguing possibilities for books, photos, comics, promotional material, and so on. But for such uses, you also need to think about how to make it obvious enough that someone can pick one up and figure out how to flip through it.

The easiest flexagon for someone unfamiliar with them to understand is the square tetraflexagon we first saw in chapter 7. The book flex is simple enough that people can do it with minimal instruction, yet it still allows you to visit multiple hidden faces.

Let's start with a square tetraflexagon where you can do a series of book flexes in a row because it's fairly straightforward. Figure 35.1 has a template for a square tetraflexagon where you can do three book flexes in a row.

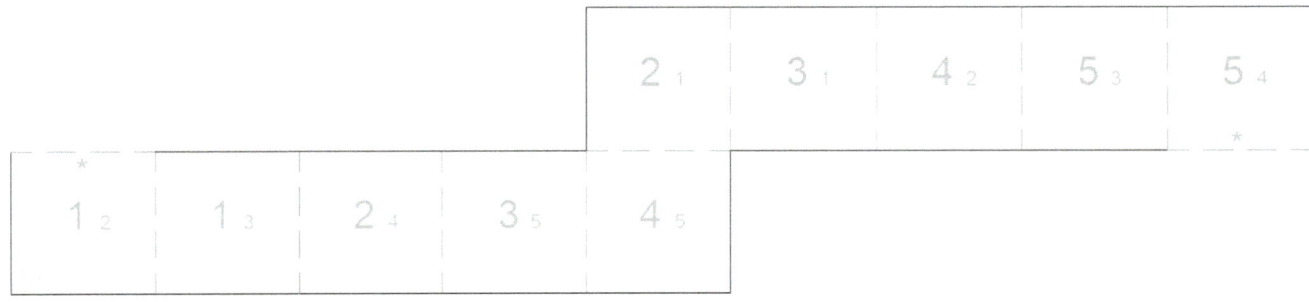

FIG. 35.1 A tetraflexagon (generator: BBB) that supports several book flexes in a row. The numbers are lighter so it's easier to add decorations.

Figure 35.2 shows a storyboard for a simple series. With this approach, you start with the intro on face 1, turn it over to face 2, then book-flex to faces 3, 4, and 5. Note that the first four pages include directions so that people understand how to walk through the pages.

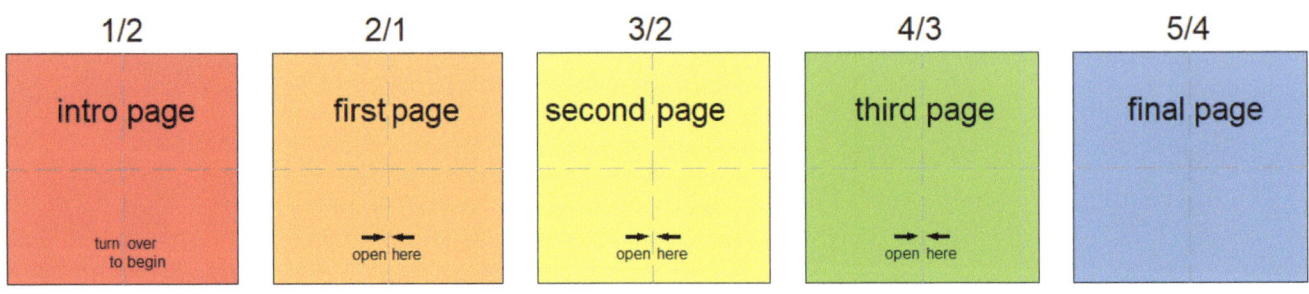

FIG. 35.2 A storyboard for the layout of five pages on a square tetraflexagon (generator: BBB).

DOI: 10.1201/9781003433538-40

Alternatively, you could have several of the pages only take up half of a face, having the story progression use how the leaves swap position after flexing. See figure 35.3 for a possible storyboard on the same template from figure 35.1. A clever design could weave together the story from different pages, while still making it clear how to flex to the next page. Perhaps pages that share a face incorporate the border into their content. Or maybe the information on the two pages interacts in some way.

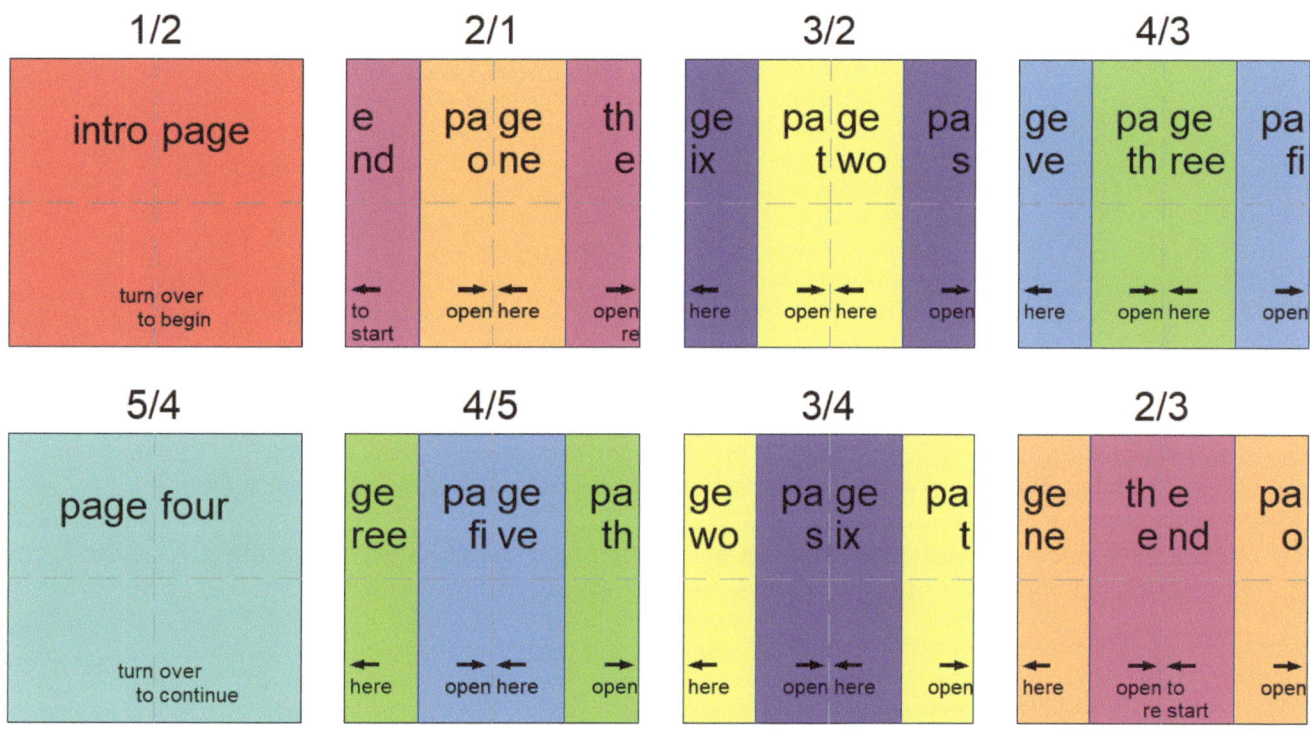

FIG. 35.3 A storyboard for the layout of eight pages on a square tetraflexagon (generator: BBB).

Chapter 7 contains other templates for square tetraflexagons that support several book flexes in a row. The basic approach in these two storyboards applies to any of those.

If a more complex flexagon is acceptable for your audience, you could use a flexagon where the faces get rotated or mixed up. The template in figure 35.4 offers one example.

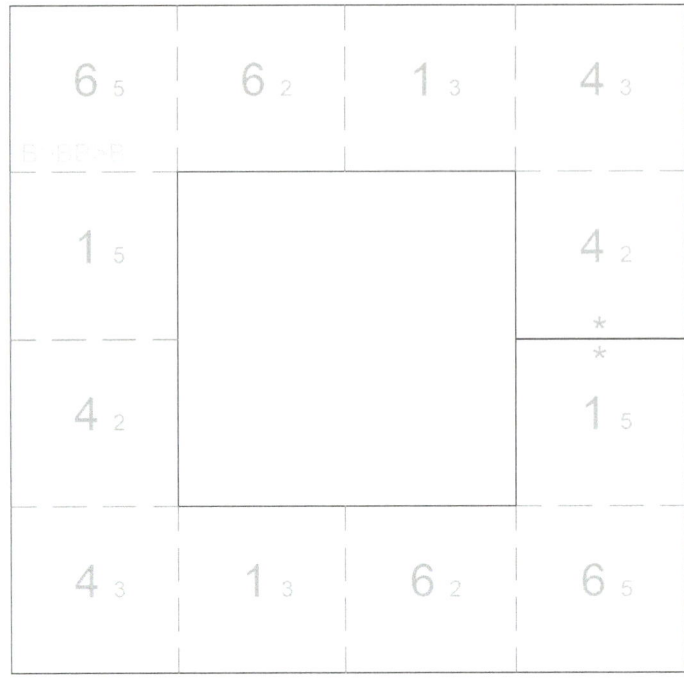

FIG. 35.4 A tetraflexagon (generator: B>BB>B) that sometimes rotates the leaves within faces. To remove the center, either cut along the solid line between the stars, or fold in half before cutting.

One possible route through the flexagon is shown in figure 35.5. Start with state 3/4, flex to faces 2, 1, 6, 5, 2, 1, and finally 4, noting that some flexes require you to first rotate the flexagon.

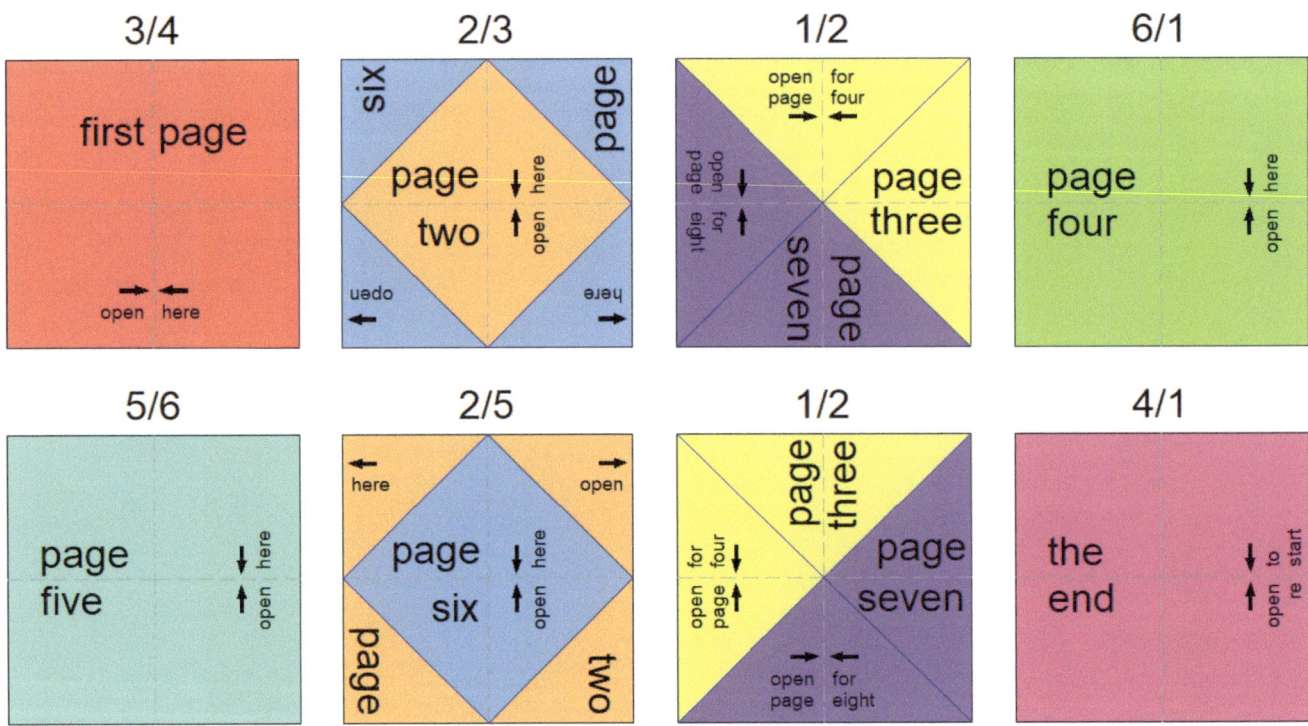

FIG. 35.5 A storyboard for the layout of eight pages on a square tetraflexagon (generator: B>BB>B).

The full instructions for flexing from page to page are written on the faces so the reader can follow them. If you first rotate the flexagon so the "open here" text is horizontal, every new page should open up in the proper orientation. But to be more explicit about the steps to follow as you're designing your story, they can be written as the following flex sequences:

1. Start on page 1, state 3/4.
2. B to page 2, state 2/3.
3. <B to page 3, state 1/2.
4. B to page 4, state 6/1.
5. <B to page 5, state 5/6.
6. <B to page 6, state 2/5.
7. <B to page 7, state 1/2.
8. B to page 8, state 4/1.
9. <B> to restart with page 1.

Faces 3, 4, 5, and 6 only appear once each, so you have full freedom to decorate those faces as a single page. However, faces 1 and 2 both appear twice, offering additional challenges and opportunities for decoration. Note that face 2 gets rearranged, while face 1 is simply rotated.

The techniques in this chapter could be used with more complex flexagons, of course: making successive pages that you flex between, providing directions on the pages, dividing faces between pages, and decorating faces that get mixed up from flexing, taking care to make sure your intended audience can navigate your design.

Summary

The book flex, being simple to explain to a flexagon novice, offers creative opportunities for using flexagons to tell stories. We presented some templates and storyboard options as a base for book-like interactive creations.

Chapter 36

Flexagon Puzzles and Mazes

You can take advantage of the way flexes shift and rotate leaves to make interesting puzzles. We'll show a couple examples and talk about how you can leverage flexagon behavior to design your own.

Clock puzzle

The goal for this first puzzle is to rearrange the picture pieces to assemble an image of an alarm clock by doing a series of pinch flexes. You can find the template in figure 36.1. Cut along the solid lines, including between the stars in the upper right corner, then copy the small numbers on each leaf onto the back. Fold together adjacent 6's, then 5's, 4's, and 3's. Finish by taping together the two edges with the stars on them.

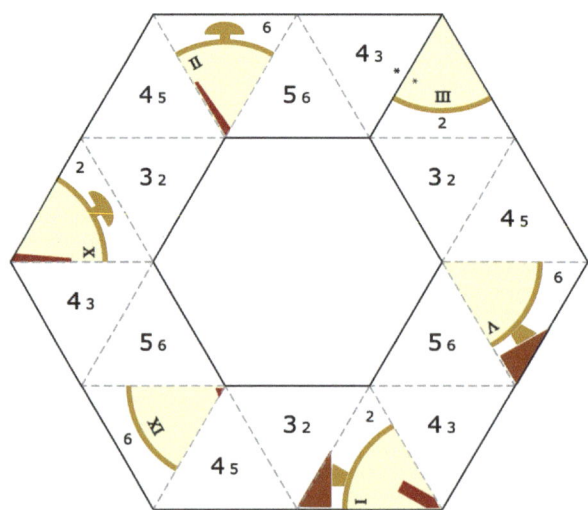

FIG. 36.1 Template for a clock puzzle on a hexa-hexaflexagon.

When you finish folding the flexagon, you should see a mixed-up picture of an alarm clock on one face, as shown in the first image in figure 36.2. The second image in that figure shows what that face looks like after doing a pinch flex. It almost looks like an alarm clock, except that the numbers are all mixed up and it has two minute hands. Can you figure out how to get the picture of the alarm clock arranged properly just by doing pinch flexes? What time does the clock say? You may want to solve the alarm clock puzzle before reading about how it was designed and how to solve it.

DOI: 10.1201/9781003433538-41

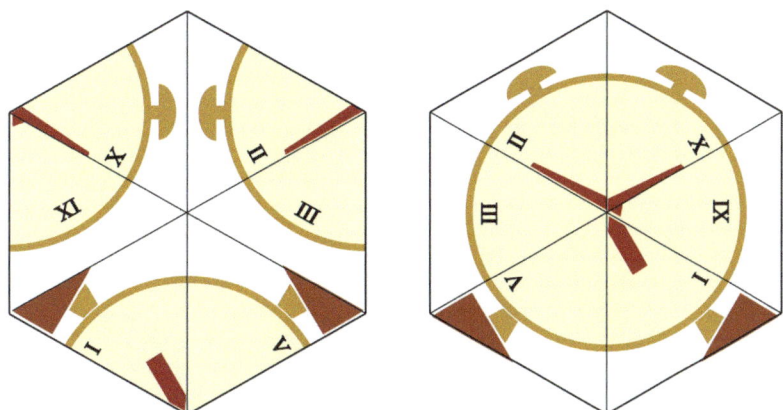

FIG. 36.2 The state when you first fold the clock puzzle (left) and a mixed-up state after one pinch flex (right).

This is the 6b variation of the hexa-hexaflexagon. We first saw it in chapter 3, then took a closer look at how it behaves in chapter 17. The numbering for this clock flexagon looks slightly different because we purposely numbered it so that it would maximize the number of flexes needed to get from the initial state after folding to the solution.

We've previously seen the Tuckerman Traverse, which is a flex sequence that visits every state in a pinch class. This will lead you to the solved state in six pinch flexes.

But you can actually get to the solution with fewer flexes if you know how. Take a look again at figure 17.6, which contains the hexa-hexaflexagon's pinch state diagram. The numbering we chose essentially starts you in state 1/6, while the correct arrangement for the alarm clock is in state 1/4. Those two states are as far apart as you can get in the diagram, and a Tuckerman Traverse is going to take six steps. But there are only four lines separating those two states, which means you can travel between them in only four pinch flexes. This route is the sequence (P^>)4.

Something else to note about this puzzle design is that the template only needs to be printed on one side because all six pieces of the alarm clock are on the same side of the template. Recall chapter 4 where we observed that templates designed for the pinch flex repeat the same pattern multiple times. If the flexagon has an even number of faces, the pattern repeats on the same side. But if the flexagon has an odd number of faces, the numbers in the pattern alternate, switching between the front and back sides. Thus, a puzzle like the clock, which only needs graphics on one pinch face, can be printed more easily if you pick a flexagon with an even number of faces. However, as we'll see when we discuss the maze later in this chapter, there are tricks you can use to print on a single side even with an odd number of faces.

Party Puzzle

Ela Schwartz, whose time-themed flexagon we saw in chapter 34, has created many interesting designs. This next one she calls the Party Puzzle, which tells a story as you puzzle out how to find the final arrangement: six pairs of identical twins go to a party and your job is to make sure everyone is talking to someone new. See figures 36.3–36.6 for the sequence of arrangements that tell the story.

FIG. 36.3 The twins first enter the room. They're busy talking, but only to their respective twin.

FIG. 36.4 Three pairs talk to their twin, but the others look like they're trying to leave!

FIG. 36.5 Progress! Three pairs of twins are meeting new people.

FIG. 36.6 Finally, everyone is mixing. To top it off, each set of twins is talking to two different pairs. For instance, the man with a beard is talking to the young woman with the Afro, while his twin is talking to the young man in the blue shirt. What a great party!

The party face has four different arrangements, which can be done on any hexaflexagon with at least five faces. One option is to start with the straight hexa-hexaflexagon template from figure 3.4. Find state 1/4 and draw the picture from figure 36.3 on face 1. Can you figure out how to quickly pinch flex between the four parts of the story? If you need a hint, the pinch state diagram in figure 17.5 can help you, or try the alternate traverse described in chapter 18.

If you want to make the puzzle more challenging, you could put the puzzle on a flexagon with even more faces, such as the dodeca-hexaflexagon from figure 3.7. It can be fun to have different puzzles on different faces.

Study aid

You can also take advantage of the interesting ways the faces get rearranged to make a combination puzzle/study aid/fidget toy. For example, the face in figure 34.7 shows important events from the thirteenth century along with the year when they occurred, but they only line up when you've flexed to the proper state.

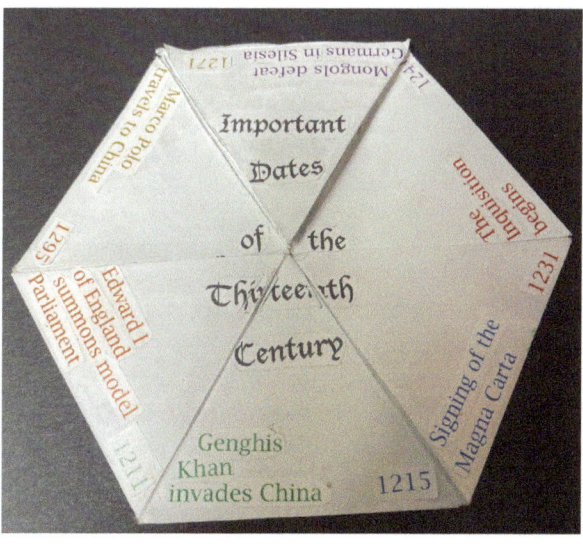

FIG. 36.7 A history study aid. Note that the dates and events are color-coded.

If you want to make a particular decorated phase appear more or less often, you can use the pinch state diagrams we saw in chapter 17 to help guide you. For example, figure 17.5, which contains the pinch state diagram for the straight hexa-hexaflexagon, shows that faces 1, 2, and 3 appear more often than faces 4, 5, and 6. So if you want your properly arranged study aid face to appear more often, pick a state where 1, 2, or 3 are on both the front and back faces.

Maze

Our next example is a maze created on a regular triangle penta-tetraflexagon, which we first saw in figure 4.3. A nice attribute of a regular triangle tetraflexagon is that it forms a cup, where one side always stays on the outside, even as flexes rearrange the leaves. This means the maze always stays on the outside, never disappearing into the flexagon, even after multiple flexes.

Figure 36.8 contains the template for this maze. Since the flexagon has an odd number of faces (five), half of the outside face ends up on the front of the template and the other half is on the back face. Since it's more convenient to print a single side, the template has two flaps for the leaf-faces that would otherwise print on the back. Fold the two flaps labeled A and B backward, and paste them against the back of the adjacent leaves. After pasting the two flaps, copy the small X and small face numbers (including the numbers in parentheses) to the back of the template. Fold 4 on 4, 3 on 3, and 2 on 2. Finally, paste the two leaves with X's on them face to face. When finished, you should have a pyramid-shaped cup with the maze on the outside and 1's on the inside.

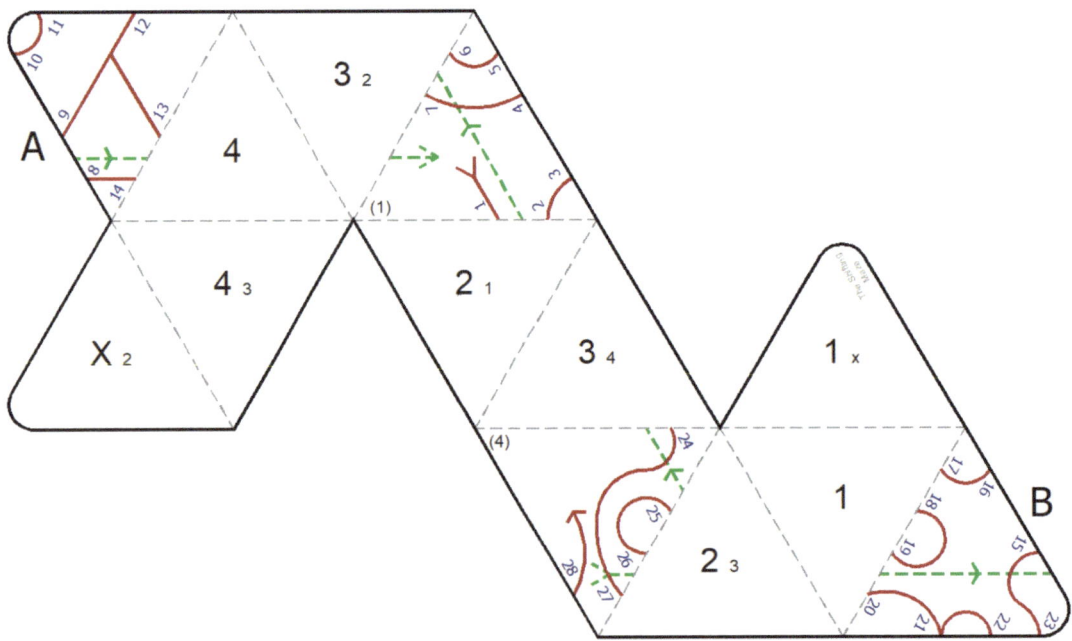

FIG. 36.8 Template for a maze on a regular penta-tetraflexagon.

The dashed green lines provide a simple example to give you a feel for how the maze works. After you fold up the flexagon, you should see 1's on the inside of the pyramid. The green maze starts next to the number 27 (the start of the green arrow) and continues across two adjacent leaves before falling off the edge, as shown in figure 36.9. You now need to use the pocket flex to rearrange the faces so that the arrow continues. After the first pocket flex, the 2's should be on the inside and the dashed green line should extend across another leaf. A second pocket flex gives you 3's on the inside, and extends the dashed green line across yet another leaf. A final pocket flex, with 4's on the inside, should lead you to the head of the green arrow. Note that you won't see the entire green

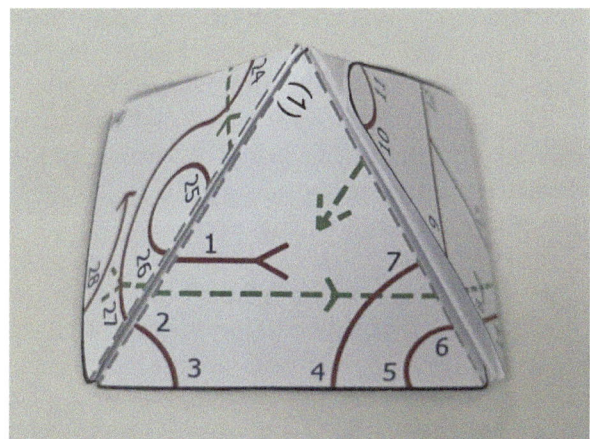

FIG. 36.9 Assembled maze on a regular penta-tetraflexagon showing the start of the dashed green maze along the bottom.

path all at once; each pocket flex removes one section from the end of the path and adds a new section to the front of the path.

Once you finish the dashed green maze, the green arrow points to the beginning of the solid red maze, which is trickier to solve. You need to flex back and forth across every state in order to find your way through the red maze. To make it harder, sometimes a single path can connect to more than one adjacent leaf, depending on which state you're in. Use the blue numbers to help keep track of where you are as you wander through the labyrinth.

Rotating leaves

We've now seen puzzles for both regular triangle hexaflexagons and tetraflexagons that utilize how the leaves rotate as you do either a series of pinch flexes (for the hexaflexagon) or pocket flexes (for the tetraflexagon). In order to design such puzzles, it's useful to understand how the leaves change. Because all triangle flexagons follow the same pattern when pinch-flexing, it's simplest to examine the regular tetraflexagon, since the same face always stays on the outside as you flex so you only need to examine four leaves. But the same principles apply to other triangle flexagons as well.

To see how this works, make a regular hepta-tetraflexagon from the template in figure 36.10. Before folding, copy the a's, b's, and c's from leaves 1B and 6C to the opposite face, being careful to keep them next to the same leaf edge. As usual, fold 6 on 6 on down to 2 on 2. When taped together, the leaves labeled A, B, C, and D and edges labeled a, b, c, and d will be on the outside, with 1's on the inside.

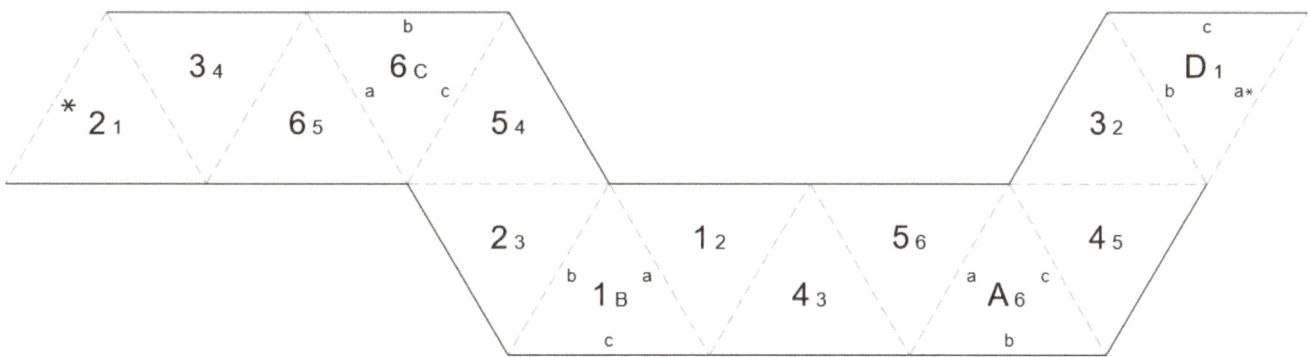

FIG. 36.10 Template for a regular hepta-tetraflexagon (generator: (P*^>)5).

As you flex this tetraflexagon, the inside will change from 1 up through 6 while the outside continues to show the letters, which get shuffled after each flex. Figure 36.11 shows how the leaves on the outside face get rearranged, where the leaves are labeled A, B, C, and D and the edges are labeled a, b, and c. Notice how each flex causes the order of the leaves to reverse direction, alternating between ABCD going clockwise around the cup (when 2, 4, or 6 is on the inside) to counterclockwise (when 1, 3, or 5 is on the inside). The edges connecting the four leaf-faces cycle between a and b (when 1 or 4 is on the inside), b and c (when 2 or 5 is on the inside), and c and a (when 3 or 6 is on the inside). This gives us a total of six different arrangements (two arrangements of the faces times three arrangements of the edges). We don't gain any additional arrangements by adding more than seven faces (one outside face, six inside faces), because the pattern repeats after this. Figure 36.12 shows the full graph of connected edges, highlighted by the inside face.

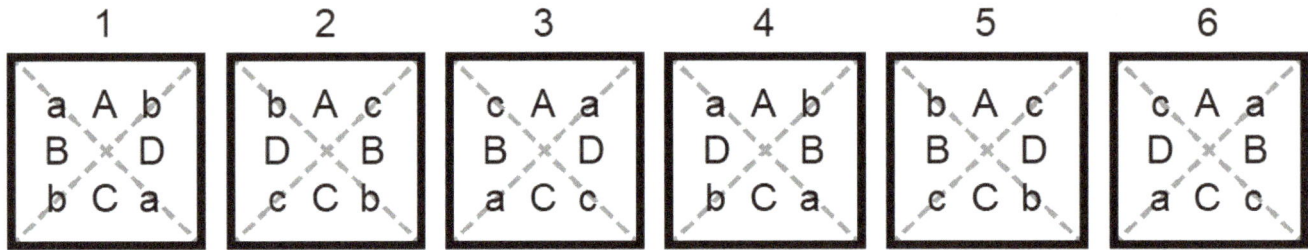

FIG. 36.11 Arrangements of leaves and edges as you flex between states. The numbers are inside the pyramid, while the lettered squares are what you see when looking down at the top of the pyramid.

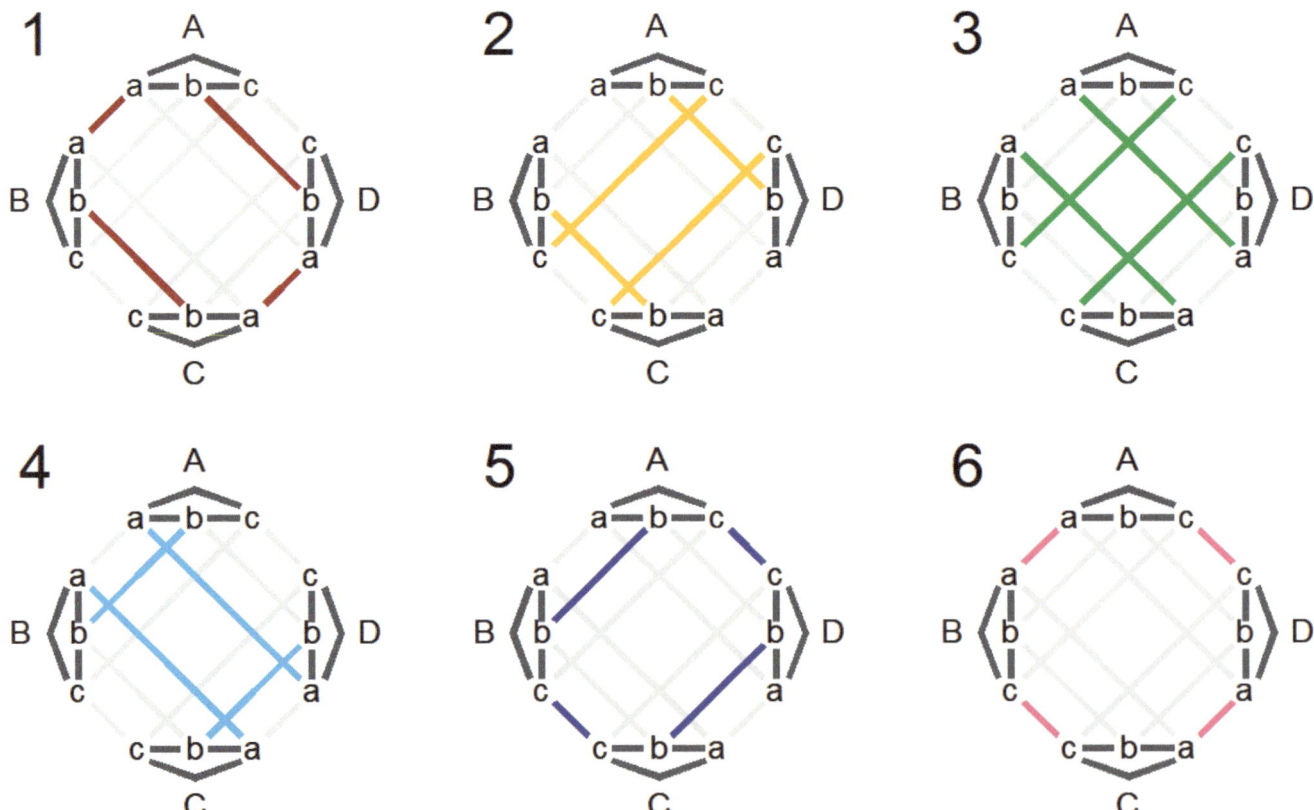

FIG. 36.12 Diagrams showing how the leaf-faces are connected across edges on the outside face given the number on the inside face. For example, when face 1 is on the inside, the leaf labeled A is connected across edge a to leaf B, and across edge b to leaf D.

If you want to design your own maze, or any picture that gets rearranged as you flex it, these illustrations can help you understand how everything gets rotated and swapped around. And, of course, you don't need to use all possible arrangements. The maze, for example, only used four of those states rather than all six.

Ela Schwartz's Path Puzzle

The path puzzle (created by Ela Schwartz) shown in figure 36.13 takes advantage of the maximum number of arrangements that we just saw. Most arrangements are mixed up in various ways, so the goal is to flex until you find the arrangements with a complete circular path. Notice how two arrangements are circles, but the images are in reverse order.

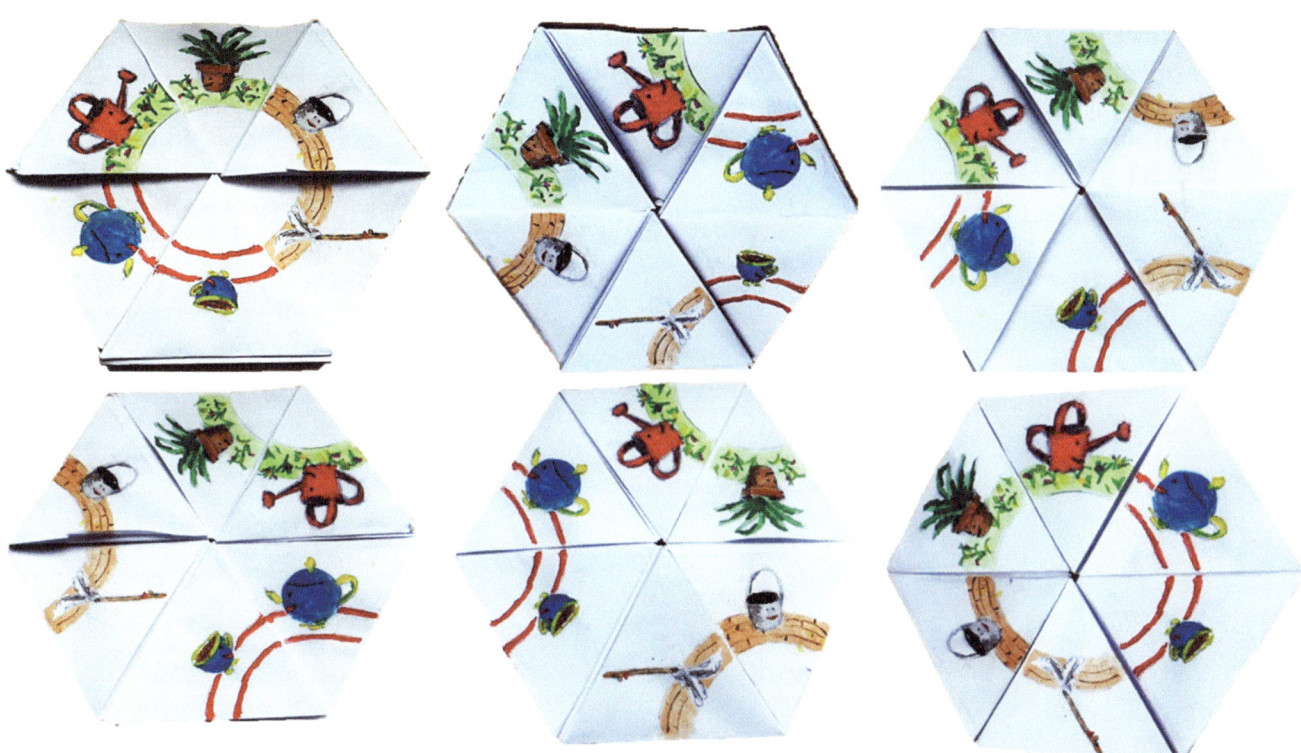

FIG. 36.13 All six ways a face in the path puzzle can be rearranged. In one of the two states with circular paths, the objects are drawn such that they're looking at the object they're most associated with.

You can design your own on face 1 of the hepta-hexaflexagon template in figure 36.14. This is the same pattern as the hepta-tetraflexagon we just saw (figure 36.10), but extended to six pats.

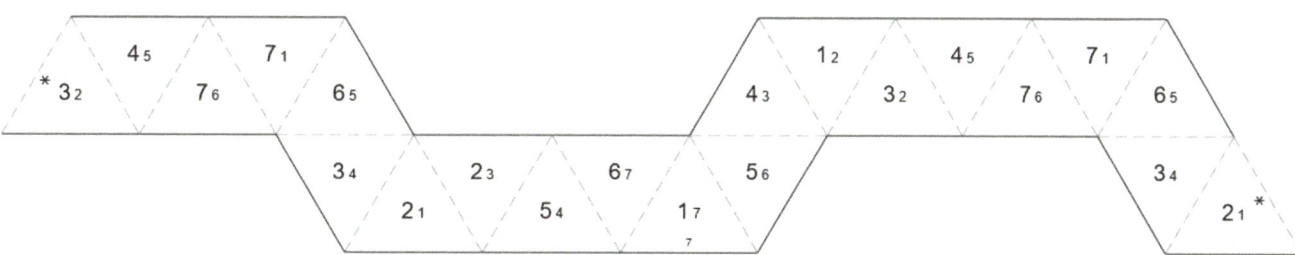

FIG. 36.14 Template for a hepta-hexaflexagon (generator: (P*^>)5) that supports the maximum number of rotations for face 1.

More ideas

Here are some additional ideas for putting designs on flexagons that go beyond simple decorations.

Word puzzles: Instead of pictures, you can put words across a face or around an edge that are only readable when flexed to the correct state.

Use other flexagons: Flexagons such as the hexagonal silver dodecaflexagon with its turn flex can provide an interesting alternative to the standard flexagons.

Make a puzzle using local flexes: Recall that local flexes, such as the tuck and pyramid shuffle, replace leaves on part of a face without rotating the leaves. You could draw a picture on a face that requires the right combination of local flexes to assemble.

More complex flexes: If you want a much harder puzzle, you could require global flexes such as the v-flex or the slot flexes. Keep in mind that such a puzzle can quickly feel impossible if you get too carried away. You probably want to limit it to a flexagon with a simple generating sequence.

Magic square: A *magic square* is a square grid of numbers arranged so that every row, column, and main diagonal adds up to the same number. Typically, they include just the sequential counting numbers starting from 1, with every number unique. Can you figure out how to put a magic square on a square tetraflexagon such that it keeps its magic properties even as you flex it? Could you adapt the concepts to another shape, such as a hexaflexagon? See Robin Moseley's flexagon.net for some examples.

Fold a picture from a template: For example, you can directly fold the template from figure 36.1 into a finished alarm clock if you fold 5 on 5, 4 on 4, 3 on 3, and 2 on 2. If you removed all the folding numbers, could you figure out how to fold up the alarm clock? You could try something similar with other flexagons: draw a picture on a face, unfold it, and try to figure out how to fold it without any labels that provide hints.

The dynamic nature of flexagons opens up lots of interesting possibilities for puzzles, mazes, games, and other clever designs.

Summary

In this chapter, we explored ways to add puzzles to flexagons to extend the challenge of flexing them. Specific examples included a clock face, partying pairs of twins, a study aid, a pyramid maze, and a circular path. We showed the six different leaf arrangements a face appears in as a result of the pinch flex. The chapter finished with additional ideas to consider for making your own puzzles.

Chapter 37

Flexagon Pop-ups

Another fascinating form of paper craft is the art of creating pop-ups, used for both books and greeting cards. A **pop-up** is a structure that goes from flat to three-dimensional when you open a hinge. And since flexagons are simply a collection of hinges, there are some intriguing possibilities in combining the two forms of paper engineering. You can imagine pop-ups appearing in all sorts of surprising places as you explore a flexagon.

However, a little bit of study of how flexagon hinges operate will quickly demonstrate that you can't add a pop-up to just any random hinge. Flexagons are so dynamic that sometimes the two leaves in a hinge travel independently, which means you can't anchor the two sides of a pop-up to them without preventing some flexes. Sometimes the place you'd like your pop-up to hide might interfere with the smooth operations of another flex. Furthermore, some hinges need to swing a full 360° as you travel between flexes, while pop-up hinges are generally designed to stop after 90° or 180°. This means you need to be strategic in where you put a pop-up construction.

There are several possible pop-up techniques you can use with flexagons that work well under different conditions:

- With the **one-piece technique**, the pop-up is created by cutting the flexagon template itself. This design needs to account for how the front and back faces of each leaf appear in different states. The pop-up appears when you open the flexagon, and it works best when the hinge only needs to open partway, such as to 90°.
- The **multi-piece technique** uses additional paper that is cut out and anchored to both sides of a given hinge. This can be nice in that the two faces of each leaf are kept independent, so that a pop-up on one face of a leaf doesn't impact its opposite face. The pop-up appears when you open the flexagon and it works when the hinge opens partway or flat, up to 180°.
- For an **embedded pop-up**, rather than attaching the multi-piece pop-up mechanism to both leaves of a hinge, attach the pop-up to a single leaf. You can then lift an extra flap to reveal the pop-up hidden inside. While not as automatic as having the pop-up appear as you flex, it has the advantage that you can add a pop-up to any leaf-face. This works even when the hinge needs to travel a full 360°.
- Make use of the **flexagon structure** itself. Flexes such as the pocket flex can turn a flexagon into an interesting three-dimensional sculpture that can be used as the scaffolding for decorations or additional pop-ups.

DOI: 10.1201/9781003433538-42

One-piece technique

Before creating a pop-up on a flexagon, it's useful to start by experimenting on stand-alone hinges. Take a piece of paper and fold it in half to make a hinge. Then try various pop-up mechanisms on that simple hinge.

Our first example makes a box that stands out from the background. Refer to figure 37.1 while following these steps:

1. Fold the paper in half.
2. Make a cut perpendicular to the hinge.
3. Fold the bottom portion forward and then back to flat. Fold the same portion backward along the same crease and then back to flat.
4. Open up the hinge 90°.
5. Push out the lower portion of the hinge to invert it, making a box. When you close the hinge, keep this box sticking out, so it automatically reappears when you open the main hinge back up.

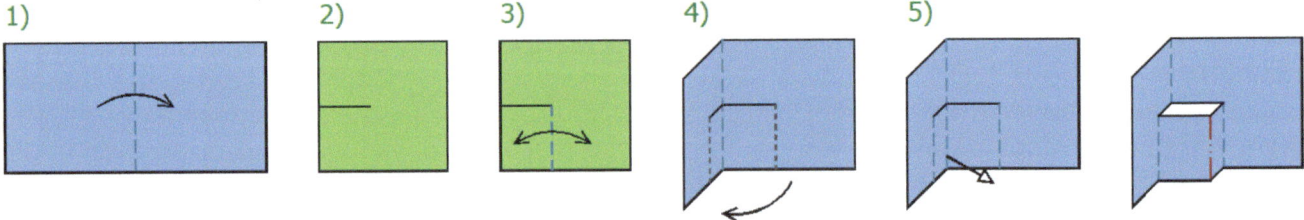

FIG. 37.1 A box from a single slit.

The second example creates a mouth extending out from a hinge. Work through the following steps using figure 37.2 as a guide:

1. Fold the paper in half.
2. Make a cut perpendicular to the hinge.
3. Fold a portion of the paper above the cut backward and forward across a diagonal and return it to the flat position. Repeat with a portion of the paper below the cut.
4. Open up the hinge 90°.
5. Push the top and bottom sections forward to invert the folds, making an open mouth shape. When you close the hinge, the mouth should continue to stick out, so it automatically reappears when you reopen the main hinge.

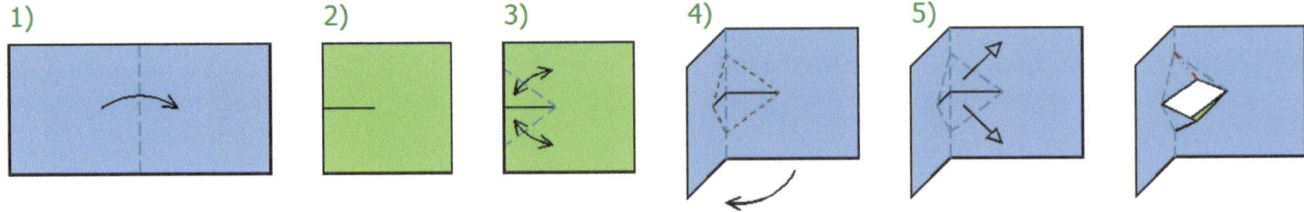

FIG. 37.2 A mouth opening from a single slit.

Both of these single-slit techniques can be used anywhere there's a crease, even at creases made by the new pop-up you just created. You can vary the angles where you cut and fold, as well as add additional pop-ups. This allows you to create a wide variety of three-dimensional shapes.

The next technique creates a bench-like shape that stands off of the leaves when the hinge is open at a 90° angle. See figure 37.3 for a sample template. The solid black lines represent where to cut, the blue dashed lines are valley folds, and the red dashed-and-dotted lines are mountain folds. But we need to be a bit more careful in how we cut and fold this template since some cuts are in the middle of leaves and the folds don't always extend across an entire hinge. When cutting the interior lines, you can be more precise if you use a craft knife instead of scissors, preferably backed by a mat to protect your furniture or floor. You can use a dull edge to score the folds, then carefully fold the mountains and valleys as marked. When the main hinge is folded at a 90° angle, the bench should stick out from the leaves, with each portion parallel to one of the leaves.

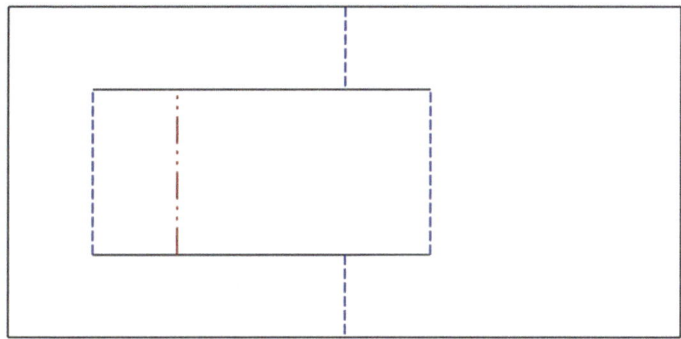

FIG. 37.3 A template for making a bench from a double slit.

If you look at the resulting pop-ups and their hinges, you can see how the cuts impact both the front and back faces of the two leaves. You can also see how they're optimized for opening up to 90°. These constraints affect where in a flexagon you can add this style of pop-up.

Square flexagon pop-ups

While any flexagon can contain a pop-up, it's natural to start with square flexagons, since they resemble the pop-up cards and books we're most familiar with.

The most basic one to start with is the square tri-tetraflexagon, using figure 7.1 from the chapter 7. Or you can use the blank template in figure 37.4, folding it the same way as the numbered template.

FIG. 37.4 Blank template for the square tri-tetraflexagon. Fold as if the leaves are numbered, from bottom left to top right, 2/1, 2/3, 1/3, 1/2, 3/2, and 3/1.

Try the one-piece techniques we just discussed at various hinges to get a feel for how they work and how they affect the faces as you flex between them. Figure 37.5 shows an example of combining several of these techniques on a single face of a tri-tetraflexagon. Notice how the cuts impact the back faces of the leaves.

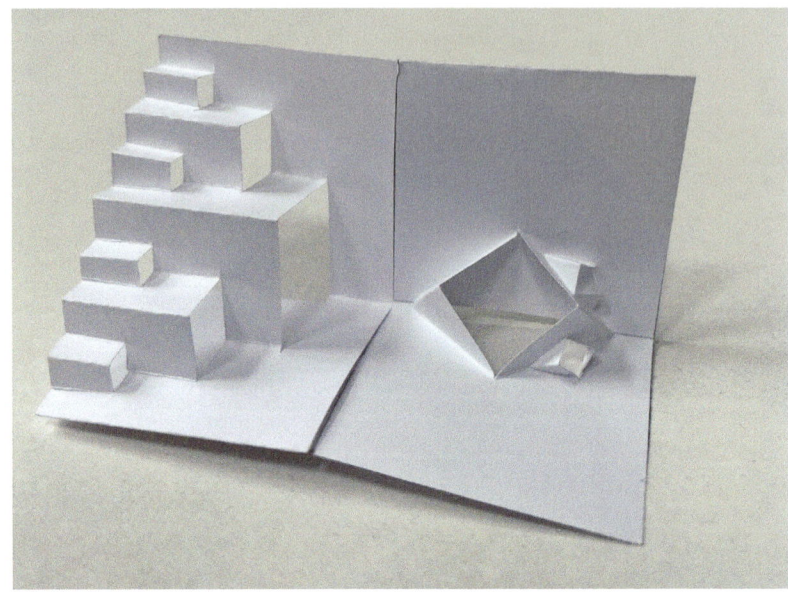

FIG. 37.5 Square tri-tetraflexagon that's been opened partway to reveal several single-slit pop-ups.

For a concrete example, we'll use a square tri-tetraflexagon to hide two different pop-ups, one showing the entrance to a castle and another showing a bedroom inside. See figure 37.6 for the template. Figure 37.7 shows a decorated and assembled version of this pop-up.

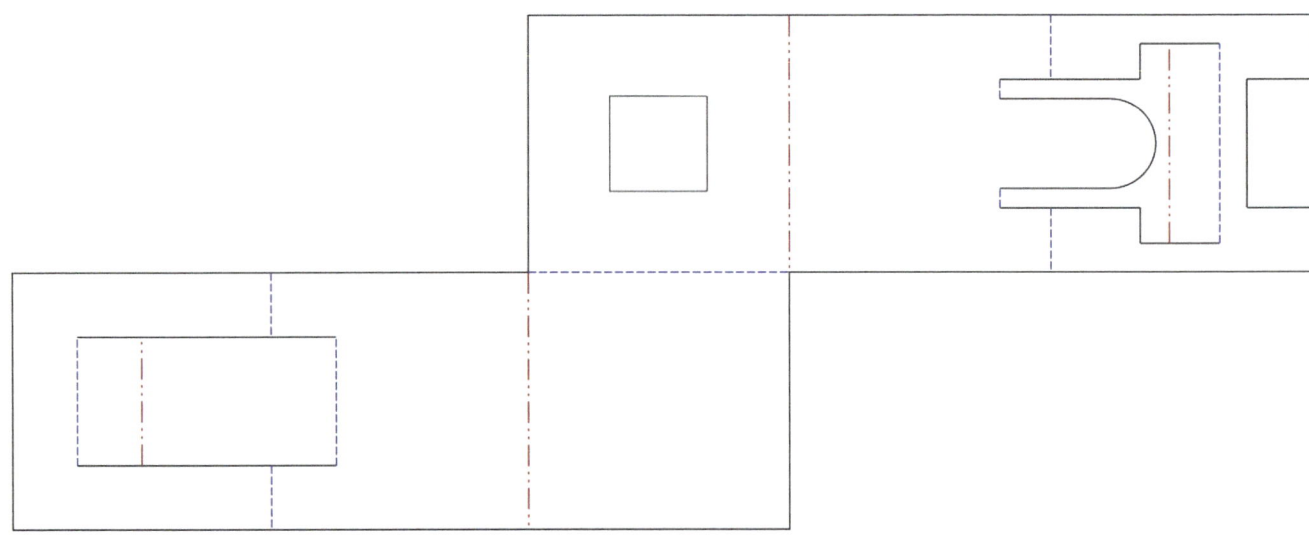

FIG. 37.6 Template for the castle and bedroom pop-up.

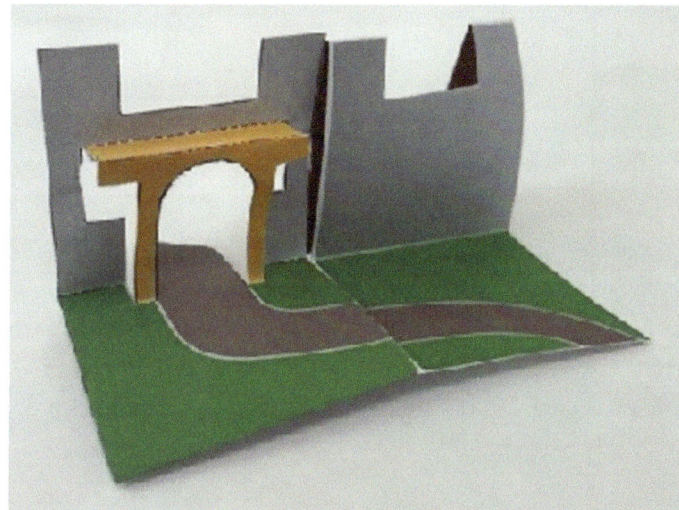 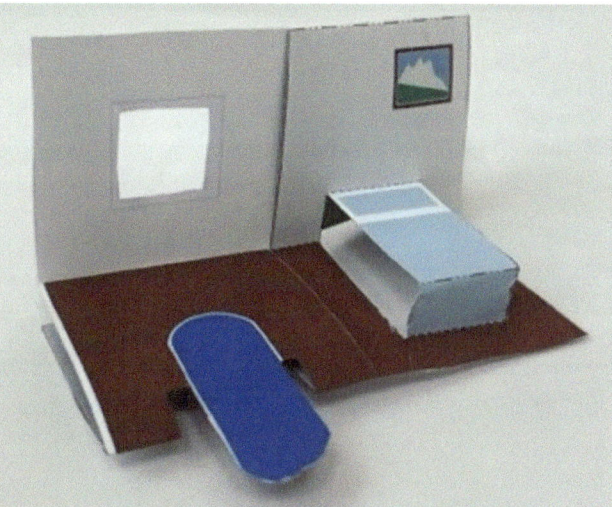

FIG. 37.7 A square tri-tetraflexagon that flexes between two different pop-ups.

This pop-up illustrates some of the tricks and challenges of creating multiple pop-ups in a single flexagon. Here are some interesting details to note:

- The bedroom window is on a leaf that's visible in the bedroom view but not visible in the castle view. That means you can cut out the window without impacting the other view.
- The bed is cut out from two adjacent leaves. The long side is on a leaf that's hidden in the castle view. But the short side causes a missing notch in the castle view. We incorporate this into the design by cutting a second notch in the adjacent leaf of the castle view to make it look like the battlement on top of a castle. This second notch is on a leaf that doesn't impact the bedroom view.
- The door to the castle accomplishes multiple goals at once. It's cut out so that the door is open. The cut-out lies flat, serving as a path into the castle. And from the bedroom view, the path becomes a rug sitting on the floor.

The square tri-tetraflexagon can be flexed back and forth, alternating between the two different pop-ups. Can we add more pop-ups if we try more complex square tetraflexagons? Yes, we can, as we'll see next.

Multi-piece technique

To demonstrate the multi-piece technique, we'll make a pop-up tent attached to a simple hinge. Cut out the two pieces in figure 37.8. Pre-crease the mountain and valley folds as marked. Finish by attaching the end flaps of the smaller piece onto the gray rectangles in the larger piece. As shown in figure 37.9, when the larger piece is opened flat, the tent will rise from the surface, and when it's folded in half, the tent will be completely tucked away inside.

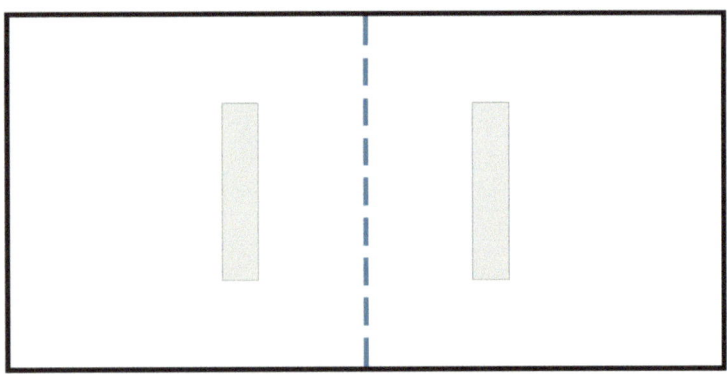
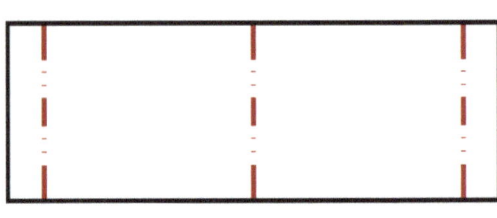

FIG. 37.8 Templates for a tent pop-up. The blue dashed line is a valley fold. The red dashed-and-dotted lines are mountain folds. Decorations can be added to the top face of both pieces.

FIG. 37.9 The tent pop-up as it appears in the open state (left) and partially closed state (right).

This basic tent construction can be varied in many ways to make different shapes: the length and width of the tent can be modified, the placement of the pop-up flaps can be moved, and the shape of the tent itself can be changed. You can also add additional multi-piece pop-ups together. Note that you typically want the pop-up to be hidden inside the flexagon when its face is closed, which puts an upper limit on the size and shape of the pop-up.

This technique works on any hinge that opens up to 180°. We could, for example, add this style of pop-up on the same square tri-tetraflexagon template we used for the castle, which would allow the flexagon to be opened flat rather than to 90°.

Or we could use the square tetra-tetraflexagon (generator: B>B) from figure 7.3, which supports a cycle of B>B>B>B>. It has four flat states where adjacent hinges always travel together without needing to go beyond 180°, making it ideal for multi-piece pop-up techniques. You could create a four-part story that optionally cycles back to the beginning. The one additional caveat, however, is that every time a flex moves a face from the front side to the back side, it also reverses the left and right sectors. You may choose to ignore this fact and just focus on one side, or figure out a way to work this detail into the story.

Another way to leverage the multi-piece technique is to make an embedded pop-up by taking an independent pop-up mechanism and pasting one side of it to a single leaf-face so that you have to manually open up the pop-up after a flex reveals it . Consider again the tent pop-up from figure 37.8. Instead of attaching the tent to two different leaves, take the whole mechanism (large rectangle with the tent attached to the inside) and paste one outside face against a single leaf-face of a square flexagon. When a flex reveals that leaf-face, the pop-up doesn't automatically open up since it's not attached to both sides of a hinge. Instead, you open up the pop-up independently of the flex. This has the benefit that you can attach it anywhere in the flexagon.

Triangle tetraflexagon pop-ups

The regular triangle tetraflexagon gives you configurations where you could open a pocket to reveal a pop-up inside. You can take advantage of the fact that you're looking at the inside of a pyramid to make pop-ups with perspective, using the inside tip of the pyramid as a vanishing point. See figure 37.10 for an example, where the view looks down a path, with a bench on one side and a hot air balloon floating above the scene.

FIG. 37.10 Regular triangle tetraflexagon with a pop-up inside.

If you want to try your hand at creating your own pop-up on this flexagon, see figure 4.1 for the template for the regular tri-tetraflexagon, which allows you to flex between two different states that can both contain a pop-up inside the pyramid. Note that if you use a regular tetraflexagon with additional inside faces (such as the penta-tetraflexagon in figure 4.3), you can't put pop-ups on those additional faces because the four inside leaves travel independently. You could, however, draw flat decorations on them.

Another interesting flexagon for a pop-up is a silver tetraflexagon. Like the square tetraflexagon, pop-ups can be designed for when the flexagon lays flat. However, it's different from the square flexagon in that the pop-up needs to collapse inside a triangle, which adds additional limits on your design so that it remains hidden when folded. It's also different in that it requires multiple flexes to get between the flat states, though these faces could still be decorated to continue your story. You can use the template in figure 5.19 for the silver penta-tetraflexagon to create a flexagon that has two flat states you can flex between. Place the pop-ups on faces 2 and 5. Note that the leaves on face 1 get rearranged after every flex.

Pentaflexagon pop-ups

The final example is unique to flexagons, using the structure of the flexagon itself to create a three-dimensional structure. It's a triangle pentaflexagon that turns into an animal-like shape after three pocket flexes.

The template in figure 37.11 folds into the pentaflexagon. A series of three pocket flexes, as shown in figure 37.12, turns it into a three-dimensional shape you can decorate, with an example shown in figure 37.13.

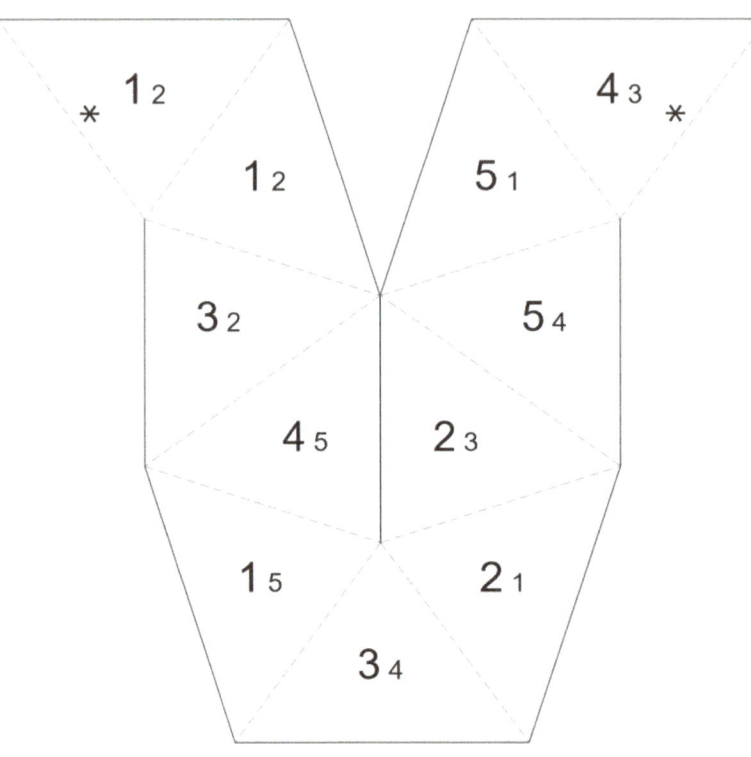

FIG. 37.11 Template for a pentaflexagon that supports a sequence of three pocket flexes leading to a three-dimensional structure.

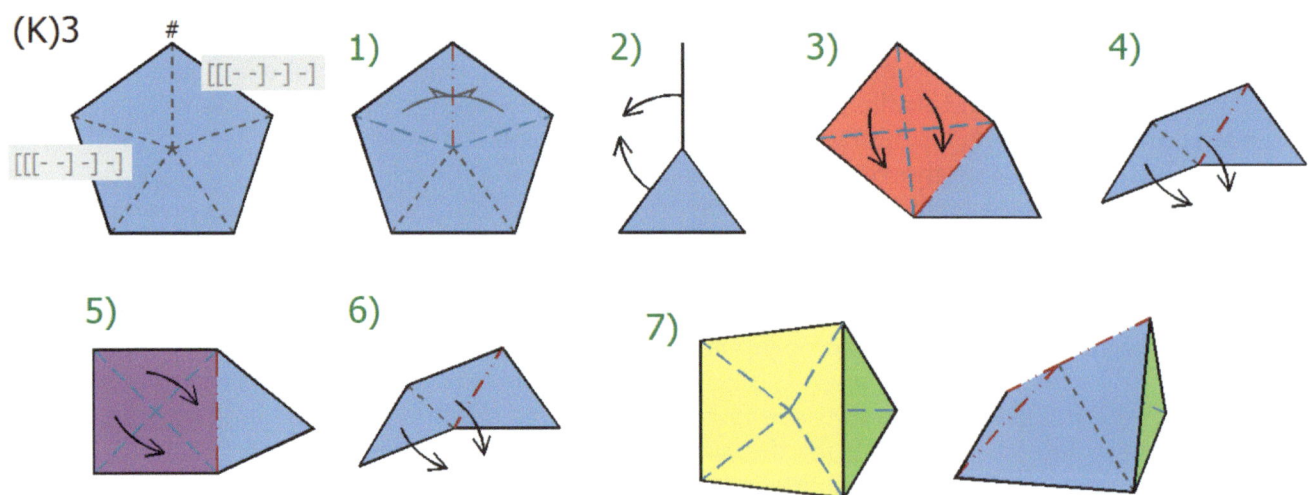

FIG. 37.12 Flex diagram for three pocket flexes, (K)3, on a pentaflexagon.

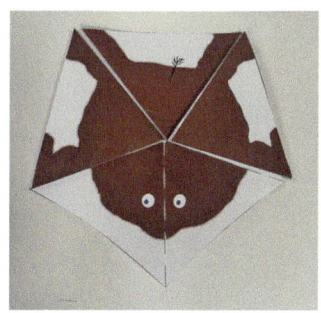 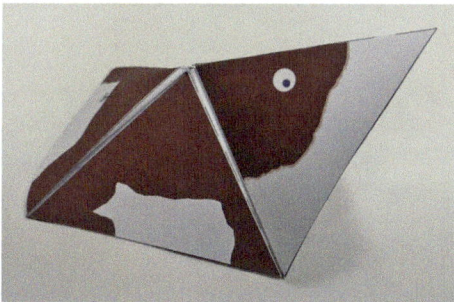

FIG. 37.13 An animal drawn on a pentaflexagon in its flat state (left) and flexed state (right). Notice how the hind legs change position between the two states.

Summary

We've experimented with several techniques for making pop-ups appear in flexagons: the one-piece technique cut into the flexagon itself, the multi-piece technique attached to both sides of a hinge, embedded pop-ups attached to a single leaf-face, and using the pocket flex to turn the flexagon into a three-dimensional sculpture.

But there are many other one-piece and multi-piece techniques beyond the ones we showed, as can be found in the many books on making pop-ups. There are also other flexagons that can be turned into interesting three-dimensional shapes through a series of pocket flexes.

What other flexagons might make for interesting pop-ups? What other pop-up techniques could work? What stories could you tell through your own unique flexagon pop-up?

Chapter 38

Cutting Flexagons

Another of the many ways to enjoy flexagons is to cut the leaves to create interesting patterns. There are a number of ways to do this.

Snipping flexagon corners and edges

The simplest option is to snip the corners of flexagons. This gives flexagons a more flowery feel. It can also make the flexagon easier to flex because the hinges work better with the extra space created in between the pats. Figure 38.1 shows an example of snipping off the corners of a hexaflexagon. First, make a hexa-hexaflexagon. Then, fold all the pats into a single stack, which will be 18 leaves thick. Cut off each of the three corners of the flexagon. When you've done this, open the flexagon and flex it to see the patterns that you get. We could call these "holey" flexagons.

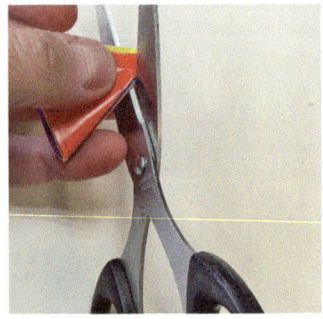 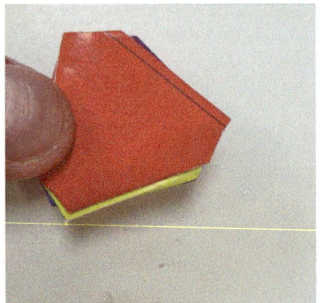 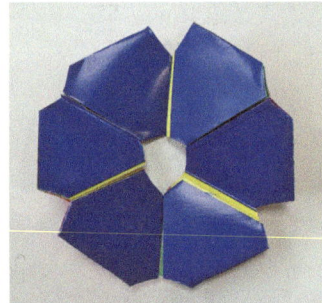 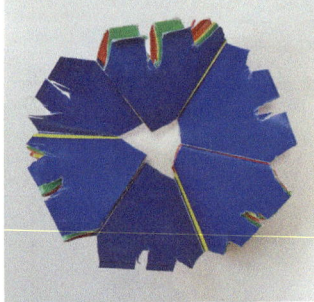

FIG. 38.1 Trimming all the leaves together in order to make patterns along the edges of a flexagon (first three images). The result of cutting a more elaborate pattern out of the outer edge (last image).

The next addition would naturally be to cut some patterns along the edges of the flexagons, as shown in the right-hand pane of Figure 38.1. To create a nice symmetrical effect, this should be done while the flexagon is still folded into one stack after the corner snipping. You can try doing these effects with other flexagons as well. Just be careful not to slice off entire hinges!

Circular cuts

For a different variation, after folding all the leaves of a flexagon into one stack, make a circular cut, completely removing several of the flexagon's corners, but leaving the working hinges intact. Figure

DOI: 10.1201/9781003433538-43

38.2 shows this method when applied to a tri-hexaflexagon along with a couple of the resulting shapes after pinch flexing. The result flexes to a clover and a "flexa-frame"! Experiment with other flexagon templates and cut patterns to get different beautiful and surprising results.

  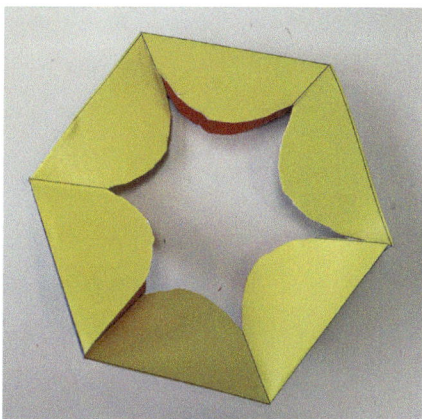

FIG. **38.2** Making a circular cut in a flexagon.

Cutting templates

A different approach to cutting flexagons is to cut the template *before* the flexagon is folded. For example, truncate a flexagon strip, such as a hexa-hexaflexagon template, by cutting off two straight strips, each a few millimeters long, from the top and the bottom of the original template. Figure 38.3 shows a truncated hexa-hexaflexagon template you can use.

FIG. **38.3** Truncated hexa-hexaflexagon template, with strips removed from the top and bottom edges. Paste the top and bottom strips back to back so the colors are on the outside.

Fold the truncated strip into a flexagon. Make sure that you still fold along the lines, even though (obviously) there will now be overlaps between different colored leaves. Note that there is an extra white trapezoid marked with a glue symbol on the end of each side of the strip. This is an alternate way to join the ends of flexagons together and appears in some flexagon publications. When you finish folding the template, the two white trapezoids should be glued face to face. The resulting flexagon can be flexed as usual. The colorful patterns are very beautiful and also provide insight into the inner structure of flexagons. Figure 38.4 shows what it looks like to fold this truncated hexa-hexaflexagon and the resulting flexagon.

FIG. 38.4 Folding the template of a hexa-hexaflexagon that's been trimmed along its edges.

Transforming the leaf shape

More satisfying truncated flexagons can be made if the templates are cut symmetrically before folding so that the shape of the leaves of the folded flexagons are identical. This method was used by Les Pook in his pioneering book *Serious Fun with Flexagons* to create silver and bronze flexagons by cutting them from square and regular triangle flexagon templates. After cutting the leaves into the desired shape, the template is folded in the same way that it would have been before the leaves were cut. It is important to make sure that the shape that is cut out still has enough edge on each hinge of the flexagon template. This preserves the dynamics of the flexagon created from the original template. If not enough edge is left at the hinges, the flexagon will, at best, turn into a point flexagon, and, at worst, fall apart completely.

In many cases the leaves will align perfectly when folded, creating a symmetrically pleasing flexagon. To do this, whatever shape you want to cut out must be reflected correctly along the fold lines of the flexagon. Whether or not leaves will perfectly align depends on the shape and folding scheme of the original template. Figure 38.5 provides the template for a heart-leaf tri-hexaflexagon, and figure 38.6 shows the resulting flexagon. Note the mirror reflection of the heart shape along the fold lines each time. Also note that the two lobes of the heart do not touch the hinge line at a single point. Rather, each lobe aligns with the hinge for a short distance so that the flexagon won't tear at these delicate junctions.

FIG. 38.5 Template for the heart-leaf tri-hexaflexagon.

FIG. 38.6 Flexing the heart-leaf flexagon.

Another example is the "hat" flexagon, inspired by an important mathematical discovery concerning shapes that tile an infinite plane. We all know that a square or rectangular tile can tile an infinite plane without there being any holes between the tiles. Square or rectangular tilings are periodic, which roughly means that all parts of the tiling look the same, with the same pattern repeating everywhere. Mathematicians have known for a few decades that there exist tilings made from two or more shapes that tile the plane aperiodically, that is, they completely cover the plane without a repeating pattern. The tiling is different everywhere you look.

In March 2023, four mathematicians – David Smith, Joseph Samuel Myers, Craig Kaplan, and Chaim Goodman-Strauss – discovered that even a single shape can tile the plane aperiodically. To put it simply, this monotile, which looks like a hat, can completely cover an infinite plane just like a rectangular tile can, yet not in any regular way. This was a major mathematical discovery, so to celebrate it, we created a "hat" flexagon. Figure 38.7 shows a section of the "hat" tiling, and figure 38.8 shows the tile superimposed on a tri-hexaflexagon template. To make the flexagon, first cut out the two strips and paste together their backs so the colored faces are both visible. Then cut along the lines of the superimposed tiling. Fold the template in the usual way. It's straightforward to fold, since every "hat" should fold neatly onto its neighboring mirror image. While the flexagon clearly isn't aperiodic, it still provides a brief introduction to this fascinating tiling.

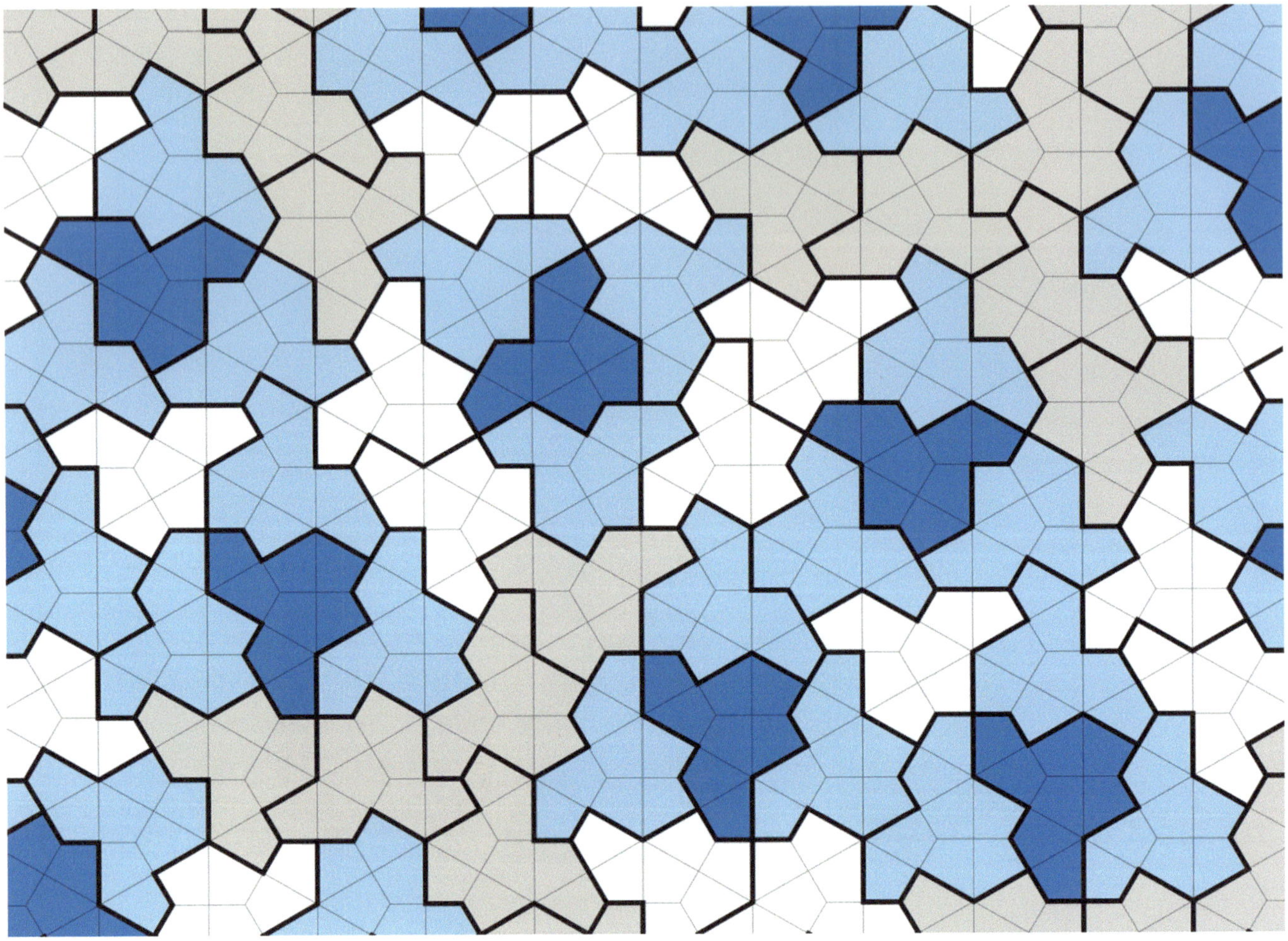

FIG. 38.7 The aperiodic "hat" tiling.

FIG. 38.8 Front and back of the template for the "hat" tri-hexaflexagon. Paste the backs together (leftmost triangles pasted together and rightmost triangles pasted together) so that the colors are on the outside.

Figure 38.9 shows the cut-out template and the resulting flexagon. It is important to note that we had to round some of the pointy corners (similar to what we did for point flexagons in chapter 21) of the original shape so that there is enough space along the fold lines to fold the leaves without them becoming completely separated from the strip.

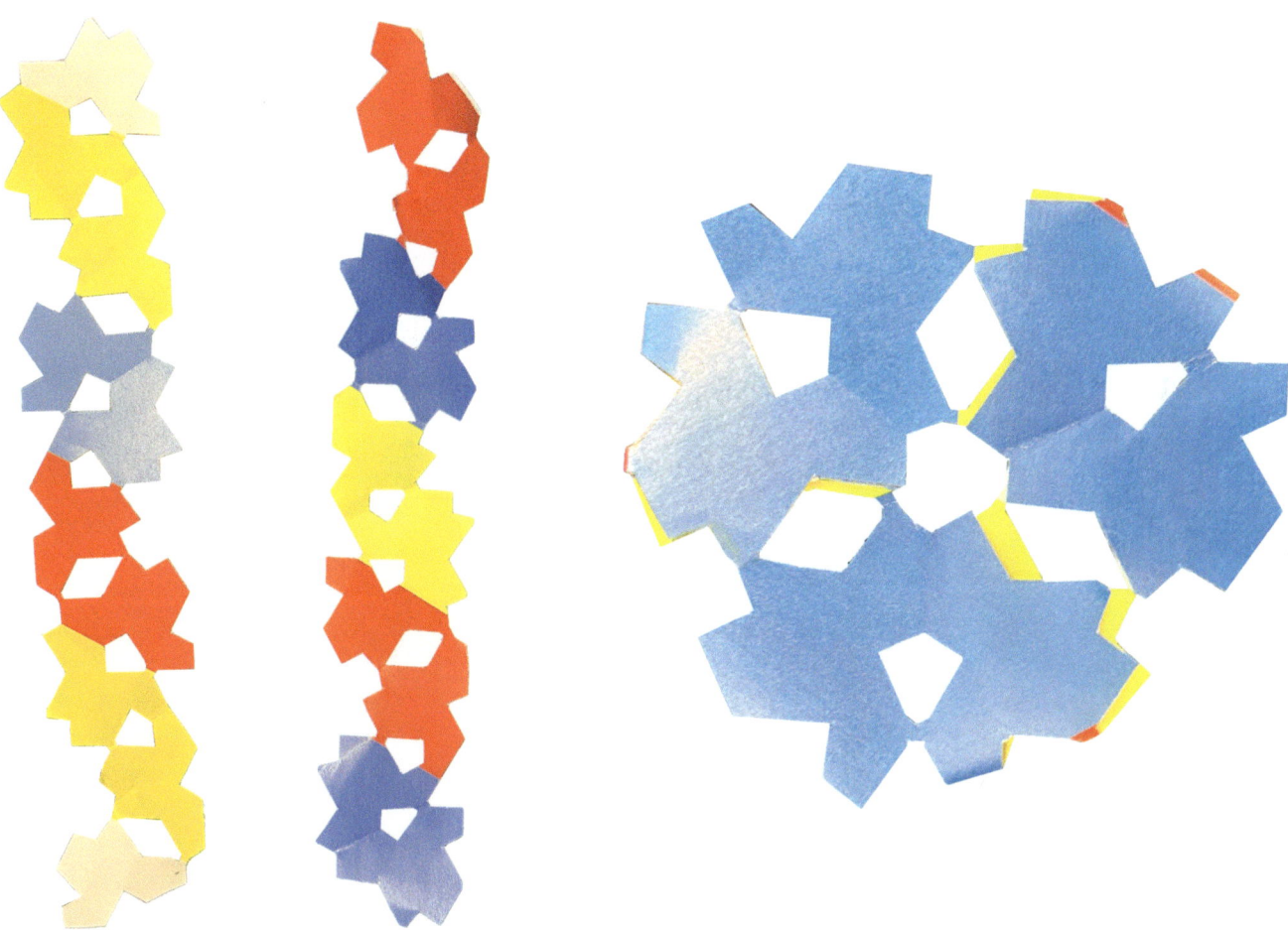

FIG. 38.9 The "hat" flexagon cut from a tri-hexaflexagon template: flat template (left) and folded flexagon (right).

Summary

We've experimented with several ways of cutting a flexagon to make interesting patterns, but there are other possibilities. For example, instead of truncating the strip, you can cut holes in the middle of some or all of the leaves, allowing you to see through to the inside of the pats and reveal their inner structure. Feel free to experiment with different ways of cutting up flexagons, combining this with other decorations, puzzles, pop-ups, etc.

Epilogue

In this book, we have explored many ways to have fun with flexagons using a wide variety of shapes and flexes, exploring the mathematics behind them, decorating them, turning them into pop-up sculptures, and creating puzzles with them.

But there is a lot more you can do and discover beyond what we've shown you. Here are some ideas for what you might want to try next:

- Pick an interesting flexagon and decorate it.
- Experiment with different flex sequences to see what patterns you can find.
- Design a puzzle that uses how leaves move around and rotate.
- Figure out how to "solve" a flexagon after mixing it up.
- Use Flexagon Inspector to create new flexagons and explore them.
- Try making up new flexes.
- Enhance atomic flex theory so it works for non-triangular leaves – squares, pentagons, hexagons, and so on. Or extend it so it can model the three-dimensional constraints of a physical flexagon.
- Make point flexagons and contrast their behavior with the corresponding edge flexagons.
- Take inspiration from the "Breaking the rules" section in chapter 21 to go beyond our definitions of *flexagon* and *flex*.

We hope all the ideas we've investigated in this book inspire you to keep exploring *The Secret World of Flexagons*.

Appendix A

Definitions

atomic flexes
The four flexes >, ^, ~, and Ur, which can be combined to create any basic flex.

basic flex
An operation on a flexagon that can be defined as a sequence of atomic flexes.

basic flexagon
A structure where every leaf is connected to exactly two mirror image copies of itself, and it's folded using the atomic flex axioms.

class
All the states accessible from a starting state using a given flex or set of flexes, e.g., a pinch class or a {P, V} class.

corner hinge
In a flexagon where each leaf contains a 90° angle, this is a hinge where the adjacent leaf angles are not 90° so they form a corner.

cycle
A flex sequence that returns to the original state.

edge flexagon
A flexagon with hinges between the edges of the leaves. See also *point flexagon*.

edge hinge
In a flexagon where each leaf contains a 90° angle, this is a hinge where there are adjacent leaf angles of 90° so they form a straight line.

face
One visible surface of a flexagon in its main position.
Note: The phrase "a flexagon has N faces" typically refers to pinch faces.

flex
A series of one or more manipulations of a flexagon that take it from one valid state to another, where the manipulations consist of folding together adjacent pats, unfolding pats, and sliding pats.

generating sequence
A flex sequence used to create the minimal flexagon that supports the sequence.

global flex
A flex that makes changes across an entire flexagon. See also *local flex*.

infinite tape flexagon
A flexagon with an infinite number of pats.

isoflexagon
A flexagon where every pat has an identical relationship to every other pat, e.g., flexagons that form a regular polygon where every pat meets in the middle.

leaf
>A single polygon in a flexagon.

leaf-face
>One of the two faces of a leaf.

local flex
>A flex that only changes a portion of a face. See also *global flex*.

main position
>The primary appearance of a flexagon when it isn't being flexed.

matryoshka
>A nested flexagon that appears when pats are flexed or folded together, which has fewer pats in its main position than the original flexagon and supports its own flexes.

minimal flexagon
>The flexagon with the fewest leaves that supports a given flex or flexes.

morph
>A position you can flex to that's a valid flexagon where the directions between pats and/or the number of pats has changed.

morphing flex
>A flex that takes you to a morph of a flexagon.

pat
>A stack of leaves in a flexagon.

pinch face
>A set of leaf-faces that travel together within a given pinch class.

point flexagon
>A flexagon with hinges between the vertices of the leaves. See also *edge flexagon*.

position
>A specific arrangement of pats as described by the directions between the pats.

reference hinge
>The hinge that a flex is defined relative to.

relatively prime flexes
>A set of flexes where none of them can be expressed as a flex sequence consisting only of the other flexes from the set.

sector
>A set of one or more pats, where each sector around the flexagon has the same internal structure. Note that this term is usually only useful for flexagons made from generating sequences that only contain pinch flexes.

state
>A unique arrangement of all the leaves and pats in a flexagon.

template
>An unfolded flexagon. Alternate terms: net, frieze, plan, strip.

thumbhole
>A gap in a pat where you can slide your thumb.

traversal
>A flex sequence that visits every state out of a class of states.

universal flexagon
>A flexagon where every pat and subpat are infinitely nested, so any flex supported by the pat directions can always be performed.

Appendix B

Flexagon Charts

This book uses the following naming convention, where each piece of the name is optional:

[overall shape] [leaf shape] [face count]-[pat count prefix]flexagon [(details)]

The flexagon name is optionally followed by additional details about the flexagon's internal structure enclosed in parentheses:

- **Generator:** The flex generating sequence, e.g., (generator: F*>>S*).
- **Pats:** The pat structure in one state of the folded flexagon, e.g., (pats: [[1,2],3,4,[[5,6],7]]).
- **Directions:** The directions from leaf to leaf in the template going clockwise, e.g., (directions: //\\//).
- **Leaves:** The total number of leaves in the template, e.g., (leaves: 15).

The following pages can be used to help identify and understand the names of a wide variety of flexagons. They show many examples for a flexagon's *overall shape*, *leaf shape*, *pat count*, and *directions between pats*. Note that some of the names can be simplified based on the context or when referring to a general class of flexagon, such as using *hexaflexagon* rather than *hexagonal regular triangle hexaflexagon*. Also note that this list emphasizes flexagons that lie flat, though many do not, and some are Möbius strips or even knots.

Unlike the other flexagon prefixes, the *face count* does not affect the flexagon's shape. It is the theoretical number of independent faces and is indicated by a Greek numeric prefix followed by a hyphen to distinguish it from the pat count. For example, a hexa-decaflexagon has six faces and ten pats.

Overall shape

An adjective that describes the shape of the flexagon in the main position is often the first part of a flexagon name: for example, triangular, quadrilateral (square, rhombic), pentagonal, hexagonal, octagonal, decagonal, dodecagonal. The overall shape may include additional modifiers such as *ring* or *star*.

3: triangular

triangular
bronze
hexaflexagon

triangular ring
dodecaflexagon

triangular
bronze
octadecaflexagon

triangular ring
trapezoid
triflexagon

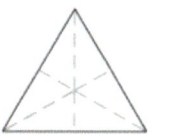 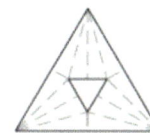 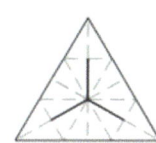 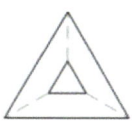

4: quadrilateral

square
silver
tetraflexagon

square
silver
octaflexagon

square ring
hexadecaflexagon

square
tetraflexagon

rhombic
bronze
tetraflexagon

rhombic
bronze
dodecaflexagon

 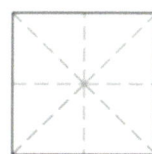 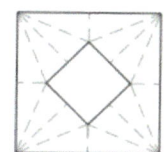 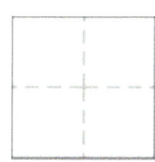 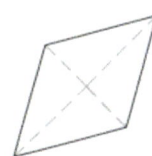 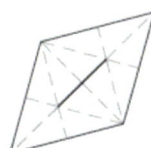

5: pentagonal

pentagonal
isosceles
pentaflexagon

pentagonal
right
decaflexagon

pentagonal ring
right
icosaflexagon

pentagonal ring
trapezoid
pentaflexagon

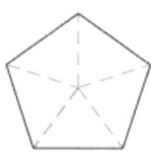 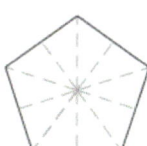 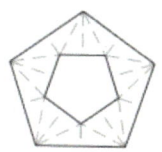 

6: hexagonal

hexagonal
regular
hexaflexagon

hexagonal
bronze
dodecaflexagon

hexagonal ring
bronze
icositetraflexagon

hexagonal ring
regular
octadecaflexagon

hexagonal
kite
hexaflexagon

hexagonal ring
trapezoid
hexaflexagon

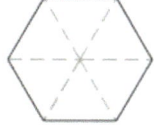 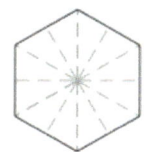 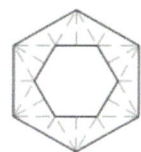 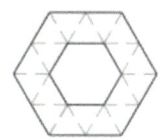 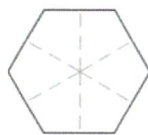 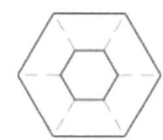

293

hexagonal ring
isosceles
enneaflexagon

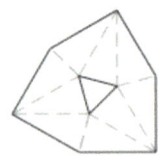

hexagonal
regular
decaflexagon

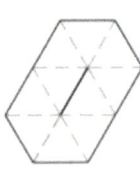

hexagonal ring
isosceles
dodecaflexagon

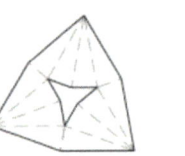

hexagonal
silver
dodecaflexagon

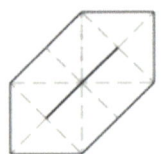

hexagonal ring
regular
dodecaflexagon

hexagonal ring
regular
tetradecaflexagon

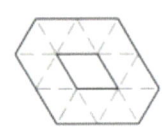

8: octagonal

octagonal
isosceles
octaflexagon

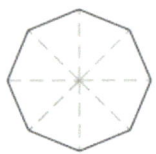

octagonal ring
isosceles
dodecaflexagon

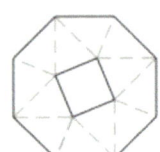

octagonal star
triangle
octaflexagon

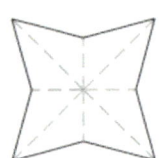

octagonal ring
isosceles
hexadecaflexagon

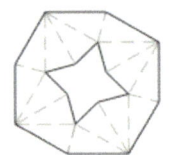

octagonal ring
isosceles
tetradecaflexagon

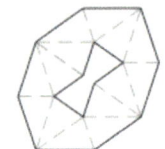

octagonal ring
trapezoid
octaflexagon

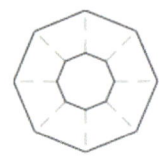

10: decagonal

decagonal
isosceles
decaflexagon

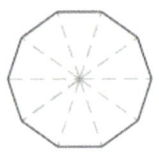

decagonal ring
isosceles
icosaflexagon

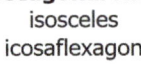

decagonal star
isosceles
decaflexagon

decagonal ring
isosceles
pentadecaflexagon

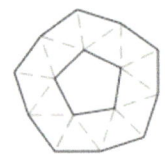

12: dodecagonal

dodecagonal
isosceles
dodecaflexagon

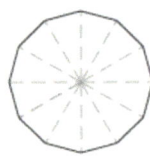

dodecagonal star
isosceles
dodecaflexagon

dodecagonal ring
isosceles
icositetraflexagon

dodecagonal
rhombus
hexaflexagon

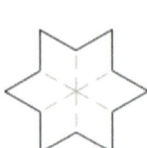

dodecagonal
pentagon
tetraflexagon

Leaf shape

The polygonal shape of the leaves can be included as a noun, not an adjective, so as not to be confused with the overall shape: for example, triangle (regular, silver, bronze, right, isosceles), quadrilateral (square, kite, rhombus, trapezoid), pentagon, heptagon, hexagon.

3: regular triangle

hexagonal **regular** hexaflexagon	hexagonal **regular** decaflexagon	hexagonal ring **regular** dodecaflexagon	hexagonal ring **regular** tetradecaflexagon	hexagonal ring **regular** octadecaflexagon

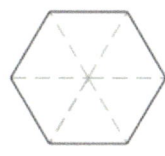  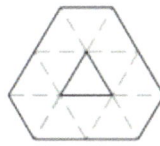 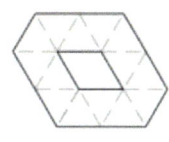 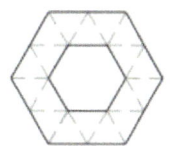

silver triangle

square **silver** tetraflexagon	square **silver** octaflexagon	hexagonal **silver** dodecaflexagon	hexagonal ring **silver** tetradecaflexagon

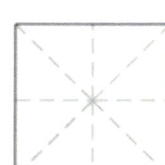  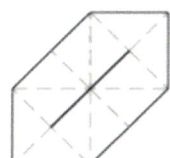 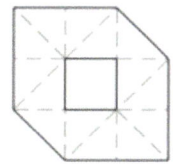

bronze triangle

rhombic **bronze** tetraflexagon	triangular **bronze** hexaflexagon	hexagonal **bronze** dodecaflexagon	rhombic **bronze** dodecaflexagon	rhombic **bronze** hexadecaflexagon	triangular **bronze** octadecaflexagon

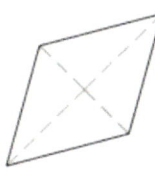 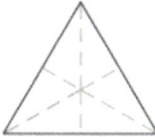 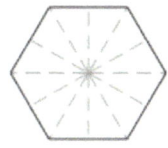 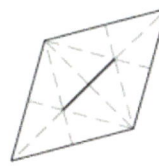 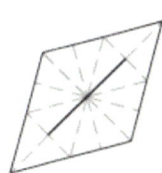 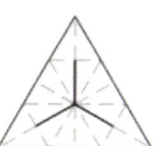

right triangle

pentagonal **right** decaflexagon	triangular ring **right** dodecaflexagon	square ring **right** hexadecaflexagon	pentagonal ring **right** icosaflexagon

isosceles triangle

pentagonal **isosceles** pentaflexagon	heptagonal **isosceles** heptaflexagon	enneagonal **isosceles** enneaflexagon	decagonal **isosceles** decaflexagon	dodecagonal **isosceles** dodecaflexagon	star **isosceles** decaflexagon

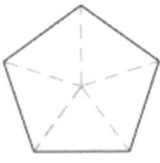  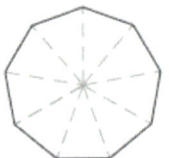 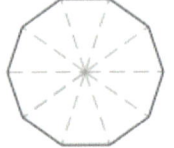  

hexagonal ring **isosceles** enneaflexagon	hexagonal ring **isosceles** dodecaflexagon	octagonal ring **isosceles** dodecaflexagon	octagonal ring **isosceles** hexadecaflexagon	octagonal ring **isosceles** tetradecaflexagon	star **isosceles** dodecaflexagon

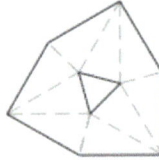 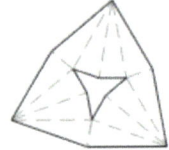 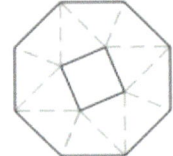  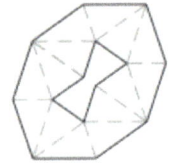 

4: quadrilateral

square tetraflexagon	hexagonal **kite** hexaflexagon	dodecagonal **rhombus** hexaflexagon	square ring **trapezoid** tetraflexagon	pentagonal ring **trapezoid** pentaflexagon	hexagonal ring **trapezoid** hexaflexagon

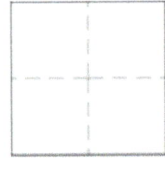 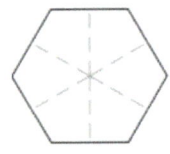 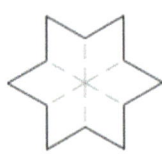 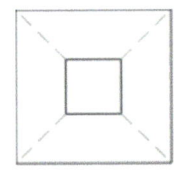 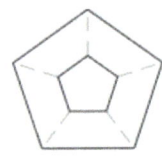 

5: pentagon　　　　　**6: hexagon**　　　　　**7: heptagon**　**8: octagon**

pentagon tetraflexagon	**pentagon** hexaflexagon	**hexagon** tetraflexagon	ring **hexagon** hexaflexagon	**heptagon** tetraflexagon	ring **octagon** tetraflexagon

 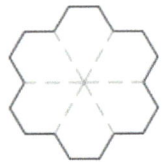 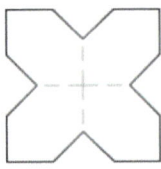  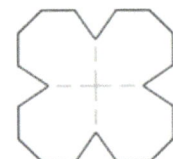 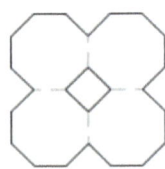

Pat count

A Greek numeric prefix indicates the number of pats in the flexagon's main position.

4: tetra

square silver **tetra**flexagon	rhombic bronze **tetra**flexagon	square **tetra**flexagon	square ring trapezoid **tetra**flexagon	pentagon **tetra**flexagon	hexagon **tetra**flexagon
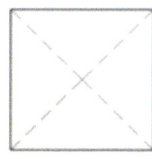	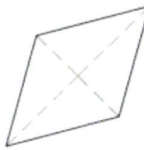	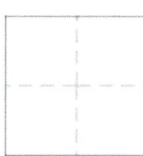	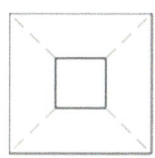	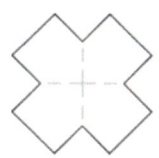	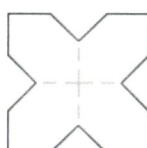

5: penta

heptagon **tetra**flexagon	ring octagon **tetra**flexagon	pentagonal isosceles **penta**flexagon	pentagonal ring trapezoid **penta**flexagon
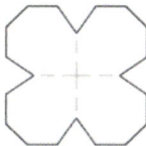	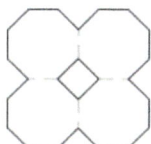	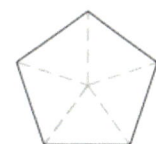	

6: hexa

hexagonal regular **hexa**flexagon	triangular bronze **hexa**flexagon	hexagonal star triangle **hexa**flexagon	hexagonal kite **hexa**flexagon	hexagonal ring trapezoid **hexa**flexagon	dodecagonal rhombus **hexa**flexagon
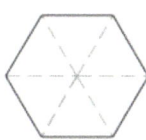			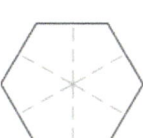	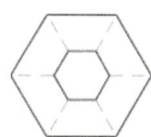	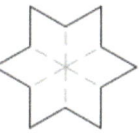

7: hepta

pentagon **hexa**flexagon	ring hexagon **hexa**flexagon	heptagonal isosceles **hepta**flexagon	heptagonal ring trapezoid **hepta**flexagon
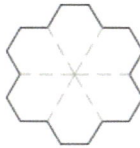		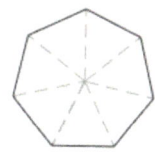	

8: octa

octagonal
isosceles
octaflexagon

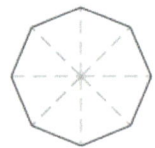

square
silver
octaflexagon

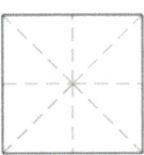

octagonal star
triangle
octaflexagon

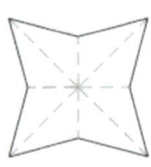

square ring
square
octaflexagon

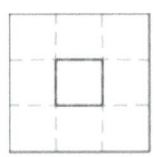

9: ennea

enneagonal
isosceles
enneaflexagon

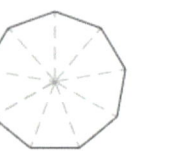

hexagonal ring
isosceles
enneaflexagon

10: deca

decagonal
isosceles
decaflexagon

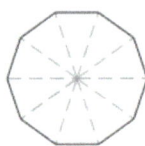

pentagonal
right
decaflexagon

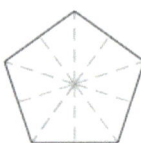

decagonal star
isosceles
decaflexagon

hexagonal
regular
decaflexagon

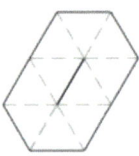

12: dodeca

dodecagonal
isosceles
dodecaflexagon

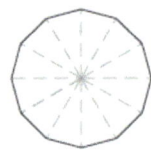

hexagonal
bronze
dodecaflexagon

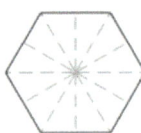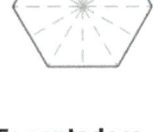

octagonal ring
isosceles
dodecaflexagon

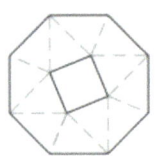

rhombic
bronze
dodecaflexagon

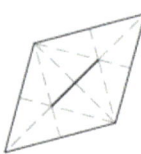

hexagonal
silver
dodecaflexagon

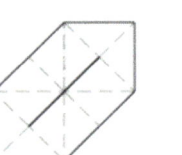

hexagonal ring
regular
dodecaflexagon

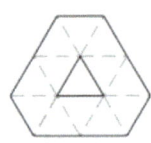

14: tetradeca

octagonal ring
isosceles
tetradecaflexagon

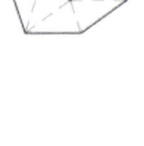

15: pentadeca

decagonal ring
isosceles
pentadecaflexagon

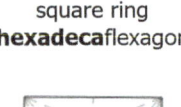

16: hexadeca

square ring
hexadecaflexagon

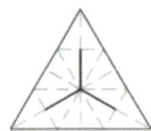

18: octadeca

triangular
bronze
octadecaflexagon

20: icosa

decagonal ring
isosceles
icosaflexagon

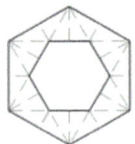

24: icositetra

hexagonal ring
bronze
icositetraflexagon

Directions between pats

The following images categorize flexagons by subsets of directions supported by common flexes. For example, some contain \//\, which supports the morph-kite inverse flexes. Note that the leaf angles may prohibit flexes that the directions between pats would otherwise allow. These directions can be used to find flexes that may work on the various flexagons.

/ × n

| silver tetraflexagon //// | isosceles pentaflexagon ///// | regular hexaflexagon existence newCustomers existed exists ////// | bronze hexaflexagon existed existence ////// |

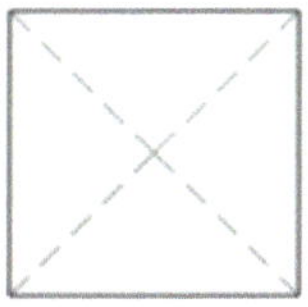

silver
tetraflexagon
////

isosceles
pentaflexagon
/////

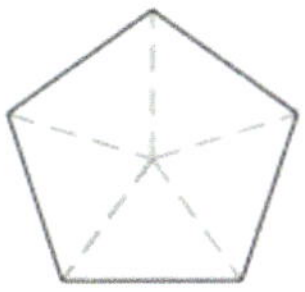

regular
hexaflexagon
//////

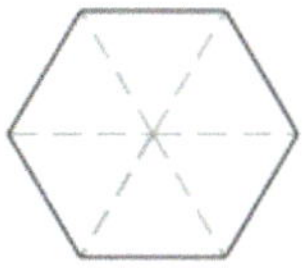

bronze
hexaflexagon
//////

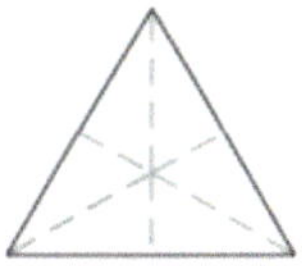

isosceles
heptaflexagon
///////

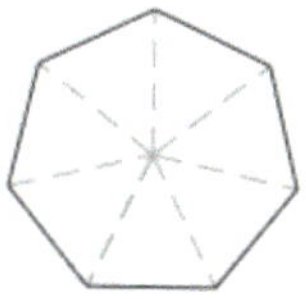

isosceles
octaflexagon
////////

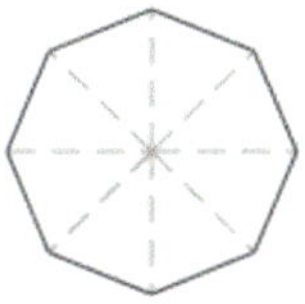

silver
octaflexagon
////////

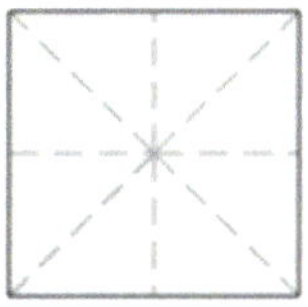

isosceles
enneaflexagon
/////////

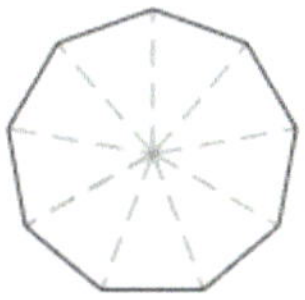

isosceles
decaflexagon
//////////

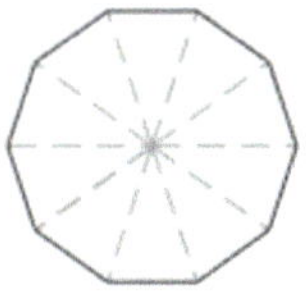

right
decaflexagon
//////////

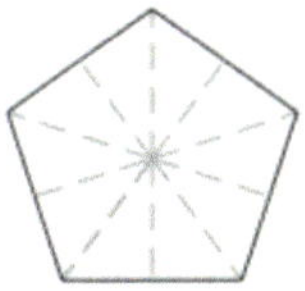

star
decaflexagon
//////////

isosceles
dodecaflexagon
////////////

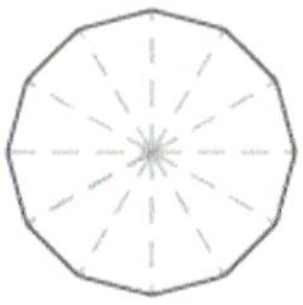

bronze
dodecaflexagon
////////////

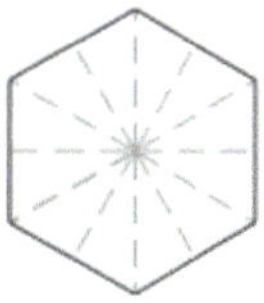

star
dodecaflexagon
////////////

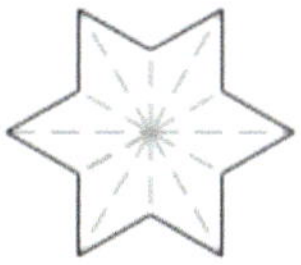

//\ × n rings

hexagonal ring
enneaflexagon
/\/\/\

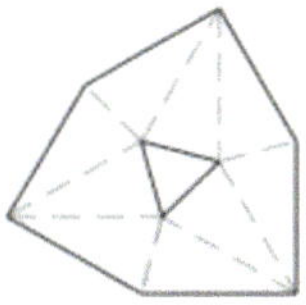

octagonal ring
dodecaflexagon
/\/\/\/\

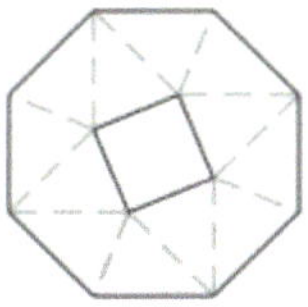

decagonal ring
pentadecaflexagon
/\/\/\/\/\

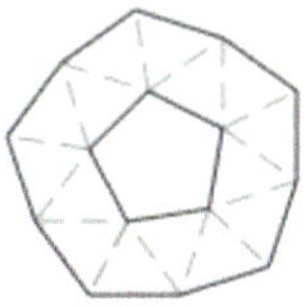

hexagonal ring
octadecaflexagon
/\/\/\/\/\/\

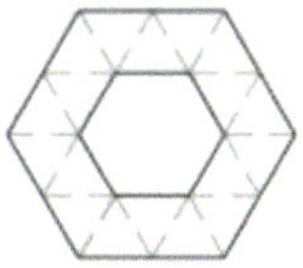

tetradecagonal ring
icosihenaflexagon
/\/\/\/\/\/\/\

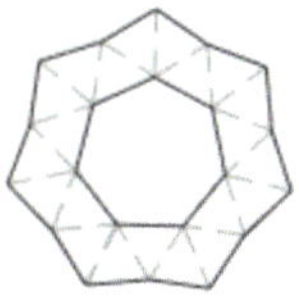

//\\ × *n* rings

triangular ring dodecaflexagon
/\\/\\/\\

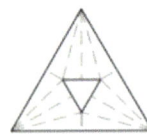

hexagonal ring isosceles dodecaflexagon
/\\/\\/\\

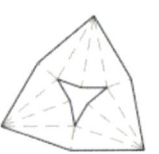

square ring hexadecaflexagon
/\\/\\/\\/\\

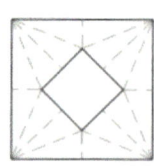

octagonal ring isosceles hexadecaflexagon
/\\/\\/\\/\\

pentagonal ring icosaflexagon
/\\/\\/\\/\\/\\

decagonal ring isosceles icosaflexagon
/\\/\\/\\/\\/\\

//\ × *n* other

hexagonal silver dodecaflexagon
/\\/\\/\\

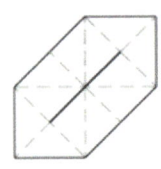

triangular bronze octadecaflexagon
/\\/\\/\\/\\/\\

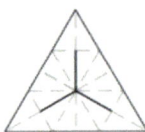

//, \//\

octagonal ring tetradecaflexagon
/\\/\\/\\

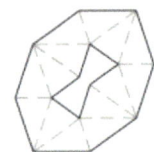

///, /\/

hexagonal ring dodecaflexagon
///\\//\\//\

///, \//\

hexagonal ring tetradecaflexagon
///\\//\\//\

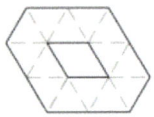

////, /\/

hexagonal regular decaflexagon
///\\///\

rhombic bronze dodecaflexagon
/\\////\\/

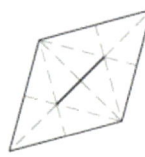

hexagonal ring silver tetradecaflexagon
///\\/\\///\

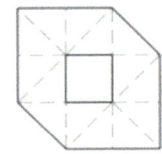

////, \//\

kite bronze octaflexagon
///\\/\

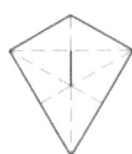

rhombic bronze hexadecaflexagon
///\\/\\///\\/\

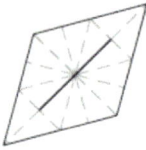

Appendix C

Flex Compendium

The tables in this appendix provide summary information for the flexes included in this book. Note that these lists are not exhaustive. There are many other possible flexes.

Common flex notation

The following shows some of the symbols used in flex notation, where A and B represent arbitrary flexes:

A'   the inverse of flex A
(A)n   repeat flex sequence A n times
{A|B}   do sequence A if supported, otherwise do sequence B
I   the identity flex, the flex that doesn't change the flexagon
A+   create minimal pat structure needed for A without doing A
A*   A+ A (create minimal pat structure and perform A)
A = B   flex sequences A and B do exactly the same thing
A = B   {when A is supported}

Useful identities

The following are identities that are useful when manipulating flex sequences:

$>' = <$ $<' = >$ $\wedge' = \wedge$
$<> = >< = \wedge\wedge = I$
$IA = AI = A$
$(AB)' = B'A'$
if $A = B$, then $CA = CB$ and $AC = BC$

Atomic flexes

The atomic flex axioms are $>$, $\wedge$, $\sim$, Ur, and their inverses. These building blocks can be used to describe all the valid operations for folding templates and doing any basic flex. The exchange flexes are often used to build other flexes. For flexes that rotate leaves, we note how the leaf immediately to the right of the reference hinge rotates given that the angle at the bottom is α, the angle counterclockwise is β, and the other angle is γ.

symbol	name	equivalence	input	output	angles
>	shift right	^<^	a # 1 / b	a 1 / # b	α γ β
			a # 1 \ b	a 1 \ # b	β α γ
<	shift left	^>^	a 1 / # b	a # 1 / b	α γ β
			a 1 \ # b	a # 1 \ b	β α γ
^	turn over	^'	a # b	$-b$ # $-a$	α β γ
~	change direction	~'	a # b	$-a$ # $-b$	γ β α
Ur	unfold right	~Ul~	a # [-2,1] / b	a # 1 \ 2 / $-b$	α β γ
Ul	unfold left	~Ur~	a # [1,-2] \ b	a # 1 / 2 \ $-b$	α β γ
Xr	exchange right	Ur < Ul' >	a 1 / # [-3,2] / b	a [1,-2] \ # -3 \ b	γ α β
Xl	exchange left	Ul < Ur' >	a 1 \ # [2,-3] \ b	a [-2,1] / # -3 / b	β γ α

Triangle flexes with decomposition

The following table lists triangle flexes. For each flex, its row shows the symbol used in flex sequences, its name, its inverse if it can be done using the flex itself, and one or two ways of decomposing the flex into simpler flexes. If no inverse is listed for a flex, you undo the flex by performing its steps in the reverse order. The number of pats is represented by p, which can be used to indicate how many times to repeat a sequence.

symbol	name	inverse	decomposition
Bf	backflip	^Bf^	Mkf' Mkb
Ds	double slide	^Ds^	(Ur>>Ur'>)4
F	flip	^F^	Mkf Mkb'
F3	flip 3	^F3^	>Xl'<<Ul' <<Ur>>>Ul'<< Ur<Ul>Ur'>
F4	flip 4	^F4^	>Xl'<<Ul' <<<Ur>>>>Ul'<<< Ur<Ul>Ur'>
Fm	Möbius flip		< Mkr Mkb' > (or <<^ T <<^ S)
K	pocket		Xr~^ > Ul^
L3	slot-triple	^L3^	(K^)3 (<)4 (K'^)3
Lbb	slot-tuck-bottom-back		Lbf < P' >
Lbf	slot-tuck-bottom-front		Mkf Lkk >>> Mkf' <<
Lh	slot-half		Ltb T'
Lk	slot pocket	^Lk^	Ltf > S <
Lkk	kite-to-kite slot	^Lkk^	> Ul Ur <<<<< Ul' < Ul' >>
Ltb	slot-tuck-top-back		Mkf Lkk >>> Xl' ^>> Ur >>>^ Ur' >
Ltf	slot-tuck-top-front	^Ltf^	Mkf Lkk Mkf' <
Mkb	morph-kite back		^Mkf^
Mkbs	morph-kite back shuffle		^Mkfs^
Mkf	morph-kite front		K > Ul' <
Mkfs	morph-kite front shuffle		> K < Ur << Ul' >> Ur'
Mkh	morph-kite-half		Xr~ >>> Xl <<<
Mkl	morph-kite left		^Mkr^
Mkr	morph-kite right		<< Ur >>>> Xl << Ur'

symbol	name	inverse	decomposition
Mkt	morph-kite-tuck		< Ur ^<<< Ur' <<^ Xl <<<
P	pinch	^P^	(Xr>>)(p/2) ~
P222	hexaflexagon pinch	^P222^	(Xr>>)3 ~
P333	pinch 333	^P333^	(Xr>>>)3 ~
P3333	pinch 3333	^P3333^	(Xr>>>)4 ~
P444	pinch 444	^P444^	(Xr>>>>)3 ~
Rhm	rhombic morph		Mkf (>)6 Mkf Xr~ (<)6 Xr
S	pyramid shuffle	^>S<^	K Xl' Ur' < Xl >
S3	pyramid shuffle 3	^>>S3<<^	< Mkr Mkl' >
St	silver tetra	^St^	Mkf Mkfs' (or >T'< <^T^>)
St2	silver tetra 2	^St2^	>>T'<< <<^T^ >>
T	tuck		Xr Xl
Tao	three-and-open	^<<Tao>>^	>> Ul' << Ur < Ul' << Ur >>
Tf	forced tuck	^Tf^	Xr Xl
Tr2	transfer 2	^Tr2^	Xl Xr
Tr3	transfer 3	^Tr3^	Ur << Ul' >> Ul << Ur' >
Tr4	transfer 4	^Tr4^	> Ul <<< Ur' >>> Ur <<< Ul' >>
Ttf	tuck-top-front		Mkh Mkf' (or T>P<)
Tu	turn	^>>Tu<<^	<Ul'~ >>Ur >>>> Ur' >>Ur >>>
Tw	twist	^Tw^	(>>Xl' >>Xr)2 ~
V	v-flex	^V^	< (Xr>>)(p/(2 − 1)) Xl'>>~ >

Local triangle flexes

The next table defines *local* triangle flexes using atomic pat notation, showing the input and output states. Flexes are organized by the number of pats modified and the required directions between the pats for both the input and output states. Note that the table includes the forced tuck (Tf), but not other tuck flexes (T, T1, T2, etc.), which vary depending on which extra flap you open up across from the reference hinge.

	input		output
	-//-		-//-
Tf	a 1 / # [[−3,4],2] / b		a [3,[1,−2]] / # 4 / b
	-\\-		-\\-
Tr2	a 1 \ # [2,[4,−3]] \ b		a [[−2,1],3] \ # 4 \ b
	-/\/-		-/\/-
Tr3	a 1 / 2 \ # [[−4,5],3] / b		a [3,[1,−2]] / # 4 \ 5 / b
	-///-		-///-
S	a [[[3,−2],−4],1] / −5 / # [7,−6] / b		a [−2,1] / −3 / # [7,[−4,[−6,5]]] / b
	-///-		-\///-
K	a [−2,1] / −3 / # [5,−4] / b		a 1 \ 2 / # [−4,3] / −5 / −b
	-////-		-////-
F	a [[3,−4],[1,−2]] / −5 / # [7,−6] / 8 / b		a 1 / [−3,2] / # −4 / [[−7,8],[−5,6]] / b
Fm	a [[3,−4],[1,−2]] / −5 / −6 / # [8,−7] / b		a 1 / [−3,2] / −4 / # [8,[−5,[−7,6]]] / b
S3	a [[[3,−2],−4],1] / −5 / −6 / # [8,−7] / b		a [−2,1] / −3 / −4 / # [8,[−5,[−7,6]]] / b
St	a [3,[1,−2]] / 4 / # [7,[5,−6]] / 8 / b		a 1 / [[−3,4],2] / # 5 / [[−7,8],6] / b

	input	output
	-/////-	-\//\-
Mkf	a [–2,1] / –3 / # [5,–4] / 6 / b	a 1 \ 2 / # [–4,3] / [–5,6] \ b
Mkb	a 1 / [–3,2] / # –4 / [6,–5] / b	a [1,–2] \ [4,–3] / # 5 / 6 \ b
Mkr	a [–2,1] / –3 / # –4 / [6,–5] / b	a 1 \ 2 / # [[–4,5],3] / 6 \ b
Mkl	a [–2,1] / –3 / # –4 / [6,–5] / b	a 1 \ [4,[2,–3]] / # 5 / 6 \ b
Mkfs	a 1 / [[–3,4],2] / # 5 / [–7,6] / b	a [1,–2] \ –3 / # [[5,–6],–4] / –7 \ b
Mkbs	a [–2,1] / –3 / # [–6,[–4,5]] / –7 / b	a 1 \ [4,[2,–3]] / # 5 / [6,–7] \ b
	-\//\-	-\//\-
Bf	a 1 \ 2 / # [5,[3,–4]] / [6,[8,–7]] \ b	a [[–2,1],3] \ [[–5,6],4] / # 7 / 8 \ b
Tr4	a 1 \ 2 / # 3 / [4,[6,–5]] \ b	a [[–2,1],3] \ 4 / # 5 / 6 \ b
	-/////-	-/////-
F3	a [[3,–4],[1,–2]] / –5 / –6 / # [8,–7] / 9 / b	a 1 / [–3,2] / # –4 / –5 / [[–8,9],[–6,7]] / b
	-///////-	-///////-
F4	a [[3,–4],[1,–2]] / –5 / –6 / –7 / # [9,–8] / 10 / b	a 1 / [–3,2] / # –4 / –5 / –6 / [[–9,10],[–7,8]] / b
St2	a [3,[1,–2]] / 4 / 5 / # 6 / [9,[7,–8]] / 10 / b	a 1 / [[–3,4],2] / 5 / # 6 / 7 / [[–9,10],8] / b
	-//\//-	-\//\//-
Lkk	a 1 / 2 / 3 \ 4 / # 5 / [[–7,6],8] \ b	a [1,[3,–2]] \ 4 / # 5 / 6 \ 7 / 8 / b

Global triangle flexes

The next table defines *global* triangle flexes using atomic pat notation, showing the input and output states. The flexes are organized by the number of pats modified and the required directions between the pats for both the input and output states. Additional requirements on *p*, the number of pats, are noted when needed. In generalized formulas, *n* refers to the number of leaves and *i* is used for referencing leaves in the middle of the template.

		input	output
		(//)p, p ≥ 2	**(//)p**
P	βαγ	1 / # [–3,2] / ... i / [–(i + 2),i + 1] / ... / n – 2 / [–n,n – 1] / n = 3p/2	[2,–1] / # 3 / ... [i + 1,–i] / i + 2 / ... / [n – 1,–(n – 2)] / n /
		(//)p, p ≥ 3	**(//)p**
V	γβα	1 / [–3,2] / ... i / [–(i + 2),i + 1] / ... [n – 4,–(n – 5)] / n – 3 / n – 2 / [–n,n – 1] n = 3p/2	[2,–1] / 3 / ... [i + 1,–i] / i + 2 / ... n – 5 / [–(n – 3),n – 4] / [n – 1,–(n – 2)] / n
		(/)p, p ≥ 5	**(/)p**
Lk	αγβ	# [[[3,–2],–4],1] / –5 / ...–i / ... [[[–(n – 2),n – 3],n – 1],–(n – 4)] / n /	# n – 2 / [–2,[n – 1,[1,–n]]] / –3 / ...–i / ... [n – 3,[–(n – 6),[–(n – 4),n – 5]]] /
Ltf	αγβ	# [[[3,–2],–4],1] / –5 / ...–i / ... [n – 1,–(n – 2)] / n	# n / [–2,–1] / –3 / ...–i / ... [n – 1,[–(n – 4),[–(n – 2),n – 3]] /
		(/)5	**(/)5**
L3	αβγ	# [[[3,–2],–4],1] / [[[–7,6],8],–5] / 9 / [[[12,–11],–13],10] / –14 /	# 7 / [–11,[8,[10,–9]]] / –12 / [–2,[–13,[1,14]]] / [6,[–3,[–5,4]]] /
		(/)6	**(/)6**
Lh	γβα	# [[–2,[–4,3]],1] / –5 / –6 / –7 / [[9,–10],–8] / –11 /	# [[–11,–1],10] / –2 / –3 / [5,–4] / [[–7,8],6] / 9 /

		input	output
Lbb	γαβ	# [[–2,3],1] / [–5,4] / –6 / –7 / [[9,–10],–8] / –11 /	# –2 / [4,–3] / 5 / [[–7,8],6] / 9 / [[–11,–1],10] /
Lbf	βαγ	# [[–2,3],1] / [–5,4] / –6 / –7 / [9,–8] / 10 /	# [–3,2] / –4 / [6,–5] / [–8,7] / –9 / [1,–10] /
Ltb	γβα	# [[–2,[–4,3]],1] / –5 / –6 / –7 / [9,–8] / 10 /	# –1 / –2 / –3 / [5,–4] / [[–7,8], 6] / [–10,9] /
Ttf	γβα	# [[-2, 3], 1] / [-5, 4] / -6 / [8, -7] / 9 / 10 / (/)6	# [4, -3] / 5 / [-7, 6] / -8 / [[10, -1], -9] / -2 / /\/\/
Mkh	βαγ	# [-2, 1] / -3 / -4 / [6, -5] / 7 / 8 /	# 2 / 3 \ [-5, 4] / -6 / -7 \ [1, -8] /
Mkt	βαγ	# 2 / 3 / 4 / [-6, 5] / -7 / [1, -8] / (/)8	# [-2, 1] / -3 \ [5, -4] / 6 / 7 \ 8 / (/)8
Tw	γβα	# [–2,1] / [4,–3] / 5 / 6 / [–8,7] / [10,–9] / 11 / 12 / (////\)2	# 2 / 3 / [–5,4] / [7,–6] / 8 / 9 / [–11,10] / [1,–12] / (////\)2
Tao	βαγ	# [–2,1] / –3 / –4 / –5 \ –6 / –7 / –8 / [10,–9] / 11 \ 12 / (/)12	# –11 / [1,–12] / 2 / 3 \ [–5,4] / –6 / –7 / –8 / –9 \ –10 / (/\\//)2
Rhm	γβα	# [–2,1] / –3 / –4 / –5 / [7,–6] / 8 / [–10,9] / –11 / –12 / –13 / [15,–14] / 16 / (\//)4	# –1 / [3,–2] / 4 / 5 \ 6 / [–8,7] / –9 / [11,–10] / 12 \ 13 \ 14 / [–16,15] / (\//)4
Ds	αβγ	# [–2,1] / –3 / –4 \ [6,–5] / 7 / 8 \ [–10,9] / –11 / –12 \ [14,–13] / 15 / 16 \	# 1 \ 2 / [–4,3] / –5 \ –6 / [8,–7] / 9 \ 10 / [–12,11] / –13 \ –14 / [16,–15] /
Tu	γβα	# 1 \ 2 / [–4,3] / –5 \ –6 / –7 / –8 \ –9 / [11,–10] / 12 \ 13 / 14 /	# –13 \ [1,–14] / 2 / 3 \ 4 / 5 / 6 \ [–8,7] / –9 / –10 \ –11 / –12 /

Pinch variations

With the standard pinch flex, you pinch every second hinge all the way around the flexagon. With the pinch variants, you may skip hinges. Since a single pinch rotates the leaves, it often changes the overall shape of the flexagon, while repeating the same pinch variant a second time at the same hinge restores it to its original shape. These flexes rotate the angles in the same way as the pinch flex: β α γ. A hinge represented as a question mark (?) means it can be either / or \. The output will contain ? if the hinge direction is preserved or ¿ if it's reversed. A pinched hinge requires //. See chapter 6 for the naming convention. The following table contains a few selected pinch variants, with similar patterns applying to other pinch variants.

	input	output
	(/?/)3	(/¿/)3
P333	# [–2,1] / –3 ? –4 / [6,–5] / 7 ? 8 / [–10,9] / –11 ? –12 /	# 2 / 3 ¿ [–5,4] / –6 / –7 ¿ [9,–8] / 10 / 11 ¿ [1,12] /
	(/?/)4	(/¿/)4
P3333	# [–2,1] / –3 ? –4 / [6,–5] / 7 ? 8 / [–10,9] / –11 ? –12 / [14,–13] / 15 ? 16 /	# –2 \ –3 ¿ [–4,5] \ 6 \ 7 ¿ [8,–9] \ –10 \ –11 ¿ [–12,13] \ 14 \ 15 ¿ [16,–1] \
	(/??/)3	(/¿¿/)3
P444	# [–2,1] / –3 ? –4 ? –5 / [7,–6] / 8 ? 9 ? 10 / [–12,11] / –13 ? –14 ? –15 /	# 2 / 3 ¿ 4 ¿ [–6,5] / –7 / –8 ¿ –9 ¿ [11,–10] / 12 / 13 ¿ 14 ¿ [1,15] /

Non-triangular flexes

Here we list some flexes that operate on non-triangular leaves such as squares, pentagons, hexagons, etc.:

symbol	name	description
B	book	analog to the pinch flex
Bl	book-left	used to clarify when book opens from the upper left corner
Br	book-right	used to clarify when book opens from the upper right corner
Bb	box-bottom	box followed by opening from the bottom
Bt	box-top	box followed by opening from the top

Bibliography

Anderson, T. et al. (2010) 'The combinatorics of all regular flexagons', *European Journal of Combinatorics*, 31(1), pp. 72–80.

Beier, J. and Yackel, C. (2015) 'Groups associated to tetraflexagons', in Beineke, J. and Rosenhouse, J. (eds.) *The Mathematics of Various Entertaining Subjects: Research in Recreational Math*. Princeton, NJ: Princeton University Press, pp. 81–94.

Berkove, E. J. and Dumont, J. P. (2004) 'It's okay to be square if you're a flexagon', *Mathematics Magazine*, 77(5), pp. 335–348.

Carter, D. A. and Diaz, J. (1999) *The Elements of Pop-up*. New York: Little Simon.

Chapman, P. B. (1961) 'Square Flexagons', *The Mathematical Gazette*, 45(353), pp. 192–194.

Conrad, A. S. (1960) *The Theory of the Flexagon*, RIAS Technical Report 60-24. Baltimore, MD: Research Institute for Advanced Studies.

Conrad, A. S. and Hartline, D. K. (1962) *Flexagons*, RIAS Technical Report 62-11. Baltimore, MD: Research Institute for Advanced Studies.

Cook, M. (2004) 'Universality in elementary cellular automata', *Complex Systems*, 15, pp. 1–40.

Elran, Y. and Schwartz, A. (2019) 'Should we call them flexa-bands?', in Beineke, J. and Rosenhouse, J. (eds.) *The Mathematics of Various Entertaining Subjects: The Magic of Mathematics*, Volume 3. Princeton, NJ: Princeton University Press, pp. 249–261.

Elran, Y., Sherman S. and Schwartz, A. (2024) 'We should call them flexa-bands! On defining flexagons and flexagon language', Proceedings of the 15th International Congress on Mathematical Education.

Gardner, M. (1956) 'Flexagons', *Scientific American*, 195(6), pp. 162–168.

Gardner, M. (1958) 'About tetraflexagons and tetraflexigation', *Scientific American*, 198(5), pp. 122–126.

Gardner, M. (1959) *The Scientific American Book of Mathematical Puzzles and Diversions*. New York: Simon and Schuster.

Gardner, M. (1961) *The 2nd Scientific American Book of Mathematical Puzzles and Diversions*. New York: Simon and Schuster.

Gilpin, M. (1976) 'Symmetries of the trihexaflexagon', *Mathematics Magazine*, 49(4), pp. 189–192.

Hilton, P. and Pedersen, J. (1994) *Build Your Own Polyhedra*. Menlo Park , CA: Addison-Wesley.

Hilton, P. and Pedersen, J. (2010) *A Mathematical Tapestry: Demonstrating the Beautiful Unity of Mathematics*. Cambridge, UK: Cambridge University Press.

Hilton, P., Pedersen, J. and Walser, H. (1997) 'The faces of the tri-hexaflexagon', *Mathematics Magazine*, 70(4), pp. 243–251.

Jackson, P. (1993) *The Pop-up Book*. New York: Henry Holt and Co.

McLean, T. B. (1979) 'V-flexing the hexahexaflexagon', *The American Mathematical Monthly*, 86(6), pp. 457–466.

Mitchell, D. (2003) *The Magic of Flexagons: Manipulative Paper Puzzles to Cut Out and Glue Together*. St Albans, UK: Tarquin Publications.

Oakley, C. O. and Wisner, R. J. (1957) 'Flexagons', *The American Mathematical Monthly*, 64(3), pp. 143–154.

O'Reilly, T. (1975) 'Classifying and counting hexaflexagrams', *Journal of Recreational Mathematics*, 8(3), pp. 182–187.

Pedersen, J. (1972) 'Sneaking up on a group', *The Two-Year College Mathematics Journal*, 3(2), pp. 9–12.

Pook, L. (2003) *Flexagons Inside Out*. Cambridge, UK: Cambridge University Press.

Pook, L. (2009) *Serious Fun with Flexagons: A Compendium and Guide*. Dordrecht: Springer.

Schwartz, A. and Rutzky, J. (2009) 'The hexa-dodeca-flexagon', in Pegg, E., Jr., Schoen, A. H. and Rodgers, T. (eds.) *Homage to a Pied Puzzler*. Wellesley, MA: A K Peters, Ltd., pp. 257–268.

Sherman, S. (2012) *Flex theory*. Available at http://loki3.com/flex/g4g10/Flex-Theory.pdf (Accessed: 4 October 2023).

Sherman, S. (2018) Explorable Flexagons. Available at http://loki3.com/flex/explore/ (Accessed: 6 June 2023).

Various authors (2007–2016) *Flexagon lovers Yahoo newsgroup*. Available at https://github.com/loki3/flexagonator/tree/master/docs/newsgroup (Accessed: 10 November 2023).

Wheeler, R. (1958) 'The flexagon family', *The Mathematical Gazette*, 42(339), pp. 1–6.

Index